UNDERSTANDING HOUSING DEFECTS

Understanding Housing Defects provides a concise, coherent and comprehensive introduction to the causes, investigation and diagnosis of defects in domestic buildings.

For this new edition, many of the chapters have been substantially updated and new photographs have been added. There are four new chapters covering:

- *how defects are defined*
- *an overview of building and architectural history*
- *external joinery and painted finishes*
- *environmental and health-related building problems.*

Many of the 21 chapters cover a specific building element and include a brief introduction setting out construction principles and the evolution of current practice. All of the chapters consider the identification, cause and diagnosis of common (and sometimes not so common) defects. This book is a must-have for all those students and practitioners who require a broad understanding of housing defects. Building surveyors, general practice surveyors, architects, estate agents, housing officers and anyone involved in the management and maintenance of property as well as its construction will benefit hugely from this highly informative full-colour text. Written by the authors of *The Construction of Houses*, the book is also the natural companion to this bestselling textbook.

DUNCAN MARSHALL, BSc, MCIOB, MIMBM
was a Chartered Builder and Senior Lecturer at the University of the West of England, UK.

DEREK WORTHING, BSc, MPhil, MRICS
is a Chartered Building Surveyor and was formerly a Principal Lecturer at the University of the West of England, UK. He now works as a consultant and a Visiting Lecturer. He is also a Visiting Professor at Uppsala University in Sweden.

ROGER HEATH, FRICS, FFB
is a Chartered Building Surveyor and was formerly a Senior Lecturer at the University of the West of England, UK. He now works as a Visiting Lecturer.

NIGEL DANN, BSc, MSc, MIOC, ICIOB
was a Senior Lecturer at the University of the West of England, UK, and now works as a consultant and Visiting Lecturer in the UK, France and Belgium (KU Leuven).

UNDERSTANDING HOUSING DEFECTS

FOURTH EDITION

Duncan Marshall,
Derek Worthing,
Roger Heath
and Nigel Dann

Routledge
Taylor & Francis Group

LONDON AND NEW YORK

First edition published 1998
by Estates Gazette

This edition published 2014
by Routledge
2 Park Square, Milton Park, Abingdon, Oxon OX14 4RN

and by Routledge
711 Third Avenue, New York, NY 10017

Routledge is an imprint of the Taylor & Francis Group, an informa business

British Library Cataloguing in Publication Data
A catalogue record for this book is available from the British Library

Library of Congress Cataloging in Publication Data
Marshall, Duncan (Building inspector)
Understanding housing defects / Duncan Marshall, Derek Worthing,
Roger Heath, and Nigel Dann. -- 4th edition.
pages cm
Includes index.
1. Dwellings--Defects. 2. Dwellings--Maintenance and repair.
3. Building inspection. I. Title.
TH4812.M2368 2014
690'.8--dc23
2013016763

ISBN: 978-0-415-62208-0 (hbk)
ISBN: 978-0-08-097112-4 (pbk)
ISBN: 978-0-08-097113-1 (ebk)

Typeset in Myriad Pro Light 10.5/14.5pt
by Fakenham Prepress Solutions, Fakenham, Norfolk NR21 8NN
Printed in Great Britain by Ashford Colour Press Ltd

MIX
Paper from
responsible sources
FSC® C011748

To Duncan Marshall

Duncan Marshall, Derek Worthing and Roger Heath were the authors of the first three editions of *Understanding Housing Defects*. Sadly, Duncan died in 2009, not long after publication of the third edition.

Duncan is still rightfully included as one of the authors of this new edition, not only as a sign of respect and affection, but also because he was such an important part of developing the character and content of the original book and all the previous editions. While the new material in the book had to be written without Duncan, the essential aims, core and feel of the book remain those which he, Derek and Roger developed over the years.

Duncan's memory lives on in this new edition which we would like to dedicate to him.

Derek Worthing and Roger Heath

Contents

Preface

Understanding Housing Defects was first published in 1998 and this is its fourth edition. In this new edition, besides reviewing and extensively revising and updating the text, we have also added new chapters on:

- *how defects are defined*
- *an overview of building and architectural history*
- *external joinery*
- *environmental and health-related building problems.*

We strongly believe that the impact of the visual image enhances and clarifies the matters being discussed in the text and to that end we have added many new photographs throughout this latest edition.

As stated in previous editions, our aim in writing this textbook has not changed from our original intention – to provide an interesting, coherent and comprehensive introduction to the causes, investigation and diagnosis of housing defects. There are a number of other excellent texts on building failure, but these tend to be either highly technical or aimed at specialist practitioners. We have always felt there was a need for a more general text, aimed at those students and practitioners who require a broad understanding of housing defects as part of a wider sphere of academic or professional activity. This group includes building surveyors, general practice surveyors, architects, maintenance inspectors, facilities managers, estate agents, housing managers, environmental health officers and builders.

This latest edition, like the first, has three specific objects:

to explain why, and how, defects occur;

to enable the reader to recognise and identify building defects;

to provide, where appropriate, guidance on the correct diagnosis of defects.

Additionally, we still hold the view that a broad knowledge of building principles and good building practice are prerequisites for a genuine understanding of building defects. To this end, we have, where appropriate, introduced each chapter with a brief summary setting out construction principles and describing the evolution of current practice.

The text does not include anything on the repair of defects and, once again, we make no apology for this. Many repairs are obvious, others require input from specialists. In addition, there is often a very wide range of repair methods and materials available and, as these are constantly changing and being developed, to do them all justice would require several volumes.

Acknowledgements

The authors would like to thank a number of individuals, companies and organisations who have contributed to this book.

Individuals:
Jeff Ayling, F. Blampied, Stephen Bond, Steve Brown, Fiona Burrows, Mike Bryant, Stuart Clarke, Ian Constantinides, Roger Cowles, A. Cox, N.J. Donnithorne, C. Edmans, Dr Amy Frost, Sally Gilbert, Azadour Guzelain, Michael Hammet, Ian Hancock, Gregg Kerry, John Lobb, Ben Locke, Susan Marshall, Johan Mattsson, David McLaughlin, John Moore, Samantha Organ, Tim Pearce, Robert Shadbolt, Peter Smith, Simon Spokes, Simon Starling, Ole Martin Stensli, Ian Stokes, Graeme Todd, Jenny Walmsley.

Organisations:
BRE, Bath & North East Somerset Council, Brick Development Association, Bristol City Council, Bristol Record Office, British Gypsum, Cheshire Record Office, Chimney Care, Comb Down Stone Mines Land Stabilisation Project, Department of Environmental Protection (Pennsylvania, USA), Environmental Agency, Get Mapping, Hall & Ensom Chartered Building Surveyors, Hepworth Clay Wavin Ltd, Ibstock, J & J Sharpe Builders Ltd (Okehamptom, Devon), Knightstone Housing Association, Liverpool Anglican Cathedral, Lyme Regis Museum, Mathew Jackson Photography, mine-explorer.co.uk, Mycoteam AS, Ordnance Survey, Peter Cox Preservation, Protimeter, Public Health England, Rentokil, Sealand Aerial Photography, Surevent, TRADA.

Understanding defects: an introduction 1

This book is concerned with explaining how a housing defect is caused, how it might be identified and, to some extent, what its consequences might be. The importance of an understanding of defects can be seen by the number of legal cases related to them. It is therefore worth considering what we mean when we say something is defective.

All building materials will have a lifespan over which they degrade and/or eventually fail. Proper and timely maintenance will help to extend the function by offering protection against degradation, or preventing failure in one element affecting another element. Notwithstanding the fact that all materials will fail at some point, and require repair or replacement, early failure may occur for various reasons including:

- *poor manufacture*
- *poor design*
- *poor specification*
- *poor construction*
- *poor maintenance*
- *inappropriate use.*

These are not necessarily 'stand-alone' reasons for early failure – sometimes they work in combination – it is not uncommon, for instance, for the failure of flat roofs to be caused by the 'unholy trinity' of an initial poor design, aggravated by a poor specification and then finished off by mistakes made at the construction stage.

Perhaps the most straightforward way of defining a defect is to suggest that something is defective when it no longer performs its functional requirements. As when, for example, a wall or its individual components no longer prevent rain from reaching the interior of the house, or perhaps, more dramatically, when a support or restraint gives way and a wall collapses. However, something could also be considered defective if it is still performing its primary function but its condition threatens to cause damage to other elements of the building, or, indeed, to other buildings.

In assessing the condition of a building, one usually attempts to make an assessment of the severity of the defect, based on an analysis, or at least an assumption, of the impact of the defect on the element itself and also on the surrounding elements. An example of this latter situation might be a rising damp problem in a wall, which is in itself a defect, but if it is not dealt with, it could lead to additional defects, such as dry or wet rot in, say, an adjoining timber floor. An assessment of potential impacts might also have to take into account a situation where there could be a relatively minor defect that, if it remains uncorrected, may lead to a more serious problem – an example of this might be a small plumbing leak that is a catalyst for dry rot. In both these cases a good knowledge and understanding of the construction and materials of the house being inspected are crucial in determining the likely severity and impact of the defect. The importance of understanding how the house being inspected was built, in general and in detail, and the materials used to do so, applies not only to older houses. For example, in a modern house the effect of damp penetration through the outer skin of an external wall will potentially have a greater significance for a timber-frame house than for a brick and block house in terms of the possibility of fungal attack and also differential movement.

It is of course possible that an element can be functioning as intended but is still considered to be defective because the requirements or expectations have changed with time – the width or steepness of stairs, for example – or that health and safety issues which were not appreciated at the time have emerged, such as the use of lead piping, lead paint, asbestos, etc. Another obvious example is that of increased environmental concerns where a lack of sufficient thermal insulation or heating controls in line with current, and indeed future, standards could be identified as a defect.

Perspectives on defects

However, it should be remembered that, while sometimes defects are a matter, more or less, of fact, they are also, sometimes, more of a matter of opinion, perception or interpretation. For example, the idea of character and patina of age may be relevant. Quite often, older houses are valued because they have character. To some people, the very fact that a building is old will be enough to imbue it with character. This connection between age and character is relevant to the identification of defects because it is clear that what might be regarded as an unacceptable physical condition in a new house might be seen as part of the character of an older house. Examples of this phenomenon might include sloping floors, or steep and/or narrow staircases.

To some extent, the issue of perception and context could also be affected by other factors close to, but not the same as, ideas of character. One of these could be that of expectation. For example, many people would accept the minor undulations or hairline cracks in the walls or ceilings of an older house that they would not in a newly built house, which they might expect to be in 'perfect' condition. Expectation also plays a part in determining the notion of something being defective in other ways. There is some evidence that the higher the purchase price of a house, the more the purchaser is likely to see minor 'blemishes' as significant defects, i.e. the owners of higher value houses perceive similar defects to be more serious than do owners of houses that cost less to purchase.

There is another aspect to the link between character, age and defects, and that is how something that might be considered defective in some circumstances – such as deterioration in the surface of stonework – might, in a house that is an historic building (i.e. listed because it is of cultural significance) be considered to be an important aspect of its 'patina of age' and therefore part of its essential character.

Obviously, in all of these cases of character there is a judgement to be made about where character ends and a defect begins. The same surface 'damage' in stonework that might be seen as part of its character will also make that stone more vulnerable to defects such as frost attack and damp penetration, which will in turn damage the historic value of the building through the loss of important historical fabric.

Design flaws

Although we do not deal explicitly in the book with defects in overall design, in the sense of, say, a house not functioning because of its juxtaposition of rooms or poor provision or use of space, design flaws can affect the functioning of an element. Examples of such flaws might be where the detailing of brickwork has been inadequate and has exposed it to excessive rain and low temperatures, leading to defects such as frost attack or sulfate attack. Or where a staircase is too narrow or too steep, making it difficult or dangerous to navigate. Design defects may also include the inadequate provision of natural ventilation via window openings. Some of these situations are covered in the Building Regulations.

Although we have not included wider issues regarding aesthetics in the book (in the sense of what a house looks like overall), aesthetic judgements are often linked to the issue of quality of workmanship and clearly something could be considered aesthetically defective in, for example, the quality of a finished surface in perhaps joinery, plaster or paint. This can be a matter of relative judgement or relative experience or knowledge: if you have never seen a high-quality emulsion paint finish on a plastered wall, you may think that the 'orange peel' effect is perfectly acceptable.

Legislation

One might consider that there is a factual basis for deciding that something is defective if it is identified as being so under one or more of the pieces of legislation which cover the physical condition of buildings but in reality this is, like much law, quite a complex area of both fact and interpretation. There are a number of statutes, as well as common law principles, which govern matters relating to buildings that are defective or 'in disrepair'. The statutes overlap to a certain extent, and they approach the problem from different perspectives. Take the example of rising damp which could, depending on the severity of the problems, come within the definitions below:
- *statutory nuisance, covered by the Environmental Protection Act 1990;*
- *unfit for human habitation, covered by the Housing Act 1985;*
- *landlord's contractual repairing covenants, covered by the Landlord and Tenant Act 1984 and common law.*

Because dampness is capable of causing injury, such as allergic reactions (in the sense that damp allows moulds to grow and their spores can create the problem), it could also be covered by the Defective Premises Act 1972. If it causes structural damage, such as rotting floor joists, which makes the building unsafe, it might be covered by the Building Act 1984. If the floor joists give way while a visitor is standing on them, the owner or occupier might be liable under the Occupiers Liability Act 1984.

Action taken under these statutes could in some instances lead to the criminal prosecution of the person with ultimate control of the building (the owner, usually). Common law and other statutes give rights to the injured party to sue the person responsible and for local authorities to take action.

There are a number of Acts of Parliament and considerable case law which can apply to housing defects, and which approach the problem from a number of different angles. Some examples of these are discussed below.

Defective Premises Act 1972

This Act, which applies to work in relation to a new dwelling (new development or conversion), imposes a duty on those who undertake work for, or in connection with, the provision of a dwelling, to:
- *do the work in a professional or workmanlike manner;*
- *use proper materials;*
- *ensure that the dwelling is fit for human habitation when the work is completed.*

Occupiers' Liability Act 1957 and 1984

The duty of care imposed under this Act is to see that any visitor is reasonably safe in using the premises for the purposes for which they were allowed in. The duty of care may vary according to such factors as whether the visitor is a child, or an independent contractor employed to carry out works of repair, or in any event knows of the danger because it has been brought to their attention. Liability is imposed on the occupier, who, for the purposes of this Act, is the person in control of the premises.

Landlord and Tenant Act 1985

This Act places substantial legal duties on landlords to repair and maintain properties which they let to tenants. When landlords fail to keep their premises in repair, their tenants may be able to sue them for breach of contract. Tenants may be able to recover damages (compensation) for the breach for a period of up to six years, depending on when the landlord first had notice of the disrepair. The tenant can also ask the court to order that the repairs be carried out.

Environmental Protection Act 1990

A house in poor condition can amount to a statutory nuisance, which is a criminal offence. The law relating to statutory nuisance is set out in the Environmental Protection Act 1990 (EPA). This largely, but not entirely, replaced the Public Health Act of 1936. Statutory nuisance under the Act could include 'any premises in such a state as to be prejudicial to health or a nuisance'. An example of a statutory nuisance might be an ill-fitting window frame, not because it needs repair, but because it allows wind and water penetration which in turn causes dampness and mould growth which are prejudicial to health.

The term 'prejudicial to health' is taken to mean 'injurious, or likely to cause injury, to health'. Therefore, both actual and potential ill-health are covered. Premises will fall within this definition if it can be shown that the conditions of the premises would cause a well person to become ill, or the health of an ill person to deteriorate. 'Health' is defined broadly to include physical and mental health. Mere interference with comfort will not count but it would include stress caused by the effects of poor housing.

Dampness is a common statutory nuisance, whether it is penetrating, rising or condensation dampness. Premises which are damp are harder to heat and can aggravate conditions such as arthritis and rheumatism. Dampness can affect the furniture in the room, making beds and bed linen damp, for example. The mould growth associated with dampness, particularly condensation dampness, is a known health hazard. Many people are allergic to the mould spores, which can cause bronchitis, asthma and other complaints which cause breathing difficulties.

Liability in tort for failure to repair

A claim may be made in the tort of negligence where the state of repair of a building causes personal injury or damage to a person's property, and the injury or damage can be said to arise out of the breach of a duty of care owed to that person. The categories of persons who are subject to legal liability for inadequate buildings have been expanded, both through common law and through statutes. Defects in the design, construction, or modification of a building may also give rise to a claim in negligence. This is a complex area of law and beyond the scope of this book.

The Building Act 1984

This is primary legislation under which various pieces of secondary legislation are made. The secondary legislation includes the Building Regulations, Dangerous Structures and Demolition of Buildings. The Act allows local authorities to take action in relation to premises which are in such a 'defective state' that they may be prejudicial to health or a nuisance. It also allows a local authority to take action where it considers that a building or structure or part of a building or structure is in a condition that renders it dangerous. This may include restrictions on the use of the building until it has been made safe.

The Act also allows the local authority to take action where a building's 'ruinous or dilapidated condition' is considered to be seriously detrimental to the amenities of the neighbourhood.

The Building Regulations

The Building Regulations are not retrospective but they are a useful reference point for considering an existing building in the light of modern requirements and standards in respect of both the health and safety of occupants and energy usage.

Town and Country Planning Acts

Legislation relating to planning is perhaps a less obvious reference point for taking action concerning defects, but there are two relevant and possibly interconnected areas which may apply. The Town and Country Planning Act 1990 gives the local planning authority the power to require that land be cleaned up if its condition 'adversely affects the amenity of the area'. Although reference in the legislation is to 'land', the term is used to cover land and buildings. Therefore, a building with a number of, presumably external, defects could be the focus of action under this legislation. The normal course of action is for the local authority to serve a notice on the owner setting out what must be done to remedy the situation. The local authority has the power under the Act to carry out the work themselves and recover the costs from the owner. There is encouragement from government for local authorities to use these powers proactively, i.e. not just in response to complaints by neighbours, and they are often used in respect of buildings in conservation areas and for listed buildings. Remedies for defects to listed buildings can also be sought using Urgent Works Notices and Repairs Notices (referred to below).

If the building is listed, action can be taken against the owner by the local authority if the building falls into a state of disrepair. The legislation in this case is the Planning (Listed Buildings and Conservation Areas) Act 1990. The primary purpose of the Act is to deal with defects that are serious enough to cause damage to the character of the listed building. There are two processes that the local authority might use to do this:

Urgent Works Notice. *This can only be served in respect of an empty building. It is used where emergency work is required to avoid further damage and deterioration of the fabric of the listed building. The work involved can only be temporary in nature and intent, i.e. it does not involve repairs. Typically it might involve using tarpaulins to prevent rain coming in through a damaged roof, or making the building secure against vandals. The owner is given a reasonable amount of time to respond to the notice. The notion of what is a reasonable time will depend upon the condition of the building but, given that the title involves the word 'urgent', then clearly it implies that the owner should act before the situation worsens.*

Repairs Notice. *This involves requiring the owner of a listed building to carry out repairs which the local authority thinks are necessary for the 'proper preservation of the building'. The work required is set out in a specification from the local authority. If, after a set period, the work has not been carried out, the local authority may begin compulsory purchase proceedings on the building.*

Fitness Standards

There also have been a number of government-led initiatives related to Fitness Standards which are usually based on risk assessments aimed at evaluating the risks to health and safety present in a dwelling. Such initiatives in the past have included the Housing Health and Safety Rating System and the Decent Homes Standard, which was aimed at social housing.

The Party Wall etc Act 1996

This Act applies where a building owner intends to carry out repair or improvement works to a party wall, a party structure (for example a floor separating flats), boundary walls, or excavations near neighbouring buildings. Under the Act the building owner must meet a number of requirements but primarily the focus is on informing the adjoining owner of the intended work.

Building history　2

Section 1: Introduction

The housing stock of England and Wales was constructed over several hundred years although the great majority of those houses were built since the middle of the nineteenth century. However, there are still many houses standing that pre-date this period.

During this period, many housing architectural styles were introduced, and many of these lasted for considerable periods of time and produced buildings that are still in use today. Many different forms of construction and types of materials have also been used, many of which are still in use, while others which are no longer favoured.

This chapter discusses the principles of house design and construction during the past 500 years or so and outlines a number of the more common defects that the houses can suffer. A more detailed discussion of the causes of the defects, their investigation and their impact on a building is set out in the chapters that follow.

This chapter is divided into four sections: Section 1 is the Introduction; Section 2 covers the period up to the First World War (up to 1914); Section 3 covers the inter-war period between the First World War and the Second World War (1918–1939); and Section 4 covers the period from the end of the Second World War until approximately the end of the twentieth century (1945–1990). This division makes sense as the First World War is a watershed in terms of house construction. Before then, house building was essentially based on traditional construction methods and materials that had developed over hundreds of years. Housing since the end of the First World War (i.e. since 1918) saw modern construction methods and materials applied, first slowly and then at an increasing pace through the remainder of the twentieth century and into the twenty-first century. This was due to the increasing demands for: (1) new housing as the population increased; (2) better-quality housing as living standards rose; and (3) more environmentally efficient buildings as the effects of climate change became apparent. Industrial progress, increasing knowledge of how buildings perform and the development of new materials have also greatly influenced building design and construction in the past 100 years or so.

In this chapter, vernacular buildings and Georgian and Regency housing are discussed in greater detail than housing during the Victorian and later periods. This is because the houses from the middle of the nineteenth century up to the present day make up the great majority of our current housing stock and the problems associated with their poor performance are really the main subject of this book.

Section 2: Housing before the First World War (up to 1914)

PART A: VERNACULAR BUILDINGS

Throughout history, different economic, social and cultural influences have impacted on the design and construction of houses. Those with more financial resources had a far greater ability to exercise influence over the conscious design aesthetic of their houses. They also had significantly greater ability to select both the material and construction form of their houses. The associated status that this provided was, and remains, desirable and generally has filtered downward in society. In pre-industrial societies this process was slow. Given the major constraints on the transmission of design and construction ideas and, more significantly, access to non-local building materials, the houses of the vast majority of the population were subject to local tradition, local craft knowledge and local materials. They were also built to respond to local environmental and cultural needs.

In pre-industrial societies, the vast majority of dwellings were not intended to be of a permanent nature and were often seasonal or short term. There is little written on or archaeological evidence of these structures. Those houses which do survive from this period are all relatively high-status buildings, even if they do not all necessarily appear to be so from our contemporary point of view. Although most of these were not designed by a professional designer (in the modern sense), they were commonly professionally built by teams of craftsmen who were working in a long and verbally transmitted tradition with local materials and who clearly had the ability and knowledge to produce permanent structures. Such buildings are often now referred to as 'vernacular buildings'.

Two typical vernacular buildings. Left: A local rubble stone cottage finished in a series of coats of coloured lime-wash; rubble stone was universally lime-rendered or lime-washed in the past. The upper storey is built within the roof void and the roof is covered with locally derived thatch. Small window openings reduce heat loss and probably date from a period where shutters, rather than the more recently installed windows, closed the openings. Right: A Wealden house, a commonly high-status timber-frame building from the south-east of England. Timber-frame construction was vernacular in that it made use of local resources, responded to local needs and was a verbally transmitted technique which flourished for 500 years after 1200.

As industrial development gathered pace throughout the seventeenth and eighteenth centuries, vernacular approaches to building were increasingly challenged because of a mixture of economic and cultural factors:

- *increasing urban development as an increasing proportion of the population moved to the cities looking for work.*

- *increasingly sophisticated and widespread speculative housing development, where aesthetic fashion played an increasingly important role in marketing.*
- *greater wealth and, to some degree, a trickle-down of such wealth, enabling a larger proportion of the population to have greater concern with and control over aesthetics and status.*
- *the breakdown of verbally transmitted tradition as printing, publishing and the transmission of new cultural ideas gathered pace.*
- *an increasing need for new forms of housing in urban areas to house large numbers of workers required for industrial production.*
- *industrial production enabling more efficient production of products and services which previously had been limited in scale, or constrained by efficiency in terms of their production and their physical transportation.*
- *associated with this, the improvements in transport technology and systems, enabling bulk and industrially produced building materials to achieve a distribution well beyond their immediate zone of production.*
- *the increasing education and professionalisation of specialist designers and craftsmen.*

Though there are numerous surviving pre-eighteenth-century buildings, it remains relatively rare to find houses pre-dating 1700 in England and Wales. Numerous examples from the post-1700 stock of historic buildings can be described as vernacular in some respects, i.e. responding to relatively simple local needs, designed by builders using local tradition and local materials. However, the upheaval of industrialisation through the eighteenth and nineteenth centuries saw vernacular building techniques being replaced by increasingly standardised approaches to both design and building. This was particularly acute in the cities. Although some local variations in materials, techniques and distinctiveness did remain, this lessens with time.

In more rural areas, particularly those areas remote from urban influence, with limited access to new design and construction ideas, vernacular traditions persisted longer, in some cases almost to the present day.

The vernacular tradition is, thus, not constrained to a house size or type and nor to a period nor even a specific location. It is possible that high-status buildings from as recently as the late nineteenth century could have some vernacular influences, in terms of the use of local materials. However, they are unlikely to be influenced by vernacular building techniques, whereas a modest rural cottage, long distant from transport and urban influences, could have been built in a vernacular tradition and materials well into the twentieth century.

The aesthetic qualities of vernacular buildings are now revered for their local distinctiveness, in terms of their form, the palette of materials employed and how they seem to 'fit' with their environment. They are also admired for the simple beauty of a fundamentally utilitarian approach to design rather than a conscious and deliberate adherence to a design theory, often emulating building designs from distant places or times.

In the past 50 years, there has been an increasing interest in and study of vernacular buildings, techniques and materials. This has led not only to a far greater understanding, appreciation and conservation of such buildings, but it has also increasingly seen attempts to revive aspects of the traditions. This has included the revival of traditional building craft, such as traditional timber framing and cob building to name two, but it has also led to a greater concern and attempt to ensure that new construction responds to the specific locality where houses are being constructed.

Occasionally, it is possible to find examples of vernacular buildings hidden behind more recent façades. It was not uncommon for façades to be replaced for fashion, as well as for technological reasons. An understanding and ability to recognise the range of vernacular materials and techniques are essential, particularly when one is concerned with defects and repairs to such buildings.

A relatively narrow range of local materials was used in vernacular buildings. This was necessitated by the lack of industrial production and highly constrained transportation opportunities. Other than brick and lime, which were both modified, vernacular materials all occurred naturally and, to many, they age attractively (if they are well maintained) and have become popular again in the past few decades.

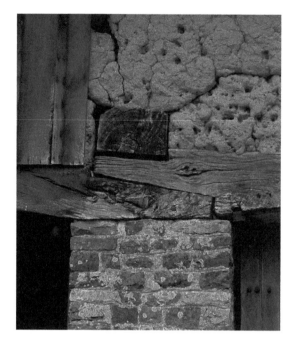

The biggest influences on the vernacular tradition are the environmental conditions of England and Wales and, more importantly, the geology. The Earth's crust provides a range of potential and frequently used building materials: soils, in particular clays and sands, stone, and timber. All these materials, in both pure and modified form, have been used for millennia in the production of buildings. Indeed, they all still continue to be used; it is the extent, frequency and geographical spread of use which have changed over time. To some extent, the increased concern about sustainability and the lack of local distinctiveness have seen an increasing focus and emphasis on the use of vernacular materials.

A range of materials, all local to the site, are employed in this extended medieval hall house. Hall houses were the preserve of the wealthy in this period. The 'hall' – a central double-height communal space in the centre of such buildings, generally with a central open hearth – can be identified externally by the double-height window. This plan form was commonly adopted in many parts of Britain. Its popularity waned in the sixteenth century when the desire for greater privacy among the upper classes reduced the need for communal space. More significantly, in the sixteenth century, masonry chimneys became widely used and provided for greater comfort, efficiency and convenience. The chimney also enabled the open double-height space to be floored over, adding valuable living space. This stone, brick and timber-framed building has been extended over many centuries.

Soils

The most widely available of all these vernacular materials are soils. Soils have been used in buildings for millennia and have commonly been a key constituent for mortars, renders and plasters. They have also been used as a mass walling material.

Burning or firing suitable 'brick earths' – predominantly clay – to form bricks is, arguably, the earliest example of a manufactured building material. It has a long history of use in England and Wales and has become the most widely and visibly used material for external load-bearing wall construction.

Buildings made of brick

The technique for brickwork was known to the Romans. Their bricks tended to be thin (25mm or so), but large on plan (300mm square) and used as a bonding course in rubble walls and for arches and barrel vaults rather than for entire walls. Such techniques and technology were widely exported and used in all parts of their empire, including the British Isles. Brickwork and its associated technologies pretty much disappeared from use once the Romans left in the fifth century, though there are several examples of salvaged and recycled Roman bricks being used throughout the medieval period.

The native brick industry really started again in the late thirteenth century in the eastern counties of England that were close to and influenced by ports which imported bricks and, more significantly, the brickmakers from the Low Countries.

The popularity and increasingly widespread use of bricks really took off in the sixteenth century for the construction of effective domestic chimney stacks. This was a largely vernacular undertaking, as transporting this bulky and heavy material for great distances was difficult and expensive. It established widespread local brick production and developed and encouraged brick-laying techniques

As a mass walling technique, the seventeenth century saw brickwork becoming increasingly more desirable following its extensive use in prestigious high-status buildings, such as Hampton Court Palace. However, the true popularity of brick did not occur until the eighteenth century. The increasingly widespread and effectively implemented regulation in favour of fireproof wall construction following the Great Fire of London and increasing demand for mass housing in industrialising cities saw brickwork become the predominant building material in urban settings. Even here, however, brick production itself remained a largely local, hand-made activity, adding a distinctiveness to the look of brick buildings and hence a vernacular feel to such buildings.

Early brickwork was hand-made and of great variability. Brick production was a local and largely hand-crafted activity, right up until the nineteenth century.

From the mid-nineteenth century, improvements in transport and the industrialisation of production meant that bricks, now in standardised sizes, but from different areas throughout the country, were increasingly easier to distribute. This, combined with increased education, training and the increasing professionalisation of the building industry significantly reduced the local distinctiveness of brick buildings in the latter half of the nineteenth century. Refer to Chapter 6 for a more detailed discussion of bricks and brickwork.

Buildings made of mud

Solid wall construction using unburnt local soils – the most universally available building material – has a long history of use. There are a range of techniques:
*   *turf (walls of stacked earth with its grass in place);*
*   *clay lump or adobe (unfired earth bricks bonded with earth mortars);*
*   *rammed earth (walls formed by ramming earth into in-situ formwork);*
*   *cob (muds mixed with straw and sometimes lime as a binder) stacked up in layers and trimmed. Cob is also known as 'clum', 'clob' and 'wichert' in various localities.*

Other than turf walls, all of these vernacular techniques are seeing a revival because of concerns about the environmental impact of more popular materials as well as the loss of local distinctiveness.

This sub-section focuses on the most commonly found technique for vernacular buildings in the UK: cob.

Cob is a simple building technique which requires a suitable earth, ideally a mixture of a clay (15–25 per cent) and sand/aggregate (75–85 per cent). Suitable soils were worked up by mixing the soil (often using animals as well as humans to 'tread' the mix) with chopped straw and by adding water to the mix. The straw is added to provide a matrix to reduce the impact of shrinkage. The water is necessary to produce a plastic, workable and consistent material for the construction of walls. On some occasions, and where available, lime or chalk was added as a binder to provide added strength.

Slightly eroded cob on a dry stone plinth wall – the typical lime-wash finish has long since disappeared. Mud buildings are still considered to be relatively inferior and this can militate against their conservation.

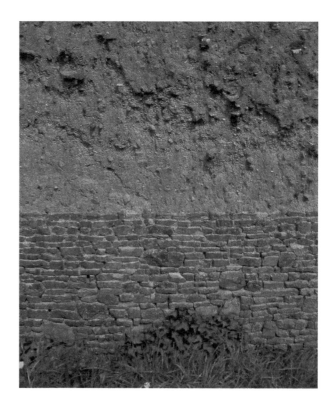

Once mixed to a workable consistency, the cob was simply piled into a layer, or 'lift', on top of a suitable stonework plinth. The plinth was necessary to minimise the contact between the cob and the damp earth. Once the lift (typically 150–300mm) had been placed and then trodden down to compact it, it would be allowed to initially dry so that it was firm enough to be trimmed to bring it to an appropriate thickness. Typically this varied between 300–1000mm. Once dried – anything up to four weeks per lift – the subsequent layer would be added and the process continued until the appropriate height was achieved.

Lintels for openings were built in as work proceeded, with the opening itself often being cut once the walls were dry and complete. Because there was no formwork and the mud was added in thick layers trimmed by hand, the corners of cob buildings tend to be rounded, with soft curving lines. This 'organic' sinuous form is characteristic of cob buildings.

On this reconstruction of a typical medieval cob building the various layers, or lifts, of cob can be seen. The whole of the outside has been lime-washed to provide additional weather protection; lime-washing also helped as an insect repellent. The idea of the importance of a 'good pair of boots and a good hat' can clearly be seen in this building – wide overhanging thatch eaves shed water away from the relatively vulnerable mud walls. At the foot of the wall, the discoloured lime-wash indicates the height of the plinth wall which is intended to provide a firm foundation as well as to avoid moisture from the ground and from rain-splashing eroding the base of the wall. Note the small window openings which were typically closed with timber shutters in lower-class buildings prior to the eighteenth century (when glass came into widespread use).

To minimise the mud wall's exposure to moisture and precipitation, they required 'a good pair of boots and a good hat'. The plinth wall, usually 300–500mm in height, reduced the chances of rain-splashing, capillary action and rising damp eroding the base of the wall. At the wall head, wide over-hanging eaves were ideal – usually in thatch. This was detailed so as to throw precipitation away from both the head and face of the wall. The walls themselves were sometimes lime-rendered, but were frequently lime-washed (a mixture of air-drying lime, water and tints) to achieve on-going moisture resistance. The use of porous lime-based coatings also encouraged the evaporation of any moisture within the body of the wall, so-called 'breathability'.

Mud buildings generally had a low-status reputation. Indeed, they still have such a reputation – this is a significant barrier to their conservation. The surviving examples tend to be relatively low-status rural workers' dwellings, agricultural buildings and garden walls. There are regional variations in vernacular mud building techniques using cob. In the northern Home Counties, cob construction is known as wychert. However, it is the south-west of England, and Devon in particular, that has the largest number of cob houses. It is in Devon, too, that there are numerous examples of relatively high-status cob buildings.

The approach described earlier is the most widely used technique, but there are variations. In the east of England, notably Leicestershire and Lincolnshire, a number of 'mud and stud' houses remain. These consist of simple, lightweight timber frames. Riven or split laths are fixed to this frame. Then, a thick cob-type mix is daubed onto both sides of the laths and the frame, and mud is used to provide enclosure and strengthen the frame by transferring part of the load of the roof (and upper floors, if there are any). Similar techniques using full or part cruck timber frames which independently support the roof have been identified in parts of Cumbria too. In other parts of the UK, similar load-bearing frames independently support the roof and lightweight wattle walls form the core of earthen wall structure. All these techniques have the advantage of being a faster building method than the multiple lift approach to cob building adopted in the south of the country.

Although mud building in the past was undoubtedly associated with poverty, its relative cheapness gave rise to a number of short-lived revivals in the nineteenth and twentieth centuries. There has been an increasing interest in the conservation of the remaining earth buildings in the UK, and Devon is the centre of a significant revival of the use of this ancient building technology – the motivation now is environmental impact, the sculptural and distinctive qualities and the motivation to support local rural enterprise.

A typical cob cottage from Devon in which thatch, a plinth wall hidden under a lime render and lime-wash complete this beautiful example of a local vernacular technique.

Buildings made of stone

In contrast to mud, stone has long been associated with durability and status. Some stones have the distinct advantage of being easily winnable from the earth, even appearing in relatively usable form at the surface. Other stones have to be worked out by quarrying or mining, which has influenced the extent to which they have been used in the vernacular tradition. The balance of durability and ease of working to enable it to be used for building purposes has had a significant influence on the extent to which particular stones have, or have not, been used in the past.

By the late sixteenth century, stone quarrying and mining had become more widespread and stone became more widely used in vernacular building. There is some evidence for the idea of a Great Rebuilding occurring across England between 1550 and 1640. Many timber-framed buildings were refaced in masonry, i.e. in brick, where it was being produced, and stone where it was easily available. Many new stone buildings were also erected.

In freestone areas, where stones are relatively easily quarried and easy to subsequently dimension, ashlar was commonly used around openings, for water-shedding elements like hood and string moulds, for the structurally important quoins and occasionally for entire façades.

Ashlar is a style of stonework in which finely cut stones of uniform height, with true squared face surfaces, are laid with very thin joints (3–4mm). Ashlar stones are expensive and used in relatively thin dimensions (100–175mm was common) and, as such, were always 'backed' with a lesser material (random rubble and later brick) to make up the finished thickness of the wall. In order to bond the ashlar to the backing, occasional bonding stones were provided. These were commonly a piece of ashlar returned at 90° to the face of the wall and running the full thickness of the finished wall. Later, bonding stones were omitted in favour of iron cramps or ties, which bonded the two halves of the wall together. Ashlar was the most skilled and expensive of stonework, demanding care in its planning and execution.

Fine-jointed (3mm) limestone laid in a lime mortar. Many limestones and some sandstones – both sedimentary rocks – are described as 'freestones' meaning that they are relatively easy to dimension and this has encouraged their use as ashlar. In this example, the entire façade is in ashlar. It is backed with rubble to make up its overall thickness. Ashlar is also commonly used in combination with less expensive masonry for the structurally significant quoins, window and door openings and for horizontal water-shedding elements, e.g. sills, string course and cornices.

The majority of vernacular stone work was of rubble construction. There were various forms of this.

Random rubble

This is the simplest of stonework techniques. Here the individual stones are laid roughly level, with little attempt to produce horizontal courses. Stones would be of irregular size, and there would have been little attempt to shape, or dress, the stones.

All rubble walls are built as two halves with the two faces being as flush as possible and with the irregularities of the stones meeting in the middle of the wall. Gaps between stones are filled with smaller stones and copious quantities of mortar. Continuous vertical joints are avoided as much as possible to try to bond the wall up its height. In order to bond the two 'halves' of the wall together through its thickness, 'through' stones (running through the thickness of the wall from face to face) and 'bonding' stones (running through at least two-thirds the thickness) are required at regular intervals. Without these, the two halves of the wall can act independently.

Random rubble is not ideal at the most structurally critical areas of stone walls: the quoins and around window and door openings. Here large, more substantial and regular masonry were often required. Bricks, ashlar or more carefully dressed and selected stones were commonly used here.

There is significant evidence that the overwhelming majority of vernacular houses with random rubble walls would have had them rendered, or at the very least lime-washed, or both, for waterproofing and decorative purposes.

Left: Random rubble walling. Right: The nature of rubble construction is revealed. The requirement for two flush faces on either side of the wall results in the core of the wall potentially containing significant voids and needing both 'sides' of the wall to be effectively bonded together. This requires occasional 'through' and 'bonding' stones to link the two halves together.

Coursed rubble

This is a more sophisticated form of rubble construction. Stones are dressed to enable greater regularity of both the horizontal courses height and vertical perpendicular joints. If there are any small gaps, these are filled with smaller stones and mortar. Again, occasional through and bonding stones link the two halves of the wall together.

Left: A coursed limestone rubble wall. Right: The interior is again revealed. Note all the issues regarding the need to tie together the 'two' halves of the wall and the requirement for smaller stones and plenty of mortar to fill the voids.

Squared or snecked rubble

In this type relatively finely cut stones are used to enable them to be laid on accurate and regular horizontal and vertical joints. The horizontal courses were not of uniform height. The term 'sneck' referred to smaller stones used to make up the height in courses using different sized stones.

Two examples of snecked sandstone walling. The example on the right has a bed mortar coloured with stone dust which has been 'tuck pointed' with a 1:1 lime putty and very fine aggregate mortar. This has been carefully added to a groove which was ruled into the unset bed mortar. This gives the impression, from a distance at least, of a finely jointed masonry wall – a desirable but expensive proposition.

Sedimentary stones

The most widespread group of stones which have been used in building are the sedimentary stones. These limestones and sandstones were formed over geological time as particles from erosion, chemical and biological processes accumulated in shallow lakes, seas, rivers and even deserts, and underwent chemical and physical changes: compressing and cementing the particles together. As they were laid down in layers (often referred to as beds) over time, these stones have a series of laminations. This has the dual effect of making them easy to split, but also makes them subject to potential defects associated with the ease of delamination.

Limestones are used to a greater extent than sandstones and this is a factor due to their geographical location, their ease of use and durability. Essentially limestones are made from calcium carbonate. Their character varies according to their origin. The organic limestones are formed from shells and other body parts of tiny sea creatures and other organic materials. The inorganic limestones are formed from chemical precipitation and have no organic matter. Jurassic limestones, which are found in a wide band across the country running from Dorset's coast (Portland being the most famous quarrying location) right up to the north-east coast in Yorkshire, are the best-known and widely used limestones for building. Chalk is also a soft form of limestone and was much used in the past (together with harder limestones where these were more available) to be kilned and slaked into lime the binder.

The commonest constituent of sandstones is quartz with little or no organic material. The quartz is held in a matrix made up of varying materials, but most commonly silica and calcium. The siliceous sandstones form the best and most durable building stones. The colour spectrum of sandstones ranges from green to purple and the texture too varies considerably. Sandstones in the west and north of the country are the most renowned. Sandstones are used in all forms of stone construction from ashlar to random rubble.

Granite is a very durable and particularly hard igneous stone. As such, it is difficult to cut and carve. It is common in the south-west of England, Leicestershire and Cumbria.

Flint is found in mostly in the south and east of the country as bands of hard nodules of irregular size and shape in chalk deposits. Here, because of the lack of better stones, it was quite widely used in building. The stone is brittle and is 'cut' by knapping or splitting. This process exposes the darker interior of this siliceous stone and the cut or split face laid flush to the face of the wall. Flints can also be used whole without knapping. Because of its relatively small size, it is laid as a form of rubble walling and will always require brick or coursed stone quoins or jambs. In better-quality work, stone or brick horizontal bands formed at regular intervals would ensure the wall was stable and effectively bonded.

Buildings made of timber

England and Wales have a fine inheritance of vernacular timber architecture. For thousands of years, indigenous timber species provided the main source of structural material for building. During this time a management system – coppicing – developed for woodland which provided a renewable and sustainable supply of timber and woodland products. Medieval carpenters were supplied with timber from a commercially managed woodland economy that was already ancient.

Timber's innate flexibility and the fact that many frames seasoned slowly (over many decades) in situ can be misinterpreted as a defect. Sometimes this is the case, but frequently the distortion is cosmetic only and adds a significant and unique character to many timber-framed buildings. Understanding the construction and the tradition, together with experience, is critical to making judgements about what is, and what is not a defect.

The typical image of a timber-framed house is a building with exposed black and white timbers. All of these buildings are timber-framed. However, timber-framed houses are often covered with a variety of materials. Occasionally, it is possible to find that what looks like a masonry building is in fact a timber frame hidden behind a retro-fitted façade.

The predominant timber species used for structural purposes in England and Wales was oak. This is an immensely strong, durable and long-grained timber. Elm, sweet chestnut, hornbeam and other timbers were also used. Typically timbers were used unseasoned, shortly after felling. 'Green' oak and other timbers were considerably easier to work and joint than seasoned timber and, given the typical cross-sections, air seasoning (the controlled evaporation of moisture) would have been impractical. The timber would season slowly in situ and quite often slowly takes up the gradual distortion caused by subsequent substructure movement.

In the twelfth century, a significant change occurred in timber construction techniques. Prior to this, builders had fixed and stabilised timber structures by sinking the posts into the ground. While these 'earth-fast' systems provided a stable structure, it was at the cost of their longevity. The new technique where posts were jointed into a sill beam, which was raised above the damp ground on a plinth wall, was developed and became widespread. This arrangement has become known as the 'timber framing'. It provided buildings of far greater longevity, but also required a higher level of technical competence.

Timber framing is a classic vernacular technique as there is no known historic text explaining the craft of timber framing and it is clear that the builders of timber-framed buildings worked on a practical basis, resolving structural problems as they arose. The majority of surviving timber-framed buildings of all periods were constructed for the upper classes. This reflects the build quality which could be afforded by this group rather than a constructional preference. There is also very significant variation in approach in different localities across the British Isles. The

The sill beam reduced the degree to which the vulnerable end grain of timber was exposed to saturation; it was a simple but revolutionary change.

builders of these structures had no recourse to theoretical calculation; measurement was based on proportion and made use of the developing knowledge of geometry. Problems were solved through a combination of experience (including failure) and knowledge passed from generation to generation orally.

Frames and bays are essential terms in understanding the general structural arrangements of timber-framed buildings. Frames can be designed in a variety of forms and are the points in the structure where loads from the roof and walls are collected and transferred to the ground. Frames are also the point at which the structure of the building is tied together across its span. Cross-frames are found within the building and end frames for the flank or gable walls. Bays are the spaces between frames. Thus, a two-frame house will have a single bay, a three-frame house will have two bays, and so on.

Essentially timber frames consist of a series of jointed vertical, horizontal and inclined timbers:

- *Vertical timbers were either* **posts** *– carrying substantial loads as part of a cross- or end frame – or* **studs** *– lighter-weight intermediate uprights supporting or connecting horizontal timbers.*
- **Horizontal timbers** *(known as beams or plates) span along both the length and across the width of the building. A* **sill plate** *sits on the plinth wall and all the lower ends of studs and posts are jointed into it. A* **wall plate** *has all the tops of studs and posts jointed into it, but it also provides the location for the feet of the rafters. A* **tie beam** *connects the cross- and end frames across the width of the building. It is frequently connected to the wall plate too and reduces the likelihood of roof spread at the feet of the rafters.*
- *Inclined timbers form both the roof structure and are also used to avoid structural deformation.* **Principal rafters** *form part of a triangular roof truss – describing the pitch and profile of the roof – the primary structure unit which forms the upper part of the cross- and end frames. Individual* **rafters** *support the roof covering and are fixed at their apex to one another or sometimes a ridge beam and, at their feet, to the wall plate; the wall plate would be connected to the top of the vertical posts and studs. An essential structural element of this form of construction was the pairs of inclined* **braces***. These stiffen the frame and provide a structural connection between the posts and horizontal timbers. This form of triangulation was essential to keep the frames plumb, level and square. They are required to be in pairs to avoid deformation in either direction. Braces are found in each frame element, i.e. cross-frames, end frames, in the roof structure and along wall frames.*

The most commonly used joint in timber framing was the mortice and tenon. The mortice is a squared slot or hole in one of the timbers while the tenon is formed by reducing the

This building is an unusual survival of an urban timber-framed terrace – all the structure described above is on display here. This is a seven-bay building. You can count these as they are the spaces between the end frames and the cross-frames. These frames 'collect' loads and transfer these vertically down to the ground. They can be identified along the upper storey of the principal façade as their posts have the pairs of curved triangulating bracing jointed into them to provide stability to the frame. The flank and gable of the building – the end frame – also has opposed triangulating bracing to provide stability across the end frame. This bracing is fixed to the two corner posts and also to the tie beam. The lowest horizontal timber is the sill beam into which the posts and studs are fixed.

Triangulating braces (which connect the horizontal and vertical) are also often found within the roofs of timber-framed (and masonry) houses. This right-hand photograph shows a series of wind braces connecting the roof trusses to the roof purlins. These have also been treated as a series of decorative 'cusped' wind braces.

end of the other timber to fit into the mortice. Like all the joints in timber framing, they were secured by tapered timber pegs. Other joinery includes scarf joints, which were used where long horizontal timbers were required, e.g. sills, purlins or wall plates.

The most complex joint in the timber-framed tradition was the structurally critical meeting of the post, wall plate, tie beam and principal rafter. This is known as the tie beam lap dovetail assembly.

The cross-frame and end frame, which collect all the building loads, have to resist significant forces. A critical point for load transfer was at the joint between the principal rafter, the inclined timber of the roof truss, the tie beam, running across the width of the building avoiding roof spread, the vertical post and the wall plate which supports the common rafter feet. This complex joinery arrangement is unique to the UK's tradition of timber framing.

It is possible to divide timber-framed buildings into four building types:

- **Box frame** *construction consists of wall frames connected at bay intervals by cross-tie beams. The roof is a separate structural element which bears onto the external walls, the roof acting as a lid on a box. Box frames have rafter roofs with no purlins supporting the rafter at mid-span. Although the tie beams helped to avoid the weight of the roof spreading the walls outward, collars were added to the pairs of rafters to lessen the spreading action of the roof. Box framing was superseded by the more structurally efficient post and truss framing.*

A rare and high-status roof – the Crown Post. This roof is only found in high-status and relatively early box frame buildings. It is not a truss, as such, and is used with rafter roofs, i.e. those where there are no purlins.

- *The **post and truss** form is by far the most common surviving timber-framed building form. They make use of the purlin roof. Roof and wall elements are structurally united within each cross-frame, which also includes tie beams, as they do in box-framed construction. The important difference in post and truss construction is that principal rafters are jointed into the tie beam and this forms roof trusses to carry purlins which in turn support the rafters and roof covering. A wide variety of truss forms exist, dealing with a variety of differing requirements and situations. The king post truss, for example, developed in the north of England where heavy stone slates used for roof covering required a structurally robust roof form.*

- **Aisled construction** *provides additional internal space by dividing the structure into a central nave and two side aisles. Wall posts become internal arcade posts, and the aisles are roofed over with lean-to roofs at an angle matching the main roof. The incidence of aisled construction is predominant in south-east England. Most examples of domestic aisled construction pre-date 1400 and were largely examples of prestigious buildings; from this point onwards, it became principally an agricultural form, but no less spectacular for that.*

- **Cruck construction.** *Crucks are cross-frames made up of pairs of inclined timbers, most frequently book-matched halves of the same tree, which meet at the apex and are tied together by some form of collar. Typically this 'A' frame supports the purlins and the wall plates. Thus, crucks transfer the load of the roof to the ground. This means the external walls are not necessarily structurally significant and frequently act simply as an enclosure. There are over 3000 examples remaining, found almost exclusively in the west and north of England and in Wales. This region contained the more marginal agricultural lands, was home to a poorer peasantry and a lower density of both occupation and building. Here larger, more mature trees and more curved timbers available from the less intensively managed woodlands played a greater part in the development of a regional style. Crucks are economic in their use of timber.*

Found in the north and particularly the west of the UK, cruck frames appear to have been developed from the mature curved trees, typically found in the less intensively managed woodland which was characteristic of this region in the Middle Ages.

The majority of timber-framed buildings were built with externally exposed timber frames. The most common form of enclosure was applied between the exposed timbers of the frame and is known as wattle and daub. It consisted of vertical timbers or staves fixed between two horizontal members of the timber frame, which supported timber laths woven between them. This element is the wattle. It supported the daub which was a rough cob-type plaster which was thrown onto the wattle and finished flush with the external and internal face of the timber frame. The most common treatment following the application of wattle and daub was to lime-wash both the panels and the frame.

Brick nogging – the use of brick between the timbers of the timber frame – was mostly a retro-fit, often replacing wattle and daub panels. This finish can cause problems – load and water retention have caused some serious defects to timber-framed houses.

In the east of the country, plaster was applied to laths covering the external surface of the timber frame and a moulded finish to the lime plaster was created. This took the form of a pattern or elaborate mouldings and is known as pargetting.

Window frames and door frames were incorporated into the timber frame itself rather than as separate components, as is the case nowadays. The glazing of windows became increasingly common in the early 1600s; until then most windows were open to the elements with protection being provided by sliding or hinged internal or external shutters.

The left-hand pair of these late timber frames has had brick noggins retro-fitted at some point in the past. On the right-hand side, the wattle and daub panels have been retained.

Timber frame pargetting.

Left: Timber frame door.
Right: Timber frame window.

In the South, close studding – a structurally unnecessary number of studs – was commonly adopted. This seems to be a 'conspicuous consumption' approach in this, the wealthiest part of the country. In the North and West, square panelling was more commonly adopted.

Top left: Close studding. Bottom left: Large square panels have been adopted. Right: Small square panels.

As towns and cities expanded, the plot size of dwellings reduced. Land was expensive and building plots were generally long and narrow. This required the maximisation of space; where there was a need for additional space, this had to be created by additional storeys. Particular solutions to these new design criteria developed. Jettying is one such solution. This is where the upper floor projects and overhangs the lower storey. It provided improved weathering characteristics and there were also positive constructional advantages in that only the internal end of the floor joists needed jointing into a beam. The projecting ends of the joists required nothing but optional decorative treatment. Jetties were also an external demonstration of the existence of an upper floor, something relatively few could afford and, therefore, there was an element of displaying the status of the owner. Indeed, many of the decorative features applied to the façades of timber-framed buildings probably originate in the towns. Although jettying developed in urban locations in response to increasing pressure on space, it was rapidly adopted in rural dwellings.

The tradition of timber framing fell into decline in the late 1600s and, by the 1700s, it became a rural tradition. Many timber-framed buildings, particularly in towns, were refaced with more fashionable (and more fireproof) materials – brick, stone tiles and render. Carpentry and the use of timber within buildings became more and more refined and largely became restricted to floor and roof framing and other elements within masonry buildings. While the same is true to an extent in other European countries, the rapid industrialisation and urbanisation of British society meant that the tradition of timber framing was almost completely lost. Other European countries have never entirely lost their tradition.

*All of these urban timber frames are highly
decorated: they are all jettied too.*

PART B: GEORGIAN AND REGENCY HOUSING (1714–1837)

The terms 'Georgian' and 'Regency' refer to specific historic periods between 1714 and 1837.
Following the death of Queen Anne in 1714, the ruler of Hanover, a small German state, become
King George I of England, and, together with subsequent similarly named monarchs (George II
(1727–1760), George III (1760–1820), George IV (1820–1830) and his brother William IV (1830–1837),
provides the name for the Georgian era of 1714 to 1837. Between 1811 and 1820, George IV, then
the Prince Regent, ruled in place of his medically indisposed father, and this specific period is known
as the Regency. It now also refers to a period of architectural and stylistic change which spans a
longer period – roughly 1790–1840 – where Georgian style and building, which had dominated for
the best part of a century, underwent some stylistic and construction changes.

The Georgian and Regency periods were times of significant change – so significant that we
now refer to the two most dramatic sets of changes as the Agrarian and Industrial Revolutions.

The Agrarian Revolution provided for a significant improvement in agricultural production.
The enclosure of common land, changes in cropping techniques and invention saw food
production rise dramatically. This changed the structure of society, which at the beginning of
the period, was predominantly rural and in many ways still medieval, i.e. farmed on a small-scale,
subsistence basis. A growing population could be fed more efficiently by a production system
which required significantly fewer labourers, thus 'freeing' them up to move to the burgeoning
centres of industrialisation. The increased financial returns accruing to landowners could be
profitably reinvested into speculative schemes such as industrial production and property
development.

The Industrial Revolution saw dramatic shifts in population as manufacturing, together with
all the associated industrial developments – science and engineering, steam power, coal mining

and iron production – drew a primarily rural population to the new centres of production for work. Trade within the country expanded as significant improvements in the transport system (turnpike and toll roads, canals, and latterly, railways) provided better communications. Colonisation and slavery boosted international trade, which in turn provided new raw materials and increased the demand for manufactured goods and a further expansion of production. Between 1700 and 1750 the population of England rose from 5.5 million to 6.5 million, but between 1750 and 1840 it grew from 6.5 to 16 million. Housing these people, including a rapidly expanding new middle class, in new and enlarged existing urban centres was a challenge that was tackled entirely by the private sector with virtually no state support.

The market, the methods and the rate of house construction significantly changed from earlier periods: speculative housing developments on large tracks of agricultural land adjacent to existing town and cities were undertaken; new design ideas and the means to convey such new ideas, new methods and new materials were used and increasingly a national style, rather than the local vernacular, gradually emerged for the first time. In contrast to the fraught and turbulent political and social times, the style emphasised proportion, harmony and good manners and was based on an increasingly disseminated understanding of classically based Palladian design principles.

Architectural styles

Palladian design principles had arrived in England in the seventeenth century with Inigo Jones. He, like many other wealthy young men at that time, travelled around the remnants of the classical world found in Italy, and brought back an interest in and desire to use the classical

The construction of these two houses is typically that from the post-Great Fire of London style, i.e. the second half of the seventeenth century and first few decades of the eighteenth century. Classical motif has been applied. For example, the columns from various 'orders', the triangular and segmental pediments above the windows in the left-hand example) and the classical proportions and symmetry in both examples. The right-hand example has a 'Shell' porch which marks this out as being Baroque-influenced. Other features of note are the dramatic and emphasised modillion cornices at the eaves and emphasis at the quoins with rusticated quoin stones. Note also the sliding sash windows, with their exposed box frames; by 1774 these would all have been set behind a half-brick-thick rebate for fire protection. All the bricks on both houses are handmade and local to the site. Both roofs are exposed (rather than hidden behind a parapet) and have exposed dormer windows. In the right-hand example, a typical narrow terraced house, the door is placed at one side of the elevation so that a two-bay room – the parlour – could be located to the side of the entrance and hallway. This was a common arrangement throughout the terraces of the period, as symmetry was not always easy to achieve.

building style in England. Ideas and designs from the classical world had been of interest in England for a century before this. The early use of classical style in England was generally arbitrarily applied, simplistic and ill-proportioned. Inigo Jones translated the works of Andreas Palladio, a sixteenth-century Italian building designer who had carefully studied and recorded the remains of many Roman buildings. Jones used this information to develop clearly expressed and accessible design principles which he applied to several buildings he was commissioned to construct for King Charles I in London – the Banqueting House in Whitehall, the Queen's House in Greenwich. Inigo Jones' career was truncated by the Civil War and his association with the monarchy. Although some aspects of Palladian design were taken up in the post-Great Fire of London reconstruction, it was not until the early part of the Georgian period that a stricter adherence to Palladian principles really took off.

Early Georgian style (1714–1760)

Palladian design principles emphasise and rely upon careful attention to proportion. Proportion was of supreme importance. This applied across the whole of the house and in its individual parts, i.e. the entire façade, the individual elements such as windows, room widths, length and heights, and so on. Palladio also promoted the use of the classical 'orders', which form a significant part of the proportional and design principles of the Roman temples he had measured and studied. Uniformity within individual buildings and through terraces of buildings, and symmetry were also significant principles. Symmetry was compromised at times – particularly when it came to where to locate the main entrance door and get a usable front room on a narrow plot. Reapplying these classically derived design principles to domestic British architecture was a challenge.

Later Georgian style (1760–1800)

The strict adherence to Palladian proportion continued to be applied to façades and all other elements. In the latter half of this period the rate of house building increased substantially. In industrial cities an even less ornate Palladian style was common. In wealthier areas, the detailing of the elements became less uniform and this allowed for greater freedom of expression. The influence of discoveries and academic interest in Greek buildings, and later Egyptian classicism, provided a wider range of design possibilities – though these tended to be quite austere in nature. What really began to free up the obsession with an adherence to accuracy and uniformity of the past was the arrival on the design scene of the Adam brothers. Robert Adam particularly became the most influential designer of the period. He produced designs that took in the entire house and, though he was primarily known for interior design for the very wealthy, his influence affected many features and aspects of all types of houses and successfully challenged the orthodoxy of strict Palladianism.

Regency style (1790–1837)

The Regency period witnessed a range of new and diverse stylistic innovations. Palladian classicism still had a significant influence on many developments in terms of proportion but increased exploration and trade brought in Chinese and Egyptian influences too. The Romantic and Picturesque movements of the early nineteenth century drew inspiration from landscape, nature and medieval gothic architecture and the gothic style began to be revived for house construction. Some wealthier landowners built in a romanticised rural style – known as Cottage

Compare the photograph on the top right to that on the top left taken from an eighteenth-century builder's guide. Note the façade's symmetry and that there is a perfect square formed on the façade around the top six windows and that this 'harmonious' shape is reproduced in the top square windows. Note too that the form of a classical column is implied by the proportions and setting out of the façade. For example, the cornice at the start of the parapet wall reproduces the cornice at the top of a column. The entablature is ended at the head of the top (square) windows. The column's shaft is implied by the piers of stonework running down to a horizontal sill/string course at the sill of the lower window – this represents the base of the column. Beneath that, the 'pedestal' of the column is implied in the façade before the rusticated ashlar. All these are typical of Palladian design principles. The buildings below do not imply a classic temple, but have applied decoration directly imitating a classical temple. The bottom left photograph shows part of John Wood's Queen Square (1729) and displays clearly Palladian design. Here the classical orders and the elements of a Roman temple have been applied directly to the façade of this domestic building. This early 'palace façade' with its symmetrical elevation, its central triangular pediment and harmonious proportions, enabled the tenants of the individual terraced houses to enjoy the grandeur of being part of a spectacular ensemble. Palace façades became a frequently used technique for Georgian and Regency terraces – the predominant urban domestic form of the era. The example bottom right is from Liverpool and shows typical Georgian features in a brick palace façade. Liverpool, like other port cities in the Georgian period, benefitted from the increasing international trade, including slavery, and wealth that flooded in to speculative housing development in the period.

Orné – on their estates. Romanticism and a somewhat nationalistic search for the true spirit of Englishness both encouraged a re-examination of medieval gothic style. In the late eighteenth and early nineteenth centuries, 'Gothik' became a fashionable design idiom.

The use of the Doric order was constrained until the later Georgian period. Top: This rather austere building uses Greek Doric order to produce a severe and archaeologically informed façade. Left: More commonly, brick-built terraces with simple detailing based on Palladian principles became an increasingly widespread urban form across the country. Right: In the country, Georgian style was also popular: in this later-Georgian, rural example the prestigious nature of brick becomes apparent – stone gable walls with brick principal façades. Note that even here there are vernacular influences— the rubble gables, the stone slate eaves courses of the roof covering. It could well be that this façade was 'added' to an existing pre-Georgian structure as a serious up-dating. Such rebuilding processes are common throughout history.

Left: A typical Cottage Orné – which became a popular Romantically inspired style for the wealthy landowner to develop for their estate workers. Right: An early nineteenth-century Gothik house.

The biggest influence on style was probably the increasing demand for house building in urban centres where trade and industrialisation very significantly boosted demand, particularly after 1815 when the end of the Napoleonic Wars witnessed a series of building booms. This prolonged period of building, together with better communications and transport, led to an increasingly nationally adopted range of materials and elements on houses.

Brickwork was increasingly hidden behind stucco, a wet-applied render that was used to emulate more expensive stone, but also used to cover cheap and poor walling materials. Stucco could have simple lines incised into the drying mix to imitate ashlar, but it could also be moulded into complex and realistic imitation stone features.

A series of classic Regency houses. Left: High-class Regency end-of-terrace house in London displaying all the characteristics of the period, i.e. white lime-washed stucco, semi-circular multi-storey bay windows, ornate cast-iron and wrought-iron work, classical and elegant proportions. Middle: A middling-quality Regency terrace with stucco, classical proportions and Chinese-inspired pent roof above the delicate cast and wrought ironwork balconette. Right: A lower-class house with far fewer features and smaller, tighter accommodation.

The Regency period also saw the development of detached and semi-detached housing.

Left: This appears to be a masonry building, but is in fact a relatively rare Regency timber frame with glazed black 'mathematical' tiles. These are hung on horizontal battens fixed to the timber frame to emulate bricks – a way to avoid the despised Brick Tax levied in 1784 and only repealed in 1850. The wide overhanging eaves and low pitch slated roof were typical of Regency semi- and detached houses. Right: This is a more substantial detached Regency villa with ornate ironwork and railings and lime-washed stucco. Note that the stucco does not extend around to the side elevations, where facing brick is used.

The other significant technological development affecting style was the use of cast-iron and wrought-iron balconies, railings, fireplace and other architectural features which were widely produced by the new metal industries. Windows and doors became even more elegant, with reduced size glazing bars and more complex fanlights.

Housing development, layout and amenity

Throughout the Georgian and Regency periods, the predominant form of housing development was speculative. As the country industrialised, more marginal agricultural land close to industry or on the edges of existing urban centres became potential development sites. Landowners (generally the aristocracy and upper middle classes) sought to realise their land assets. Throughout the period, there was an obsession with speculation: housing and urban development provided many potentially lucrative possibilities. This was not a constantly expanding and rewarding proposition, however. Many fortunes were made and many were lost as the market took violent swings of fortune – sometimes very rapid swings.

With its roots in methods developed in the seventeenth-century housing market in London, speculative development in the Georgian period was generally undertaken by a principal developer who would take a 99-year lease from a landowner. The principal developer would agree a scheme (fashionable and likely to be desirable) with the landowner and would then sub-lease individual plots to smaller developers – often tradesmen – who would individually borrow to build to the agreed scheme. They would have a two-year peppercorn rent on their individual plots during which time they would hope to construct, complete and sell on their house – at a profit. It is astonishing to consider that many of the most uniform terraces from this period were built by several different builders – frequently a different builder for each separate house. The principal developer would then collect the ground rent from all the occupiers, paying the landowner his annual dues, and would hope to make a substantial annual profit over and above this. After 99 years, the landowner's descendants would own the freehold of the completed development.

The 'architect' (there was no official definition of this until the nineteenth century) and the builders were strongly supported by published and widely distributed building guides or 'pattern books' which provided drawn examples of appropriate and fashionable plans, elevations and a range of building elements. This affordable and easy-to-use dissemination of contemporary design ideas provided a very significant push towards the development of the first modern national, non-vernacular building style.

Although a range of different house types throughout the Georgian and Regency period existed, the commonest urban form of construction, and one encouraged by speculative development, was the terraced house. The terrace maximised the efficiency of the development process, maximised the profitability of a site and encouraged a greater concern with urban design to flourish. The Building Act 1774 set out national standards for the first time (previous Acts really applied to, or were only effectively taken up, in London). This Act established four classes of house and stipulated the minimum quality for construction: it led to standardised plan and construction forms.

At the front of the house, typical construction would include a front retaining wall topped with iron railings, forming a void below street level in front of the basement. This area often contained steps which provided a tradesmen's entrance directly into the kitchen, located commonly in the front basement room with a scullery at the rear. The subsequent plan of the house would be relatively simple, with one room at the back and one at the front on each floor with a passage and staircase at one side. There were variations, but this plan form was very common. In middle- and upper-class houses, the ground floor was reserved for reception of the public. The first floor, referred to as the Piano Nobile, was intended for formal and private socialising and had the highest ceilings and tallest windows. Upper floors were for bedrooms and the storey within the roof void was the servants' accommodation. The party walls usually contained the chimney flues – these added strength to the structure, as well as serving at least one fireplace in each room.

Left: A model of a substantial late Georgian townhouse. The slender masonry is buttressed by the timber floors, roofs and partitions. The original ground level was somewhere within the height of the basement level. This provided a substantial excavation in order to complete the basement structure which in turn provides a substantial masonry foundation for the subsequent superstructure. Right: A typical construction at the lower front of a Georgian townhouse. The excavated basement is protected by iron railings and forms an open area enabling light to penetrate into the basement rooms. These usually contained the service areas, kitchen, scullery, etc. Under the pavement, substantial vaulted masonry structures were commonly provided for each house. These enabled storage for a variety of household requirements – most commonly coal and other non-perishable items.

Construction

Construction during the Georgian period was primarily hand-made and, even on some of the larger schemes such as in London and Bath, was produced by small-scale developer/ builders who would have employed separate trades – and they were frequently a tradesman themselves. As industrialisation gathered pace in the latter part of the period, particularly in the Regency period, greater industrial production and improved transportation began to impact on both materials and building techniques. Larger-scale development (particularly in London as well as various spa towns, seaside resorts and port cities), the increased professionalisation and education of the building professions and trades, also provoked the growth of large-scale building contractors – or Master Builders – such as Thomas Cubitt, whose firm employed a full range of building trades and were thus not reliant on individual subcontractors.

Roof structure

Houses during this period had timber pitched roofs, generally of softwood imported from the Baltic. For some bigger structures, and where local supply was good, oak was also used. Shorter spans were formed with common rafters and ceiling joists, while longer spans incorporated purlins, which spanned between load-bearing walls, or timber trusses (which were installed at regular intervals where there were no internal walls to provide support to a roof). Trusses could be found in a variety of designs, but King Post trusses dominate. In larger town house terraces, it was common to find mansard roofs: these have two different pitches within the roof structure in order to provide accommodation, generally for servants. In order to span the depth of the houses and to avoid the need for a very tall pitched roof, thus negating the classical 'look', a double-pile roof was common, this was partially hidden behind a parapet wall to reduce its visual impact. In smaller poorer houses, a duo-pitched 'Butterfly roof' was adopted – again, this was designed to minimise the impact of having to have a pitched roof.

Roof coverings

At the start of the Georgian period, local vernacular roofing materials were common – clay tiles in the east, stone slates in limestone and sandstone areas, and so on. Natural slate became more and more available and in many urban settlements by the end of the period, it was dominant. In poorer housing, clay tiles (pantiles and plain tiles) were common and thatch continued in some country areas.

Roof drainage

Although some roofs drained into lead or cast-iron gutters and downpipes, many of the roofs of the period were concealed behind parapet walls that drained into lead-lined parapet gutters supported by softwood boarding, Valley gutters, which were formed on double-pile roofs, were also common. This central valley was frequently drained via a 'secret gutter' – an open lead-lined channel through one of the roof voids to the parapet gutter.

Double-pile roofs. The double roof and parapet walls create significant rainwater drainage problems and require frequent attention.

External walls

These were of solid construction. The most common material was brick. Facing bricks for areas on the principal façades, lesser-quality facing bricks for rear walls and often poor-quality 'place' bricks where they were hidden. Thickness ranged from 337 to 450mm for the lower storeys of multi-storey houses with 225mm thickness at higher levels or for houses of two or fewer storeys and also for cheaper houses. A 112.5mm (or half-brick) thickness was common for single-storey annexes. In gauged arches, 'rubbers', i.e. soft bricks which are easy to shape by rubbing in order to form purpose-made voussoirs, were relatively common.

In the latter half of the period, stucco – a wet-applied render – was commonly adopted for brick buildings. It became widespread during the Regency period where it was used to form mouldings and other architectural features reminiscent of carved stone.

External walls were constructed of a mixture of brickwork and stone, Sometimes, stone details were provided around door and window openings and for stringcourses, plinths, quoins and other details.

Timber lintels were used behind masonry arches to provide structural support above openings. Window sills were commonly formed in stone.

Many better-class houses had basements. Some had barrel vaults which extended out under the pavement in front of the house for the storage of coal and other household goods. These were constructed with thicker walls, at least 337mm if of brickwork and 500mm if of stonework.

Bricks

The industrialisation process, and hence its standardisation, started in this period. Brick sizes were relatively similar in the eighteenth century, but not standardised nationally. As a largely hand-made and localised activity, there could be significant variability in size, shape and quality.

Flemish bond was the most commonly adopted bond. 'Snapped headers' were commonly employed. Here the expensive facing bricks were either cut in half, or were deliberately made as half-bricks, and bonded into the wall to give the appearance, but not the strength, of Flemish bond. Garden wall bonds were used in hidden areas and on cheaper construction.

Left: Early eighteenth-century brickwork; note the variable bonding pattern – predominantly Flemish bond but with some variation too. Middle: Tuck pointing. To reduce the visual impact of variable brick sizing, the bricks were laid in a mortar whose aggregate was coloured to match the bricks. While still soft, the mortar had precise narrow lines incised into the bed and perp joints into which a thin band of pure lime putty was carefully pointed in. The result was brickwork which looked more precise and more finely wrought than it was. Right: A gauged arch. Each individual voussoir or arch-brick is purpose-made. This skilled work required soft bricks which could be carved, or rubbed-up, on site. This highly skilled work uses 'red rubbers' to create a fine flat arch with a minimal mortar joint.

Stonework

Local stone was often used where available and cost-effective, e.g. limestone in Bath, limestone and sandstone in Bristol, sandstone in Cumbria and Yorkshire, granite in Cornwall and Devon, flints in Sussex, rag in Kent. At the start of the period, it was only ever imported from one district to another where it was used for very prestigious and expensive housing; at the end of the period, transport improvements meant that stone was being transported more frequently.

Buildings could be constructed wholly of stone or, with, for example, ashlar veneer (150–200mm thick) and backed with either rubble or brick. Rubble walls were either solid stone through the thickness of the wall or, particularly in the latter half of the period, the rubble was backed brickwork. Rubble walls would have been finished in a lime-render and lime-wash or a series of lime-wash coats.

Mortars

Throughout the period, masonry, whether brick or stone, was bedded in lime mortar. Lime mortar hardens slowly and this meant that any initial settlement and much subsequent settlement of a building was accommodated by the walls. Most aggregates would be locally sourced and thus of variable quality, particularly, in poorer-quality houses.

Internal walls

Internal walls were normally formed with common bricks, sometimes in stonework. Non-load-bearing walls were frequently formed with 'place' bricks (i.e. improperly fired/soft bricks). Often, internal walls were only a half-brick thick, even for structural walls of tall buildings of several storeys. All walls were faced with lime plaster.

Internal partitions of timber studding covered in lath and plaster were common. These were braced with extra timbers where required to act structurally. Stud partitions were sometimes infilled with bricks-on-edge to provide sound insulation.

Foundations

These originally depended on the size and quality of the building being constructed. Cheaper housing had either no foundation or a shallow foundation, often consisting of a layer of flat stones, sometimes a bed of clinker. More expensive houses normally had some form of deeper foundation. Often this was a substantial basement and sub-basement structure. It was common for the sub-basement to be excavated, with the basement above that being built at the original ground level. This means that often the ground-floor levels of many Georgian and Regency houses is a storey or so above the original ground level of the site. This effectively means that the basement or sub-basement acted as a giant masonry foundation.

Windows, doors and joinery

Box-framed sliding sash windows in softwood were the standard approach in the period. Glass technology (essentially hand-blown 'Crown' glass) limited the size of glass panes available. The division of the two sliding sashes by glazing bars resulted and was the norm throughout the entire period.

The double-hung sliding sash window was commonly adopted and developed over the Georgian and Regency periods. Imported as an idea from the Netherlands in the seventeenth century, early Georgian sashes had thick glazing bars with quite crude features. It was not uncommon to find that just the bottom sash opened. Through the later Georgian period, sashes became lighter, more refined and the glazing bars finer and finer until the Regency period where daringly thin glazing bars are common. In the earlier period, the box frames of these windows were exposed and nearly flush with the façade. Over the period, and pretty universally after the 1774 Building Act, the boxes were hidden behind masonry rebates.

Doors throughout most of the period were of either four or six panels, comprising two square panels (or glazing sometimes) above two longer panels, above two shorter panels. Doors were usually of softwood and were intended to be painted. More expensive hardwood joinery of oak or mahogany was polished or varnished.

The images above show windows from the Georgian and Regency period. Top left: An oak non-opening window from a rural cottage. Top right: An exposed box sash with thick glazing bars from the early Georgian period. Bottom left: A post-1774 late Georgian sash behind a brick rebate. Bottom right: A Regency sash window behind an ashlar rebate, which has been chamfered to improve light; also note the very thin glazing bars, with a meeting rail (between the top and bottom sash) as narrow as the glazing bar itself.

Upper floors

These were constructed with timber joists (usually of softwood), herringbone strutting and square-edged softwood floorboards. In larger houses, double 'framed' floors were used. Here the floor joists do not span the whole distance, but are jointed into a larger section beam running at right angles to the joists – this enabled large open spans in the rooms below. Ceilings were formed with lime plaster over softwood laths – additional plaster mouldings – cornices, friezes and other decorative mouldings were commonly added to enhance the spaces.

Ground and basement floors

Floors at ground level were commonly of suspended timber construction with softwood joists and square-edged softwood floorboards. Solid floors with flagstones, brick pavers or ceramic tiles over compacted earth were often used for rear extensions and in basements, where the kitchen and scullery in houses of all classes were found. In more prestigious developments, the floor of the main entrance could be finished in marble, Purbeck stone or ceramic tiles.

Services

Services and amenities were crude by today's standards, but very advanced compared to what had preceded them. Running water was very rare – collection and storage of rainwater, from springs and streams was very common. Drainage was primarily concerned with surface and rainwater disposal/storage. Sanitation was generally very crude with a privy or bog house covering a communal cesspit in the rear courtyard – shared with many people in working-class districts. Some wealthy owners had flushing toilets and WCs by the end of the period, but these were relatively rare. Coal and timber burning in chimneys and, latterly, in cast-iron chimney pieces and ranges provided heat, cooking, light and hot water production in all houses.

Typical defects of Georgian and Regency houses

These are discussed in Part D below.

PART C: VICTORIAN AND EDWARDIAN HOUSING (1837–1914)

The period between the accession of Queen Victoria and the start of the First World War falls into two distinct architectural periods: Victorian (1837–1901), covering the reign of Queen Victoria; and Edwardian (1901–1914), the latter also covering the first four years of the reign of King George V as well as the reign of King Edward VII.

When Victoria became Queen in 1837, the United Kingdom was still essentially a rural economy, but by 1914, it had become a fully industrialised nation. The Industrial Revolution of the late eighteenth century and the nineteenth century led to the mass movement of the population from the countryside to the urban centres where manufacturing took place. It created the need for huge numbers of houses to be constructed for workers, managers and employers. The total population roughly doubled between 1801 and 1851 and did so again between 1851 and 1911 to approximately 43 million people.

Much of the demand for new housing was achieved by speculative development. Landowners and builders often undertook this in a piecemeal fashion and this can be seen today where Victorian streets often follow original field patterns. Housing for workers was rapidly erected as cheaply as possible in streets of small and medium-sized terraced housing, generally of two to four storeys and sometimes five-storey buildings, many being subdivided to house several families. Separate streets of higher-quality and more spacious terraced houses together with semi-detached and detached villas were also constructed for managers and employers. The total number of houses rose by about five-fold between 1801 and 1911 (from approximately 1.6 million to 7.6 million). House building had mini-peaks in the 1830s, 1840s and 1850s. These were followed by major building booms in the late 1860s and the late 1870s as well as at the turn of the twentieth century.

Some of this Victorian property expansion did provide houses and suburbs of acceptable quality in terms of appearance, internal design and external space. Unfortunately, the cheapest housing, especially in the first half of the period, was of extremely poor quality and this resulted in many thousands of sub-standard houses constructed with questionable materials and design, e.g. the use in some areas of sun-baked bricks, the lack of a damp-proof course or a foundation. Such houses were also adversely affected by rising and penetrating damp, had little or no proper natural light and ventilation and normally had no water or toilet facilities, the occupants having to make use of communal standpipes and toilets. The extremely high density of much housing development was another problem and resulted in overcrowding and conditions like the back-to-back houses of many northern towns.

Part of the underlying causes for these types of development was that, before the First World War, 90 per cent of housing was rented. Landlords (not just speculators, but also factory owners) often had little or no incentive to provide buildings of better quality, while they had every

Terraced Victorian middle-class housing in Bristol showing that money was spent on the street elevation, with rubble and freestone details evident. However, cheaper brick was used for the flank and the (unseen) rear elevations.

Note the high density and the lack of off-street amenity in this example of back-to-back housing in Leeds.

incentive for high-density developments, as both factors directly affected the return on their property investment.

From the middle of the nineteenth century onwards, there was a gradual realisation and acceptance by various leading authorities and, finally, by Parliament, of the link between poor and overcrowded housing and ill-health. This resulted in various Acts of Parliament including the Public Health Act 1875. This, together with the introduction of various regulations and the 1877 model by-laws, set minimum basic standards of design and construction as well as layout for new housing. Minimum requirements and standards were also set for sanitation, water supply, natural light and ventilation as well as the prevention of damp. In modern terms, these standards were relatively low and they were not applied retrospectively to houses already in existence. Subsequent legislation by central and local government in the remainder of the nineteenth century as well as the twentieth and early twenty-first centuries has periodically raised these early standards to the current level of building standards and regulatory control.

A terrace of Victorian working-class houses in south-east London.

Although speculative building at the lower end of the housing market still took place, the new regulations led to better standards of construction generally, especially at the upper end of the housing market. Many houses built in the last decade of the nineteenth century and the first decade or so of the twentieth century (the Edwardian Period) are considered by construction experts to be of the highest quality in terms of workmanship and in materials, in particular, the timber and craftsmanship used for carpentry and joinery.

These two Edwardian houses show the extravagant level of detail that is typical of many houses of the period. The influence of the Arts and Craft Movement can be clearly seen.

Architectural styles

A number of different architectural styles were developed over the period for the appearance of better-quality housing for the middle and upper classes throughout the whole period. These were rarely reflected in the design of cheaper housing, which was often of unadorned brickwork or render.

The relatively unadorned flank wall of a worker's house constructed in 1884 – very few builders are as helpful in dating their work.

In brief, the range of architectural styles included:

- **Classical.** *Inspired by classical Greece and Rome, the houses had symmetrical design features, their façades including columns and pediments and often involved the use of stucco.*

Left: A terrace in Clifton, Bristol, was constructed in freestone in the classical style in 1875.
Right: A classica-style terrace with stucco finish.

- **Gothic revival.** *Based on medieval buildings, the heyday of gothic revival was between 1855 and 1885. Such houses often had an extravagance of detail to their roofs and elevations, asymmetrical features and pointed arches above external openings.*

Left: The Foster's Almshouses in the centre of Bristol were constructed between 1861 and 1883 and are an example of French Burgundian Gothic Revival Style, being modelled on the Hôtel Dieu in Beaune, France. In 2007, they were converted into private residences. Right: A slightly more restrained example of a Gothic Revival house.

- **Old English.** *More likely to be used for rural housing, this style had steeply pitched roofs, pointed gables with finials and much fancy detailing as well as false half-timbering, usually painted black.*
- **The Arts and Craft Movement.** *An influential, but to some degree peripheral style, this movement, led by William Morris, flourished from the 1880s until the 1930s and was inspired by the writing of Augustus Pugin and John Ruskin. It was arguably an anti-industrial movement and was based on the use of local building materials by traditional craftsmen and a return to medieval, Romantic or folk styles of decoration. It had considerable influence on the Garden City Movement and on the designers of Garden Suburbs.*

Examples of old English-style architecture, the houses in the right-hand photograph being rather late versions and dating from about 1895.

Arts and Craft houses; the right-hand pair of semi-detached houses were built in suburban Bristol in 1908.

Housing layout, landscape and amenity

From the 1860s onward, the quality of the environment within which a house stood was also recognised as an important factor of housing quality. Dense housing layouts plus the lack of natural landscape as well as the lack of amenities in terms of private gardens, public open spaces, trees and the like were identified as being detrimental to both the health of the occupants and the quality of life generally.

The initial impact of this enlightenment was on middle-class suburban development. In 1875, Bedford Park in Chiswick, West London, was developed. Whilst not a true **Garden Suburb**, it was an early forerunner as it included houses of different designs and appearances and, although the layout was rather too cramped, the houses did have the benefit of being set in an informal and unregimented environment that retained existing trees. Bedford Park also had some influence on future house appearance, especially in the period 1890–1910, with its use of gables of red brickwork and/or tile-hanging and multi-paned leaded windows. However, most developments continued to be on the basis of the grid-iron layout of straight and parallel 'by-law' streets beloved by local authorities of the time, with the focus on terraced housing (which comprised almost 90 per cent of all housing by 1911).

Bedford Park had a considerable influence on a number of enlightened employers who sought to improve the living conditions of their employees. As a result, a number of industrial **model villages** were created, including Port Sunlight, Cheshire, in 1886 (Lever Bros), Bourneville,

Birmingham, in 1894 (Cadbury) and New Earswick, York (Rowntree) in 1904. Such villages ranged in size from a few hundred inhabitants to the 25,000 of Bourneville, where the houses were of traditional design, but with modern interiors and large gardens and set among parks and recreation areas.

Typical cramped Victorian working-class housing: a terrace with rear annexes and small back yards.

These two photographs show houses and streets in Bourneville. Note how green and leafy it is.

Another scheme for creating a higher-quality residential environment in terms of housing and an attractive man-made landscape was **Hampstead Garden Suburb** in North London. This development started in 1906 and was based on a planning design by Raymond Unwin, the joint designer of Letchworth Garden City. It was aimed at all classes of people and income and involved low-density housing with natural boundaries of hedging rather than walls, and wide, tree-lined roads with woods and public gardens.

In 1898, Sir Ebenezer Howard inspired the **Garden City Movement** with his book *Tomorrow: A Peaceful Path to Real Reform*. A social and political reformer, he argued for a self-contained city for 32,000 people planned on a concentric pattern of open spaces, public parks and greenbelt, plus six radial boulevards and better-quality housing for the working classes. From 1904, he developed **Letchworth Garden City** in Hertfordshire, although the final design was more 'organic' and less symmetrical than Howard's original proposals. The next Garden City was not developed until after the First World War (Welwyn Garden City, again in Hertfordshire, in 1919) and further Garden Cities have subsequently been developed in the UK and elsewhere in the world.

Construction

Construction during this period was based on a mixture of traditional and newly introduced building techniques and materials. The industrialisation of manufacturing processes linked to mass production and efficient transportation by the increasingly extensive railway system reduced the cost of materials and components. Research and development of chemicals, metals and other base materials led to both new building materials and new ways of using existing materials. However, the basic principles of domestic construction did not change a great deal from those used in the seventeenth and eighteenth centuries.

Top left: An example of cheaper, but reasonably comfortable, Victorian housing. Top right: Better quality housing with added constructional details such as the parapet walls over the party wall line. Bottom left: A long stepped terrace (to take account of the sloping ground) that has butterfly roofs behind the front parapet wall. Bottom right: Butterfly roof construction and the variety of typical extensions that were frequently constructed behind Victorian terraced houses.

Roof structure
Timber pitched roofs, generally of softwood, were formed with common rafters and often incorporated purlins. Trusses, including King Post trusses and Queen Post trusses, were commonly used. Mansard roofs were sometimes formed to provide accommodation or storage. Terraced houses often had a butterfly roof construction, the roof slopes draining to zinc or lead valley gutters that ran from front to back.

In speculative housing, flat roofs were only constructed for small projections, such as above bay windows. They were formed with softwood joists and boarding and covered with lead or zinc sheeting.

Roof coverings

These were normally of clay tiles or slates although the use of thatch or stone slates continued in some country areas. Clay tiles were supplied in a number of profiles including pantiles and plain tiles. Both of these had been in use for some time, but fell out of general use in the nineteenth century, although clay plain tiles became popular again in the Edwardian period. On the other hand, single and double roman clay tiles, first produced in the nineteenth century, were extremely popular roof coverings. Slates were produced in their millions during this period in Cornwall, Wales and the Lake District, and because of cheap rail transport were used as a roof covering throughout England and Wales.

Roof drainage

The roofs of cheaper houses drained to cast-iron ogee gutters fixed to fascia boards secured directly to the face of external walls. Parapet walls were commonly formed around the roofs of more expensive Victorian houses (and over party walls more generally) in the nineteenth century, but fell out of favour in the Edwardian period. Roofs behind parapet walls drained into parapet gutters of lead or zinc over softwood boarding, zinc being used for cheaper construction. Valley gutters, which were formed between two or more adjoining roof slopes, and hidden box gutters (which drained through a roof void) as well as flashings and soakers were also formed in lead or zinc.

External walls

Normally of solid construction and in a range of thicknesses similar to those set down for Georgian houses (see Part B). Early forms of cavity wall construction (known as hollow walls) were sometimes built from the 1830s onward, initially with vitrified clay wall ties, latterly ties of cast or wrought iron; sometimes, these metal ties were dipped in tar.

External walls were constructed of brickwork or stonework, sometimes a mixture of both. Frequently, stone details (or bricks of a different type) were provided around door and window openings and for stringcourses, plinths, quoins and other details. Other materials, such as terracotta (fine-grained and very durable unglazed fired clay of various colours) or faience (glazed fired clay) were also used for moulded facings and details. Vertical tile hanging became popular in the Edwardian period.

Brick, stone or terracotta arches provided structural support above door and window openings. Window sills were commonly formed in stone or terracotta. Timber lintels were commonly provided behind the masonry or brick external arches.

Many houses had basements, which were formed with thicker walls, as with Georgian houses (see Part B). Sometimes, external walls would be coated with asphalt externally (as an early form of vertical damp-proof membrane) before being covered up.

Bricks

Brick sizes gradually became standardised from 1850 onwards and, in 1859, mass production of bricks commenced. This enabled many varieties of machine-made brick to become available, e.g. Flettons from the 1870s.

Flemish bond was often used for more expensive property, while English bond was used for cheaper, speculative construction. Snapped header bond was also used where more expensive bricks with snapped headers (sometimes called blind headers) were used for the outer course, secured every so often to a solid inner wall so that it replicated Flemish bond in appearance. Another form of brick construction was rat-trap bond, where bricks were laid on edge in two skins and bonded by a through brick, i.e. a form of early cavity; it formed a cheaper, but not strong, wall.

Stonework

As in earlier periods, local stone was often used, especially where it was a readily available and cheap material. It was also transported to other parts of the country for more expensive housing and, as the century progressed and transportation costs dropped, for cheaper housing.

Buildings could be constructed wholly of stone or, frequently for reasons of cost, were built of brick with stone details. Often, a stone face was applied to a brick or cheaper stone backing. The stone could be finished in various ways, e.g. ashlar (smooth-faced) or rubble (crudely finished) and built in courses or in a random fashion.

Mortars

Throughout the nineteenth century, bricks were bedded in lime mortar, which was also used for stonework. Lime mortar sets slowly and because of its inherent flexibility can accommodate reasonable levels of initial settlement and other movement within a building.

Portland cement (PC) was invented in 1824 and by the end of the nineteenth century the use of PC mortar for brickwork was becoming popular. However, it was not in widespread use for residential buildings until after the First World War. However, while it was cheaper than lime mortar and had a fast set, it is a less flexible mortar, cracks more easily and is less tolerant of building movement.

Internal walls

Construction was similar to that used in the Georgian period, i.e. internal walls of common bricks or stonework; non-load-bearing walls of place bricks, i.e. improperly fired/soft bricks; walls normally only a half-brick thick, even for structural walls of tall buildings of several storeys, and faced with lime plaster. Internal partitions of timber studding, sometimes braced to act structurally, sometimes infilled with brick to give sound insulation, and all covered in lath and plaster were common.

Damp-proof course (DPC)

The Public Health Act 1875 made DPCs compulsory for new houses. A variety of DPCs were installed, including lead-cored felt, bituminous felt, two courses of dense (engineering) bricks or slates laid in cement mortar to break joints. Unfortunately, the (hessian-based) felts were often not installed with proper laps and joints and have degraded with age. Brick or slate DPCs laid in a cement mortar, being formed in rigid materials, were adversely affected by building movement and often failed to prevent rising damp due to the resultant cracks in the DPC.

Foundations

In the earlier years of the period, more expensive houses had some form of foundation which might be of stone slabs or, as the nineteenth century progressed, corbelled brickwork became more frequently used. Cheaper housing often had no foundation or merely a simple shallow layer of flat stones or of clinker. By the end of the century, concrete foundations (often with corbelled brickwork as well) were required for all buildings unless built on rock.

Windows, doors and joinery

Fully rebated timber sash windows were the norm throughout most of the period, although timber and metal casement windows did appear in the late Victorian period and the Edwardian period due to the influence of the Arts and Craft Movement and the development of Edwardian mock Tudor housing. Removal of the special taxes on glass (in 1845) and windows (in 1851) plus the development of plate glass in 1832 meant that, by the mid-1800s, the number of windows in a building was no longer restrained and larger panes of glass could be used. This also resulted in fewer glazing bars, although horns were needed to strengthen the sashes. Cords were used for smaller sashes and metal chains for larger windows.

Doors throughout most of the period were of four panels comprising two long panels above two short panels, usually of pine (i.e. softwood), sometimes of pitch pine or oak, and meant to be painted.

Softwood joinery in cheaper housing was painted or, occasionally, stained. More expensive hardwood joinery of oak or mahogany was polished or varnished.

Upper floors

These were constructed with timber joists (usually of softwood), herringbone strutting and softwood floorboards, plain-edged for cheaper housing and tongued and grooved (T&G) for more expensive houses. Ceilings were formed with lime plaster over softwood laths. Moulded details, e.g. cornices, friezes, were often incorporated in better-quality houses.

Ground and basement floors

Ground floors of suspended timber construction with softwood joists and pine floorboards were the norm. The floorboards were plain-edged in cheaper housing, but tongued and grooved for better-quality work. Rear extensions, which often housed a kitchen or scullery, had flagstones or brick paviors directly over compacted earth. At the basement level, a solid floor was formed of flagstones or, sometimes (for cheaper housing) of beaten earth.

The floor of the main entrance vestibule (and, sometimes, the hallway) was often finished with fired clay floor tiles, also known as pammets. These could be encaustic tiles (coloured and patterned tiles that were normally unglazed), geometric clay tiles (smaller shaped tiles, often triangles or lozenges), or quarry tiles (plain and square, often red or brown), or a mixture. Quarry tiles were frequently used for floors in kitchens and outhouses.

Services

In the first half of the nineteenth century, the provision of services generally had remained little changed from the previous century. From the 1870s onwards, cheaper houses had a sink with cold water and a coal-fired range for cooking and hot water as well as an outside water closet at the rear or an earth closet in rural areas. No bathroom was provided except in more expensive houses, some of which also had a back boiler, often behind a range, to provide hot water from a direct cylinder.

All houses had coal fires for heating the living rooms and the bedrooms. Mains services and internal carcassing for water and gas were via lead pipes. From about 1880, a few houses had electricity, but only for lighting. Lighting was normally by gas (from about 1840) or oil lamps. Urban houses were connected to combined storm and foul water drains formed in salt-glazed clay pipes from about 1860 onwards. However, rural houses continued to use traditional cesspits throughout the period.

PART D: TYPICAL DEFECTS OF GEORGIAN, REGENCY, VICTORIAN AND EDWARDIAN HOUSES

This period of 200 years included a variety of architectural styles and basic floor plans. However, basic construction changed relatively little and the defects that the houses of this long period suffer from are generally very similar. Typical defects (many of which are discussed in detail in the following chapters) include:

- **Arch failure.** *Brick and stone arches were built over door and window openings externally with timber backing lintels internally. Frequently, the external arch can fail, leaving only the internal lintel to support the full structure above or to fail as well.*

- **Bonding failure.** *Bonding between adjoining walls was haphazard. In the Georgian period, especially, it was not uncommon for front walls at street level to be built at the end of the construction phase – the large opening enabled access during the building phase. Bonding through the thickness of the wall often consisted of the occasional brick course acting as a connection between walls. Snapped headers were fairly common. This lack of physical connection often results in lateral movement at the junction between external walls – front, flank and rear – and at the junctions with between external walls and party walls. Similarly, flank walls were often inadequately bonded to internal spine walls, or sometimes even not bonded at all.*
- **Defective lime mortar.** *Lime mortar was commonly used for brick and stonework, but it tends to weaken in an acidic atmosphere (as found in many urban areas). It then requires repointing. Lime render is similarly affected.*
- **Drainage problems.** *Salt-glazed pipes were used for below-ground drainage during the nineteenth century, but are now between 100 and 150 years old. Many have cracks and/or porous joints. Above-ground drainage relied on cast-iron pipes and gutters and these tend to corrode (i.e. rust) on the face nearest the wall.*
- **Inappropriate use of cement mortar.** *These mortars are strong and dense, but also inflexible and prone to cracking. Repointing or rendering a previously lime-mortared wall with cement mortar can lead to spalling of the face of the brickwork. Being denser than lime mortar, it also prevents a wall from 'breathing', i.e. letting penetrating moisture out.*
- **Inadequate foundations.** *Many earlier houses in the period, particularly cheaper houses, lacked a foundation or had only a shallow foundation. Movement cracks (and occasionally collapse) due to settlement and/or subsidence are a common problem.*
- **Lack of a damp-proof course.** *Georgian and Regency houses were rarely built with damp-proof courses. Similarly, in the nineteenth century, most houses built before the last decade had no damp-proof course and, where installed, early damp-proof courses were often formed in materials that have subsequently perished.*
- **Over-stressed timber floors.** *Joist sizes were generally far smaller than modern joists, especially on upper floors. This under-sizing often results in excessive floor deflection and movement due, in part, to the increased amount of furniture to be found in modern houses.*
- **Penetrating damp.** *Beam and joist ends were normally built into solid external walls by about the depth of a half-brick (112.5mm), often over a coursing timber. Where an external wall is 225mm thick or thinner, the amount of solid cover was insufficient to prevent penetrating damp adversely affecting the end of the timber or the coursing timber. Timber backing lintels over doors and windows are also often affected by wet or dry rot due to penetrating damp as the external brick/masonry arch was normally only 100–112.5mm thick, i.e. a half-brick.*
- **Poor-quality materials.** *Speculative builders would often use whatever materials were available for purchase on the day of construction. Frequently, this resulted in sub-standard materials being used, e.g. sun-baked bricks, soft stonework and inappropriately sized timber scantlings. Speculative developments were often built with an eye to the length of the lease – 40, 50 and 99 years were not uncommon leasehold periods and the quality reflected this.*
- **Poorly ventilated timber ground floors.** *The number of air bricks provided to ventilate suspended timber ground floors was often inadequate, leading to unventilated areas that are prone to damp and fungal attack. Frequently, this was exacerbated when the under-floor void had a restricted airflow because it was very narrow, blocked with builders' debris, had inadequately ventilated sleeper walls or abutted the solid floor of an adjoining part of the building (e.g. a rear extension).*
- **Roof coverings.** *Failure due to the rotting of roofing battens as well as to nail sickness (due to rust) of both slate and tiled roofs is a common problem. Slates can suffer from delamination due to frost attack. Welsh slates are more prone to this problem due to their smooth surface, while English*

slates are rougher and less prone to the capillary action which leads to the problem. Clay tiles can also suffer from pitting as well as delamination.

- **Roof timbers.** *These often suffer from the effects of wet rot as originally no felt was installed and/or parapet and valley gutters have failed.*
- **Rotten coursing timbers.** *It was common practice to level up brick courses at frequent intervals (because of their unevenness) with a softwood coursing (or levelling) timber. In external walls, these are prone to the effects of damp penetration leading to wet or dry rot, especially where they are of full wall thickness and 'protected' by only a render finish.*
- **Windows.** *Timber windows (both sash and casement) often suffer from wet rot especially to sills as a result of a lack of regular decoration externally. Metal windows are often affected by rust for the same reason. Sash cords and chains can become defective.*
- **Worm attack.** *Softwood timbers in many houses are likely to be affected to a greater or lesser degree by woodworm; it would be unusual to find a house that is not. Hardwood scantlings such as oak floor and roof timbers affected by damp may be attacked by death-watch beetle.*

Section 3: Housing between the wars, 1918–1939

The period between the end of the First World War and the beginning of the Second World War saw a surge in the construction of new houses and the development of new housing styles.

In the immediate aftermath of the First World War there was a great need for the provision of new houses. The problem was two-fold: one was that there was a housing shortage (which had been acknowledged as an issue before the war) and the other was the poor state of much of the existing stock. This meant that a major housebuilding programme was necessary, and the urgency of this was underpinned by the idea that the soldiers returning from the trenches would require 'Homes fit for Heroes', a phrase that encapsulated a national sense of obligation to those who had fought in the war.

At the time the overwhelming majority of housing was provided by the private rented sector. There was, however, a general consensus that the number and type of houses required could not be provided by the private sector and that the government and local authorities would have to play a major role. This led to the development of a national housing strategy, which in turn led to the Housing and Planning Act of 1919 which gave local authorities a responsibility to determine housing need in their area and to provide low-rent housing. Housing Committees were set up and they were encouraged to build through the provision of generous subsidies. The subsidy arrangements shared the costs of this new housing between the tenants, local ratepayers and the Treasury. Councils in areas of high housing need could apply for these subsidies. The London County Council (the predecessor to the Greater London Council) also raised money through selling London housing bonds which promised investors a 6 per cent return and this raised £4 million during the 1920s.

Subsidies were also available to private developers (through a subsequent Housing Act) but, certainly in the immediate post-war period, most of the new houses built were local authority 'council houses' – what today we would call social housing. Changes in governments in the next two decades meant that different emphasis and importance were given to either private or local authority provision, but, essentially, a pattern was set which meant a move away from the dominance of the private rented sector to one where private developers built houses for sale and local authorities built houses for (relatively) low-cost rent.

The most significant factor affecting the design and layout of local authority housing in this period was the Tudor Walters Report which was published by the Local Government Board at the end of the war. This report was heavily influenced by the Garden City Movement and the model village developments such as Port Sunlight, Bourneville, etc. (see previous section). Experts contributing to the report included women's organisations which had a strong influence on the recommendations for design and internal layout. Architecturally it was heavily influenced by the Arts and Crafts Movement with its strong rural references (the type of design encouraged by the report is sometimes referred to as cottage-style vernacular or garden-suburb vernacular).

Under the influence of the report, planners promoted the construction of new suburban 'garden' estates, situated on the outskirts of cities. Mainly consisting of three-bedroom houses for families, the design of the estates aimed to create self-contained communities of low density, often with no more than 12 houses per acre. Facilities, including churches, schools and shops, were provided; public houses were initially excluded from the plans. On most estates, the houses were provided with a generous sized garden to encourage the tenants to grow their own vegetables. The interiors varied, but generally there were significant improvements compared to (lower-class) Victorian housing. There was more space generally but also more rooms with defined functions. They tended to have three rooms downstairs and three bedrooms and usually a larder and a separate bathroom, as well as an indoor WC. For most new tenants these new conditions were a huge improvement on their previous slum housing where they had experienced overcrowding and often were without even basic facilities. Due to rising costs, some of the high standards of local authority housing seen immediately after the war were scaled back in later years but overall the standards remained relatively high.

Architecturally the idea was to create a garden village – or garden suburb – and so many 1920s council houses were built in a simple cottage style with gabled, red tiled roofs, brick walls combined sometimes with white render or pebble dash and horizontal casement windows. Houses were built in pairs or in short terraces (usually of up to about five houses). They were generally low and wide, roofs were hipped and chimneys low and squat. As well as the Arts and Crafts influence, a significant number of local authority estates were built in a neo-Georgian style.

Many of the council houses from the inter-war period were influenced by the Arts and Crafts architecture.

Both the local authority and the private sector houses built in this period signalled a move away from the emphasis on narrow front and rear elevations, and the back additions, that were common in a typical Victorian house. The houses of this period tended to be square or at least rectangular in order to let in more daylight. This emphasis on the provision of daylight and sunlight was also to be seen in the way the housing of this period was laid out, with more external space provision and attention to orientation.

After this initial burst of building activity targeted at reducing the post-war housing shortage, local councils began to tackle the problem of existing slum housing. The Housing Act of 1930 encouraged mass slum clearance and councils set to work to demolish poor-quality housing and replace it with new build. Unlike the garden estates built directly after the First World War, much of the slum clearance in major cities was replaced with flats, mostly three to five storeys high. They were often modelled on schemes in continental Europe. The construction of flats as a response to the housing shortage was aided by the developments in, and understanding of, concrete technology. This approach to housing was supported by architects from the Modern Movement who saw in flats an opportunity to work with the ethos of this architectural style and who were influenced by the approaches adopted in other parts of Europe, and in Vienna particularly. Gradually the idea of living in multi-storey accommodation began to gain acceptance and, in London, flat building exceeded house building for the first time in 1936.

Although the Garden City Movement was influential, its proposals for estate layouts, with its emphasis on avoiding linear development and its encouragement of cul-de-sacs, were by no means the universal defining approach of the period. This was because the private sector was involved in significant amounts of house building alongside the new arterial roads that were spreading out from towns and cities in what became known later as ribbon development. This expansion into the suburbs (effectively the creation of suburbs) was greatly aided by increases in bus and train travel which allowed people to commute to work much more easily.

Although the building of houses by the local authorities was a dominant force just after the war, the development of building societies as sources of mortgage lending meant that private sector housing also grew substantially from the early 1920s onwards. This resulted in a significant increase in home ownership in this period. By 1939, 1.1 million council houses had been built, but this figure was outstripped by the 2.8 million middle-class private homes built by speculative builders from 1923. The rate of building increased in the early 1930s, reaching a peak in 1936 when 370,000 houses were completed. The commonest house type was the three-bedroom semi-detached house though developments often included detached houses and bungalows.

These two houses are probably from the late 1920s/early 1930s. Note the brickwork to the ground floor and the render to the top floor which are quite typical of houses from the early part of the inter-war period. The roofs are dual pitched and the bays are squared.

The bungalow came into its own between the wars. While some were built in pairs, the detached bungalow was common and this provided an affordable way of achieving the goal of living in a detached home. They were also claimed to be less expensive to furnish and cheaper to run than a conventional house.

An inter-war period bungalow.

Architectural styles

Although as mentioned, much of (though not all) local authority housing was influenced by an Arts and Crafts aesthetic, private sector housing tended to develop a different architectural style. What we tend to refer to now as a '1930s semi' has come to define what many would think of as the typical house style of the period. In fact, of course, terraced houses continued to be built, though the terraces were shorter than the long terraces of the Victorian period. The 1930s house tended to be less ornate than previous styles partly because of the voguish emphasis on functionalism. In this they reflected some of the ideas of the Modern Movement, without in any real sense embracing it, just as some Art Nouveau and Art Deco touches could also sometimes be discerned. But perhaps the overriding style of the typical 1930s semi can be seen to be a retrospective borrowing from traditional vernacular motifs. The dominant feature was the front bay – usually of two storeys – which was surmounted by a large prominent gable typically dressed with barge boards and often fake timber framing in imitation of sixteenth- and seventeenth-century vernacular buildings. The bays were sometimes square-edged or set at an angle (canted). Curved bays were also popular – either tucked under deep projecting eaves or surmounted by a projecting gable supported from below by open timber brackets. Another recognisable stylistic component of the privately built 1930s house was the round-arched, open porch leading to the front door. One other significant feature of the 1930s semi-detached house is that, apart from the relatively few houses built in local stone, it was impossible to discern any distinct regional patterns to the style.

A 1930s house. Note the hipped roof, the canted bay and the leaded lights to the casement windows.

A typical 1930s semi-detached house. Note the oval bay, the profiled tiles to the roof – which is hipped – and the plain tiles to the gable over the bay. The windows are timber casement.

True Modern Movement architecture which had such a significant impact in the rest of Europe, and which continues to this day to influence architectural thinking, had very little impact on mass housing in this country. Where Modern Movement houses were built, they tended to be single commissions for individual clients.

These three Modern Movement houses represent perhaps the most important international architectural style of the twentieth century but one which did not find favour in this country.

The flat roofs, the smooth-faced render and the suntrap metal windows are all typical motifs of houses built under the influence of the Modern Movement.

Like the council house, the typical privately built semi adopted a rectangular plan with a small kitchen, often called a 'kitchenette'. A serving hatch was another innovation of the time, linking the kitchen with the dining room, while the latter usually had French windows opening to the back garden. Generally the houses were low and wide and of just two storeys: the first floor contained three bedrooms and a bathroom with the WC (often located in a separate room).

Another innovative feature of the private sector 1930s semi was the provision of a garage for the motor car – something that was not generally a feature of local authority housing.

Construction in the inter-war period

Although there will always be exceptions, and of course regional variations, the common construction features of buildings from this era, and particularly the 1930s include the following.

Roofs

Houses of the period mainly had pitched roofs except for the relatively small number that adopted the Modern Movement style, which emphasised the use of flat roofs. For semi-detached houses (and the end houses of rows of terraces), hipped roofs were the dominant style. The roof structure was essentially based on the same principles as Victorian roofs but adapted for a different shape. They consisted of softwood rafters supported by purlins. In larger roofs the purlins would be strutted off load-bearing brick internal walls and in some case trusses would be used.

'Typical' 1930s houses with hipped roofs and round arched open porches (two of which have subsequently been enclosed).

Roof coverings

These included slate and clay. The full range of clay tile profiles were used, with plain clay tiles continuing the renewed popularity that they had gained in the Edwardian period. Coloured tiles enjoyed some popularity, with green being a favoured colour.

Gutters and drainage

These would be cast iron for the rainwater gutters and downpipes and for the soil and vent pipe.

External walls

Brickwork tended to be the most common material for the construction of external walls at least for the 'typical' house. The use of cavity walls started to become more common throughout this period. These would be built with brickwork for both the inner and outer skin. The cavity wall ties used during the early part of the period could well be wrought iron or steel but later on galvanised (steel) ties would be commonly used. It is not possible to date a changeover from solid walls to cavity, partly because there was such a geographical difference in the way that cavity walls were adopted; for example, they were often seen first in areas exposed to high levels of driving rain.

The use of cement mortars gradually replaced lime mortars throughout the 1930s – though lime was still often added to the mix to improve workability.

A particular defining feature of houses of the inter-war period was the use of render. Often houses from the 1920s had the ground-floor external wall as exposed brickwork with the first floor rendered, but by the 1930s the tendency was to render the whole of the façade. The render was usually roughcast or pebble dash. The exceptions tended to be those houses which were more influenced by a Modern Movement aesthetic where a smooth render would probably be used.

External render was a common material in this period. Again note the hipped roof and the bay windows.

Bay windows

Bay windows were common in houses built in the 1920s but they were perhaps a more dominant feature in the 1930s. The ground floor of the bay was usually brick but the first floor would often be constructed in a timber framework. The bay would typically be rendered or covered in plain tile-hanging.

*Bays were a dominant feature
of the period.*

*Bays were usually rendered or covered, as here, in
plain tiles. Note the timber casement windows.*

*Non-structural timbers were often used on bays and the façades in general. They were purely decorative
and were meant to evoke the imagery of traditional timber-frame houses of the past ('Mock Tudor' is a
common pejorative term to describe this feature).*

These roofs sweeping down to ground level were a relatively common feature of the period.

*Open porches leading to the front door were another typical feature of the
1930s house.*

Note the leaded glass in windows and the sunrise motif on the timbers (also used in glazing).

Internal walls

It was usual to build load-bearing and non-load-bearing internal walls in a single skin of common brickwork, or sometimes clinker blocks.

Damp-proof course (DPC)

A variety of DPCs were used, including slate and engineering bricks in the 1920s, but it was quite common, particularly in the 1930s, to use bitumen impregnated felt. The bitumen felt DPC had the advantage of being relatively flexible at first but over time they tended to become embrittled and therefore susceptible to damage through building movement.

Foundations

There was a general requirement in model by-laws for concrete foundations with stepped brick footings above. By the end of the period it was often sufficient just to have the concrete strip (though brick footings were still acceptable).

Windows, doors and joinery

Although sash windows were still quite common in the early part of the period, casement windows with top opening lights grew in popularity and became the dominant style particularly for the 1930s semi. Timber was still a significant material although steel windows gained in popularity. Initially these were in plain mild steel but later they were galvanised. Although not universal, another feature of the period (particularly the 1930s) was the use of windows which were glazed, or partially glazed, with leaded lights. Stained glass was a popular feature with a design of a red or yellow rising sun or a Tudor galleon being favoured.

French doors leading onto the rear garden were another feature typically associated with 1930s housing.

Upper floors

These were constructed with softwood timber joists, with herringbone strutting which tended to be more favoured than solid strutting. The floorboards were usually square-edged softwood or, sometimes, tongued and grooved. There was a general awareness of the fire issues associated with building joists into party walls and so joist hangers or steel corbels were sometimes used.

Ground floors

Although ground-bearing in-situ concrete floors were used in the 1930s, ground floors in the period were more typically still of the timber suspended variety. From the 1920s onwards there were improvements in the design of timber suspended ground floors, particularly in relation to protection against damp and its associated problems (i.e. wood-rotting fungi).The floors would typically be softwood timber joists supported on brick sleeper walls, laid usually in a honeycomb pattern (which produced gaps through which air could flow). The rectangular shape of typical 1930s housing, the lack of a back addition which might block through air flow, and the general use of honeycombed sleeper walls all meant that cross-ventilation tended to be more efficient than it was in many Victorian houses. The sleeper walls usually had a damp-proof course to protect the joists against rising damp. Where solid external walls were used, the joists would often rest on a sleeper wall built close to that wall so that the joists were not built into it. Where a cavity wall was used, then the floor joists would be built into the inner skin. In both cases, this, and the use of DPCs, helped protect the timber against rising and penetrating damp.

The floorboards might be square-edged but tongued and grooved was quite common. Also parquet block flooring was popular and quite often found in entrance areas and hallways.

Services

Space heating throughout the inter-war period was provided by solid fuel appliances, mostly in the form of coal fires. Hot water supply to the bathroom became a more common provision. Early in the period there were still examples of hot water coppers being used, with the water heated by a fire, and circulated by a pump. The most common approach that developed through the period, however, was with water being heated by a back boiler (situated at the rear of one of the fireplaces) and feeding a (direct) water tank of galvanised steel. An alternative to this, usually later in the period, was a gas-fired hot water geyser located in the bathroom. Electric immersion heaters were also available at the latter end of the period. Pipework was usually formed in lead.

Although some new houses in the immediate period after the First World War were still being built with gas lighting, electric lighting quickly became the universal means of artificial lighting provision.

New technology

Although the typical house construction described above can be seen as the continued use or the development of existing materials and forms of construction, there were examples of relatively radical developments in technology being used during this period. The impetus for this was the problems of material and skilled labour shortages which led the government, in the early 1920s, to encourage experiments in approaches to construction. This included the development and use of what we would now refer to as system-built housing. These systems included timber frame, steel frame, concrete frame/slab and in-situ concrete wall slabs.

Common defects

Housing built in any period will suffer from defects if poorly designed and constructed. In fact, houses from this period were generally well built – in part because of the growing influence of model by-laws which set or influenced standards in the design and construction of the main elements of the house. So, for example, the depth and size of foundations were set out in these documents. Also lessons had been learnt from previous mistakes, an example being, as previously mentioned, the development of better practice in the design and construction of suspended timber floors. However, some defects are particularly associated with housing from this period and these include the following:

• **Cavity wall tie failure.** *Where houses from this period were constructed with cavity walls, then* *the failure of the wall ties through rust is common. Sometimes the rusting will have occurred because the wall ties were not protected, or it may be that, where they were galvanised, this (essentially sacrificial) protection has corroded over time.*

- **Condensation.** *Condensation problems occur in houses from all periods, but in 1930s semi-detached houses it is quite common in bathrooms. This is mainly because they tend to be situated at the rear corner of the house and therefore have two exposed external wall surfaces which can be very cold if the cavity has not been (retro)fitted with thermal insulation.*
- **Rising damp.** *Where the damp-proof course is of the bitumen felt variety, this has often become embrittled and cracked, thus allowing the passage of water. Also, although concrete floors were not particularly common, the damp-proof membrane was sometimes missing.*
- **Rusting metal windows.** *The metal windows of the period became problematic because of rusting due to inadequate protection. Often the source of the moisture was condensation on the windows, caused because they readily transmitted heat and therefore were cold.*

Section 4: Post-Second World War (1945 until the 1990s)

The final section of this chapter looks briefly at the period stretching from the end of the Second World War until just before the end of the twentieth century, as this seems a suitable end point to a chapter on building history.

The end of the Second World War in 1945 heralded significant social changes in British society, particularly with the introduction of the welfare state in the years immediately following the end of hostilities. This social revolution came about at the same time as a programme of nationalisation of key industries and services, and the need for industrial reconstruction – at a time when the country was effectively bankrupt as a result of the war. Economically the years of relative austerity in the late 1940s gradually gave way through the 1950s and 1960s to growing prosperity, and with it changes to the way people lived and the material possessions they expected to accumulate – not least in their expectations of the houses they lived in. The growth of the private motor car had a significant effect not only on where people lived but also on the design and layout of the housing which was built at this time. Changes in the make-up of families also gradually had an effect on the provision of housing.

As with the 1914–1918 conflict, the Second World War and the virtual standstill in house building brought another watershed in house design. One difference between the two world wars, however, was the number of houses that were destroyed. In the Second World War, about 475,000 houses had been destroyed or made uninhabitable through enemy action. Another factor in housing demand after the war was the high birth rate – the creation of the 'baby boomer generation'. Also slum housing remained a problem in many large towns and cities. Clearly, as with the situation at the end of the First World War, urgent action was needed to address the housing crisis. One of the responses was to provide temporary accommodation by constructing pre-fabricated houses.

Altogether, 156,000 'pre-fabs' were assembled using innovative materials such as steel and aluminium. It was intended that these dwellings should only have a short life (of five years or so) but they proved to be a successful and popular house type and many well outlived their life expectancy (some of those remaining have been listed as 'historic buildings'). However, as these structures were only intended as a temporary measure, the provision of new council housing became an urgent priority for the new post-war government. This need signalled another 'boom' in council house building. Local authority house building resumed in 1946 and of the 2.5 million new houses and flats built up to 1957, 75 per cent were local authority-owned.

The various system-built houses were another response to the housing crisis – they were of course intended to be a permanent response – unlike the short-life 'pre-fabs'.

Examples of non-traditional precast concrete houses.

The building of council houses in the post-war era was shaped to a large extent by a new approach to town planning enshrined in the 'Greater London Plan' of 1944: a blueprint for post-war reconstruction by Professor Patrick Abercrombie. Out of this came the idea of neighbourhood units, and the new town movement which revived the idea of the Garden City. In 1945, a New Towns Committee created government-sponsored corporations which were given power to acquire land within a defined, designated area, and to establish new towns. The New Towns Act passed the following year provided the government with the power to implement these plans. The result was the creation of 22 new towns between 1946 and 1972, many serving as satellite towns to Greater London.

Post-war house construction was also shaped by two housing reports: the Dudley Report of 1944; and the Parker Morris report, Homes for Today and Tomorrow, published in 1961.

As well as the surge in council housing, the post-war era also saw a sharp rise in property ownership, rising from 26 per cent of all households in England and Wales in 1945 to 49 per cent by 1970, and about 63 per cent by the mid-1980s. Part of the (further) growth in home ownership in the 1980s was due to the 'right to buy' legislation which allowed public sector tenants to buy the house that they occupied. By the mid-1990s about one and a third million houses had passed into private ownership through this route.

Bungalows remained popular in the private sector and to some extent they came to typify post-war suburbia in some of the dormitory areas of major cities.

Stylistically, there were still differences between the public and privately built houses. Inevitably, a greater variety of styles and types of dwelling were to be found in private developments. Now, some 60 years on, houses of the 1950s and 1960s are beginning to acquire a period character of their own and are sought after as being distinct from more modern housing.

The post-war estate layout was founded on the principle of the 'neighbourhood unit' – a planning concept which promoted the development of self-contained communities which were intended to incorporate a wider social mix and a greater variety of house types. Some neighbourhood units were built 'phoenix-like' out of the slums they replaced as part of urban regeneration schemes, while others – as with many pre-war council estates – were built on new greenfield sites on the edges of towns. This approach, often informed by official design manuals, was popular until the 1980s. A greater variety of house types typified the neighbourhood unit and included blocks of flats as well as the three-bedroom semi-detached house.

In some of the new towns and council estates built in the 1950s, a new type of house layout, known as the Radburn layout, was introduced, which aimed to separate vehicular and pedestrian access (something that was also encouraged in the 1960s by Parker Morris). The orthodox street frontage was abandoned in favour of the use of road access by cul-de-sacs, with access to the front door by a pedestrian footpath across a small open grassed area with no obvious boundaries between individual properties.

During the 1970s there was a lot of emphasis on high-density low-rise developments based on design manuals (such as *The Essex Design Guide* and the Greater London Council's (GLC) *Preferred Dwellings Guide*). In the 1980s, the cul-de-sac regained popularity, roads were generally wider (partly due to the increasing size of motor cars) and there was an increase in the amount of grassed and planted areas. In later years, providing parking spaces on new developments continued to be a significant issue. In the twenty-first century, concerns about sustainability gradually came to dominate housing development in terms of estate layout as well as the individual house.

Through the recommendations contained in the Dudley Report (which was seen as the equivalent of the Tudor Walters Report) post-war council houses were provided with more space and better services, including improved storage and kitchen facilities. The Parker Morris report of 1961 (entitled *Homes for Today and Tomorrow*) recommended standards for all new homes, both public and private, which reflected changing patterns of living (for example, more informality in the way space in the house was used) and the growing use of household appliances. The main recommendations were for more living and circulation space and better heating throughout the house so that all spaces could be used freely. The kitchen was to be extensively fitted and provided with plenty of space for storage and the use of electric domestic appliances, such as washing machines and refrigerators. A typical design of a Parker Morris house would have been a three-bedroom terrace with a through living room and a kitchen/diner with a room divider, and large windows to provide high levels of daylight. The recommendations in the report were made mandatory for public sector housing but not for the private sector. The Parker Morris standards were later (early 1980s) dropped, mainly because of the costs involved in implementing the space standards.

The 1950s saw an increase in the numbers of flats and maisonettes built by local authorities. Many of these blocks of flats were relatively low rise (perhaps three or four storeys high).

These three-storey flats were built in the early 1950s. The garden fencing is a later addition as all external areas were intended to be communal.

Although high-rise flats tend to be associated with the 1960s and 1970s, the first high-rise block was the ten-storey 'The Lawn' at Harlow in Essex, which was completed in 1951. High-rise flats represented an architectural ideal for some of the architects who were influenced by the Modern Movement, but they also seemed to offer a solution to housing problems by the provision of high-density housing, which also served as a counter to the outward expansion of cities, and the subsequent threats to the surrounding countryside. By the mid-1950s, government policy to extend the green belt nationwide, and a new subsidy arrangement under which, for the first time, grants increased with blocks of more than six storeys, led to the adoption of high-rise blocks by many local authorities. The drive towards the building of tower blocks was to some extent a response to the problems of poor housing conditions and homelessness that still existed in the 1960s. The creation of the Housing Corporation and Housing Associations was also a result of this continuing housing crisis.

Examples of medium- and high-rise flats.

Many of the high-rise blocks were poorly designed but others, particularly those that tried to create a sense of community through their layout, design and services, were arguably more successful. However, in general, the attempts at 'social engineering' were not successful and mounting criticism of their social and visual short-comings, and the structural failure of one particular block in dramatic circumstances – Ronan Point, a 23-storey prefabricated concrete tower block in Newham in London – caused a strong reaction against the building of more high-rise blocks. The Ronan Point disaster occurred in 1968 and, after 1970, the concrete tower block was no longer seen as a workable model for urban regeneration.

Both of these high-rise developments have been listed as 'historic buildings' (Park Hill in Sheffield, left, and Trellick Tower in London).

Most of the tower blocks were effectively system-built. The houses that were system-built in this period also fell into disrepute – though to some extent they were unfairly maligned. The technical issues relating to pre-cast concrete and timber-framed houses in particular are discussed in Chapter 17.

Architectural styles

In a sense, and particularly in the period after the 1950s, it can be said that there was little in the way of coherent architectural styles that distinguished the post-war period. What there were instead were various fashions for the treatment of materials and elements.

Typical houses from the 1950s.

It has been suggested that a typical 1950s house was a development of the Arts and Crafts 'cottage' but the influence of Modernism can be seen in the lack of decoration (which can also be explained by materials and skills shortages of the time). The external walls of the 1950s house were usually finished in plain brickwork, often with a rendered panel around or next to the main entrance door. Another break with the typical 1930s house was a return to gable roofs rather than hipped.

However, some 1930s influences still pertained in the style of the post-war housing.

Some 1950s houses still retained features of the 1930s house.

The large number of non-traditional system houses built in the period immediately following the Second World War had their own distinct style which was determined to some extent by the materials used, for instance, steel cladding sheets or pre-cast concrete and the stylistic treatment of the major elements (such as in the use of mansard roofs on some systems).

During the 1960s there was, in the private sector, a marked growth in the size of building firms who used their own architects. Exterior styling varied but there were, nevertheless, some common characteristics which give 1960s housing an identity of its own. Plain, flat wall surfaces with large oblong 'picture' windows were typical. The windows usually had robust wooden frames with opening top lights. Front doors were usually glazed in small glazed panels, usually with rippled glass while the woodwork was painted a light colour. Some roofs were flat but the typical 1960s house has a low pitched roof with an end gable finished with a prominent but unadorned barge board painted white. Roof tiles were generally of brown or grey concrete and although red brick walls were found on 1960s housing, light brown, grey or buff coloured bricks were widely favoured. Claddings of tiles (usually of concrete) and white painted boarding applied between the ground and first windows typified the 1960s house and continued to be popular into the 1970s.

In the private sector, the inclusion, in the late 1980s and early 1990s of such features as utility rooms, en-suite bathrooms, and integral or linked garages had an effect on the layout and design of many new houses. The issues of parking and garage space in new housing developments – and the attempt to reduce off-street parking – were, and remain, a major problem. Another characteristic of 1980s private housing, particularly following the waning influence of Parker Morris (which was abandoned in the early 1980s) was the small size of many of the rooms. Indeed, the show houses used to sell the design on many developments had to have the internal doors removed, otherwise the normal expected furniture layout could not be achieved. In private sector estate development in particular, although the semi-detached house continued to be popular, there was a growth in the number of detached houses. Three-storey town houses have been popular at various times. These often had the kitchen at ground-floor level, and the living rooms at the first-floor level.

These photographs show houses from the early 1960s (left) and the late 1960s (right).

Although some houses from the 1960s were influenced by the architecture of other countries, in particular Scandinavia, an important stylistic approach of the post-war period was the increasing use of pastiche, i.e. the reproduction of past architectural styles. This perhaps began in the 1960s with neo-Georgian style – characterised by red brick walls, pedimented porches, small paned windows and fake louvred shutters. Another style was sometimes called the 'Colonial' with double garages under a colonnaded façade incorporating bow windows and concrete urns. This pastiche approach was also often used internally with, for example, the past being mimicked by ornate plasterwork details in fibreglass. Neo-cottage detailing with ledged and braced doors, prominent ironmongery and mock timber beams (again, often in fibreglass) was also common. Such mimicry would often be inappropriate both in detail and scale, and sat in contrast to the other more modern details in the house. This reproduction of past styles, or at least the, often random, incorporation of elements of the past, became particularly prominent in the 1980s and 1990s, but it is a trend that continues to this day with Georgian, Victorian and even 1930s references being incorporated in new houses. The development of often large private estates by national contractors, particularly from the 1980s onwards, has led to, in many people's eyes, a lamentable loss of character and the obliteration of vernacular traditions in housing. This stylistic trend was often later followed by the public sector – in the guise of housing associations who were becoming the main providers of social houses for rent.

One stylistic break with the past which began in the 1960s was that, for the first time in five hundred years, the roof line of many new houses was unbroken by chimney stacks. Following the recommendations of the Parker Morris Report, most new houses were built with full or partial central heating; moreover, the use of open coal fires had declined since the passing of the Clean Air Act in 1956.

Many houses on estate development attempted to evoke images of the past in the way that individual elements were designed. Often, however, this stylistic 'cherry picking' resulted in a poor pastiche which had no architectural coherence.

A development from 2012 – still mimicking the past.

Some examples of the design of individual elements in the post-war period

1960s and 1970s details

- *Generally low-pitched gabled roofs, but also some examples of flat roofs. Mono-pitched roofs are often seen on 1970s developments.*
- *Clean simple lines on elevation but rendered wall panels were quite common.*
- *The use of tile and artificial slate hanging (including on cross-wall housing) and painted or stained timber boarding.*
- *Relatively large (picture) windows unobstructed by glazing bars.*
- *Entrance doors often fully glazed (increasing energy losses).*
- *Roof tiles were often of brown or grey concrete.*
- *Although red brick walls were found on 1960s housing, light brown, grey or buff coloured bricks were widely favoured.*

These examples of flat roofs to houses are from the 1950s (left) and the early 1960s (right).

Mono-pitched roofs were popular in the 1970s.

1980s details

- *The standard roof trusses from the 1960s to the 1980s produced low pitches and were appropriate for the large concrete tiles of the time, but the development of more innovative truss designs from the 1980s onwards allowed for a greater variety of roof slopes, including hips and steeper pitches, which also saw an increase in the popularity of dormer windows.*
- *Bay windows (and, to an extent, oriel windows) became fashionable again.*
- *From the 1980s onwards, window sizes reduced, compared to the picture windows of the 1960s, partly as a response to energy issues. Sliding sash windows regained some of their popularity.*
- *Return to chimneys, often with a gas fire.*

Oriel windows became more fashionable from the 1980s onwards (as did bay windows). Developments in roof truss technology allowed for a greater variety of roof shapes.

Construction

As always, developments in technology and expected performance standards were significant in the way new houses were built but legislation was also important. The various model by-laws gave way to the Building Regulations in 1965. These were applicable nationally throughout England and Wales, with the exception of the Inner London Boroughs (the area of the former London County Council) where the London Building Acts continued to prevail for some time. The Building Act 1984 consolidated the various amendments to the regulations and resulted in the introduction of the Building Regulations 1985. This had a significant effect on the standards and national uniformity of the technical aspects of house construction, including in later years issues of energy efficiency and water usage. In recent years the Code for Sustainable Homes has led the way in promoting a range of sustainable practices in the design and construction of houses.

Roof structure

Although traditional 'cut' roofs were still used in the 1950s, there were restrictions on timber imports and so new techniques were developed. One of these was the TRADA truss. The TRADA truss was essentially a lightweight version of the traditional roof truss. It did away with the need for internal load-bearing walls upstairs and allowed for smaller section rafters, often at slightly wider centres. However, TRADA trusses were relatively short-lived in terms of popularity and by the mid-1960s most developers were using trussed rafters, a situation that continues to this day. Flat roofs were not common but there were examples of them being used on some system-built houses as well as medium- and high-rise flats.

Roof coverings

Some houses and low-rise flats, particularly in the 1950s and early 1960s, used metals such as copper as a roof covering. Clay tiles were also used but most houses used the increasingly popular concrete tiles which were becoming available in various profiles, again a situation that continues to the present time.

Drainage and guttering

This was usually cast iron, asbestos cement or pitch fibre in the 1950s (and some houses had concrete gutters) but almost universally plastic was the material used by the late 1960s/early 1970s, a situation that continues to the present day. The two-pipe system of above-ground drainage was still in use at the start of the period but a single-stack system was the usual approach from the end of the 1960s onwards.

Walls

Most (traditional) houses from the mid-1950s onwards were built in cavity construction with brickwork in both leaves or with a brick outer leaf and block inner one. By the 1960s, blockwork was almost universal for the inner leaf. Cavity wall ties were usually galvanised steel, twist or wire type. By the 1980s, cavity wall construction was virtually universal with brick, stone or artificial stone in the outer leaf and aerated blocks in the inner leaf. Cavity wall ties were mostly galvanised twisted steel, galvanised wire butterfly or plastic. Copper was also used. Later on, wall ties would be mostly stainless steel (or galvanised ties with an Agrement certificate). Modern wall ties were mostly stainless steel.

Mortars were cement-based ordinary Portland (cement or masonry cement) and often mixed on site. Some developers used a partly pre-mixed mortar (containing sand, lime and pigment) with the cement being added on site. In the 1960s and 1970s walls were mostly finished with two coats of Carlite lightweight plaster. In later years, dry lining would become the standard finish.

Some houses in the 1960s and 1970s were built with brick and block cross-walls, i.e. the gable and party walls, with timber studding to front and rear walls. To some extent, these were a hybrid of traditional and non-traditional forms of construction.

These three houses all use cross-wall construction.

Damp-proof courses (DPCs)

These were mostly lead-cored bitumen coated, or bitumen felt in the 1950s. These materials were replaced by polythene DPCs over time.

Internal walls

Earlier on in the period these might be half-brick, timber studding or blockwork for load-bearing partitions. Non-load-bearing partitions included clay blocks, timber studding, lightweight concrete blocks, and stramit (straw-based walling panels). By the 1980s, 100mm dense blocks were generally used for load-bearing partitions, and 75mm blocks for those that were non-load-bearing. Non-load-bearing partitions might also be constructed of one of a number of proprietary systems which could be plastered or self-finished (laminated and Paramount were the two most common).

Windows

Galvanised metal windows were still quite common in the 1950s although by the 1960s, if metal was used, it would usually be aluminium. Single-glazed timber casement windows were still dominant in the 1960s and 1970s with double glazing starting to become more common in the 1980s. Different designs, such as pivoting windows (as used on tower blocks), became more common in houses. By the 1980s, plastic windows were available but most developers used standard metric timber windows (usually of casement design and often with no top lights). These could be single or double-glazed. In the 1980s, low energy (low e) glazing started to be used.

Ground floors

Although timber suspended ground floors were still used in the 1950s, ground-bearing in-situ concrete floors with a screed became increasingly popular (in part because of timber shortages). They often used a granolithic screed covered with thermoplastic tiles bedded in bitumen (often with no separate damp-proof membrane). Thermoplastic tiles were first produced in the UK just after the Second World War. The tiles were made from a mixture of resin binders, mineral fillers, asbestos and pigments. The early tiles were quite brittle. Asbestos vinyl tiles were introduced in the mid-1950s; they were made in much the same way but they were more flexible. As the dangers from asbestos were realised, its use in vinyl tiles was stopped, and other materials were used. An alternative that was popular in the 1950s was where the floor would be finished with wooden blocks laid in bitumen.

It was often the case that the bitumen used to lay the tiles or the wooden blocks would act as the damp-proof membrane (DPM). In other cases, a separate pitch or bitumen damp-proof membrane would be painted onto the concrete slab although sometimes no DPM at all would be used. In some areas of the UK the screed and waterproofing function were provided by 20mm asphalt laid on the slab. The asphalt could be trowelled to a smooth finish and so obviated the need for a separate screed. Polythene DPMs started to be used in the early 1970s and were the usual solution by the 1980s (they were laid on a sand blinding to protect against the danger of being punctured).

Pre-cast concrete floors were available by the 1980s and were sometime used on shrinkable clays. Ground floors in the 1980s would sometimes have included thermal insulation, often with a tongued and grooved chipboard finish instead of a screed, and this later became a common approach for the floor finish to ground-bearing slabs.

In recent times pre-cast beam and block ground floors have become the norm and ground-bearing slabs are fairly rare.

Upper floors

These normally consisted of softwood joists built into the inner leaf of the cavity wall or supported on joist hangers. They would be stiffened by herringbone struts – and later solid joist off-cuts might be used. In more recent times metal web and I joists have replaced traditional cut joists. Floor covering was softwood boards (often, tongued and grooved in the 1950s and 1960s) but by the 1980s tongued and grooved chipboard had become more or less the norm – and this continues to be the case (though oriented strand board (OSB) is also used). Ceilings were often asbestos composition boards or fibreboard in the 1950s, mostly plasterboard in the 1960s.

Foundations

The most common form of foundation from the 1950s onwards has been concrete strip foundations, though some of the system-built houses had raft foundations. Later on in the period, (concrete) trench fill came to supplant the more traditional strip. Piling was also used from the early post-war period onwards, usually in the form of short, bored piles and mostly on shrinkable clays. Later on in the period, where developments took place on 'brownfield' sites (and some other types of land), piled foundations would have been used. Also, in former mining areas for example, raft foundations may have been used.

Services

There were a variety of heating systems in the 1950s and 1960s, including coal fires, gas fires, electric storage radiators and warm air systems with ducting. Solid (smokeless fuel) boilers were sometimes incorporated into the fireplace and could provide a low level of space heating to perhaps two or three radiators. Gas-fired central heating as we know it today was the exception rather than the rule, but from the 1970s onwards it became, over time, the usual means of space heating. By the 1980s, most houses had central heating boilers, steel radiators and copper pipework. Room temperature would have been controlled by a room thermostat and thermostatic radiator valves may have been incorporated.

Defects in post-war houses

Some of the defects that are particularly associated with housing from this period include:
- *Failure of wall ties, due to inadequate (or no) rust protection or the corrosion of the galvanising.*
- *Sulfate attack in floors – where poor or no DPM was combined with a poor selection of hardcore, which contains sulfates.*
- *Cracking caused by unconsolidated hardcore in early examples of ground-bearing concrete floors.*
- *Rising damp in early concrete floors due to missing DPM.*
- *Lack of stiffening in 1950s upper floors and undersized floors due to materials shortages.*
- *Fungal attack on external woodwork (including windows and doors) due to poor quality of timber.*

Building movement: foundations   3

Both these problems have been caused by mining. Left: The Crooked House pub in Dudley, West Midlands, used to be in a coal-mining area. Right: Northwich in Cheshire is famous for its salt mines (photo courtesy of Cheshire Record Office).

Some key terminology

There are some basic definitions that need to be understood before considering foundation failure:

- **Cohesive soils.** *These are soils that have a natural tendency to 'ball' together when squeezed firmly in the palm of the hand.*
- **Consolidation.** *The action of increasing the density and strength of the soil by the expulsion of water and air under self-weight and constant external loading.*
- **Desiccation.** *The process of the drying out of a soil or sub-soil.*
- **Differential settlement.** *This occurs where settlement (see below) takes place at differing rates under different parts of a building or structure as a result of variations in loading and/or ground conditions.*
- **Heave.** *The upward or lateral expansion of the sub-soil or fill.*
- **Non-cohesive soils.** *These are granular soils which range from compact gravel and sand to loose sand.*
- **Settlement.** *The downward movement of the ground, or any structure on it, which is due to the load applied by the structure. (Note: this definition is the one accepted by surveyors and engineers, but the legal definition differs.)*
- **Subsidence.** *The vertical and downward movement of the ground which is not caused by the imposition of the building and/or foundation loads, i.e. loss of support from the sub-soil. The two main causes are trees and defective drains; less common are other causes such as mining subsidence and made ground/landfill.*

Introduction

All buildings move. In the majority of cases, this will be for quite acceptable reasons and the amount of movement will not adversely affect the building's performance.

This is an example of extreme movement in a building. However, the materials used in its construction are more flexible than modern materials and it has not collapsed. Perhaps the newer brick building is providing support.

Initially, when a building is constructed, movement downwards will be caused by the new loads imposed on the sub-soil beneath. Unless that sub-soil is particularly dense and strong, e.g. a rock such as granite, a newly constructed building will suffer from initial settlement as the sub-soil beneath is consolidated. Normally, the downward movement of the sub-soil and any resultant cracking of the building are both of a minor nature. With non-cohesive granular

Simplified diagram illustrating loading

Loads transferred downwards from roof and floors into walls and, ultimately, foundations.

Upper levels of sub-soil may be affected by seasonal change.

Ground floors can bear directly on the ground or be carried by walls.

Weaker sub-soils initially tend to consolidate under the newly applied loads of the building.

sub-soils, such as sand, this initial settlement occurs quickly and has often ceased by the time the construction has been completed. However, in cohesive clay sub-soils, settlement may occur over quite a long period, perhaps lasting 20 years or more from construction until the sub-soil is fully consolidated.

Buildings are also subject to seasonal movement and this will continue throughout the building's life. This is a result of certain sub-soils, in particular, clay, being affected by climatic changes, e.g. seasonal change as summer changes to winter and back again. Expansion of these sub-soils occurs in winter as water content increases and, conversely, contraction occurs as they dry out in summer.

It should be obvious that cracks in buildings caused by these processes will open and close seasonally, but in normal British weather conditions such movement should be within acceptable limits. Unfortunately, for a number of reasons, seasonal movement in water-susceptible sub-soils, such as shrinkable clay, can become excessive and lead to an unacceptable level of cracking and movement.

While accepting that much building movement will be of a minor nature in both extent and effect, there will still be occasions when an inspection of a building will reveal evidence of movement that causes concern. Often, the movement will manifest itself along lines of 'natural' weakness within the above-ground structure, e.g. windows, doors or junctions between separate elements of construction. It is then necessary to determine the cause(s) of the problem and its implications. This may involve assessing a wider range of defects other than just cracking and movement, e.g. cracking of an external wall due to foundation failure may also allow water penetration.

Left: The movement in this stone wall has taken place along a line of structural weakness, i.e. around an opening. Right: On the other hand, many older buildings are capable of withstanding substantial movement without cracking because they were built in brickwork (i.e. smaller structural units which can better distribute load) laid in a lime mortar.

Foundations: construction summary

Transference of load

The purpose of the foundation is to fully and efficiently transfer the load from the building to the sub-soil beneath. The exact type and performance requirements of a foundation will depend upon a number of factors including building type, use, load, etc. and sub-soil type, formation, condition.

The principle of the foundation

Ground level

The width of a simple foundation varies according to the bearing capacity of the sub-soil.

The building load, including the structure and all imposed loads (furniture, snow, etc.), exerts downward pressure on any footing/foundation.

Resistance of the sub-soil needs to be equal to or greater than the total applied load from the foundation.

The foundation is an essential element of construction in any modern building, although many older buildings have no formal foundations as such. It should be designed specifically for a particular building and the sub-soil beneath. For many buildings, especially low-rise housing, its design will be relatively simple although due attention must be given to any peculiarities of the structure and sub-soil. For other buildings which, perhaps, involve unusual construction, layout or use, or where sub-soil conditions indicate poor bearing capacity, a more sophisticated approach to both design and construction of the foundation will be required.

The pier of the brickwork in between the windows has created excess loading on the foundation/ ground. Cracking has been avoided because the walls were built in lime mortar, i.e. a flexible material.

Sub-soils: a brief outline

Consideration must be given to the type of sub-soil being built upon, both when dealing with original design and, subsequently, when trying to understand why a building is suffering movement.

Broadly speaking, there are five main types of sub-soil. These are set out below in descending order of bearing capacity:

- **Rock such as granite, limestone and sandstone.** *Strictly speaking, rock is not a soil. This should bear loads without problems. However, rock may contain faults or fissures that can collapse under load and result in settlement.*
- **Granular soils such as sands and gravels.** *These should bear loads without problems, but sands can be affected by flooding and loose sand is inherently unstable unless contained.*
- **Cohesive soils such as clays.** *These include firm/stiff clays, sandy clays, soft silt and soft clay. This group includes shrinkable clay. They frequently suffer problems, normally due to the presence of water or changes in water content. Approximately 75 per cent of UK building insurance claims are for subsidence in clay sub-soils.*
- **Organic soils such as peat and topsoil.** *These should never be built upon (unless piles are used) because they contain organic material and a high proportion of air and water.*

- **Made ground.** *Normally referred to as landfill, this is a separate category. In theory, if it is formed with non-organic material, it is possible to build directly on it once it has fully settled and consolidated. In practice, it is difficult to guarantee its content and it may contain a high proportion of vegetable matter and other deleterious materials, such as tin cans, old refrigerators, etc., which consolidate in an unpredictable manner.*

Left: An example of rock, in this case, sandstone clearly showing the bed layers. Right: A cohesive soil, in this case, clay, and its shrinkable nature can be seen.

Sub-soils will originally have been laid down in a series of layers or beds of varying thickness. Weaker sub-soils may lie over, or under, stronger ones, or there may be a multi-layered mixture. Good design must ensure that the foundations of a building are taken down to a suitable depth to avoid the effects of adverse climatic factors and, also, to bear on a sub-soil capable of providing adequate support. This may involve building down through weaker layers, such as peat or made ground, until a firmer, more stable sub-soil is reached. Alternatively, the design may involve a different approach and make use of a raft foundation that floats on, or just below, the surface of the sub-soil. Unfortunately, building movement often occurs because insufficient thought has been given to the type of sub-soil beneath the proposed building.

Foundations: a brief history

Historically, buildings were constructed directly on the surface of the ground, until it was realised that removal of at least the topsoil normally provided a firmer base. This was considered sufficient for cheaper brick and stone buildings, although timber-framed structures were often built on a stone plinth. In the nineteenth century, a thin layer of fire ash or furnace clinker was commonly spread over the exposed sub-soil to provide a level surface on which to build.

This load-bearing wall has a crude foundation composing of a shallow row of boulders rammed into the earth. The circled brush gives some idea of the scale.

Footings in poorer quality buildings up to late nineteenth century

Ground level

Depth could be as little as depth of topsoil

Topsoil removed and sub-soil surface levelled

25-50mm of ash or clinker

By 1900, new houses were normally required to have concrete or brick foundations (unless on rock). However, they were often less substantial than those used today. In addition, Building Control was far from rigorous.

Footings for larger, better quality buildings

Eighteenth century – stone slabs beneath wall

Nineteenth century – corbelled brickwork

Detail from about 1900 showing foundation for large house

Larger buildings were found to require a more sophisticated approach and this often resulted in slightly deeper foundations and an attempt to spread the greater loads involved over a larger area of sub-soil. Originally, this was by the use of stone slabs, but as brickwork became more frequently used in the nineteenth century, corbelled brick footings were developed. These were often used in better-quality housing especially in the latter half of the Victorian period.

The nineteenth century saw great advances in the knowledge of how buildings and materials perform and, at the same time, there was a realisation of the need for minimum construction standards. Although there had been previous attempts to legislate for building standards, these were neither of great detail nor necessarily applied uniformly in all parts of the country. The Public Health Act of 1875 was the first of a number of pieces of legislation that attempted to introduce minimum building standards on a national basis in England and Wales (except for London, which had its own rules and regulatory structure). Subsequent Public Health Acts, Building Acts and, since 1965, the Building Regulations, have set increasingly higher standards of construction which, in part, reflect greater understanding of how buildings perform.

Left: A section through a proposed house of the 1890s showing a shallow strip foundation. Right: Although of a later date (1920), this design shows strip foundation under stepped footings (drawings courtesy of Bristol Record Office).

London has, quite separately, followed the same raising of standards through its own building legislation, though since 1987 the Building Regulations have applied there as well.

Strict application of minimum standards by local authorities under the powers given to them under legislation resulted in both deeper and more sophisticated footings in the late nineteenth and early twentieth centuries. In particular, corbelled brick footings began to be used for all types of construction. Eventually, these were superseded by concrete strip foundations in the early part of the twentieth century.

Concrete strip foundation

It is unusual to find concrete strip foundations in nineteenth-century buildings and they only became common from the 1920s onwards, at the same time as external cavity walls became the accepted standard. Strip foundations are designed to spread building load in a similar, but cheaper, way to corbelled brickwork. By this time, it was also recognised that seasonal movement in certain types of sub-soil, in particular, shrinkable clay, required a foundation depth of at least 750mm below ground level. Subsequent knowledge, in part as a result of investigations into building movement caused by a series of extremely hot and long dry periods in the mid-1970s and the 1980s, has resulted in recommended foundation depth in climatically affected sub-soils, e.g. shrinkable clays, being increased to a minimum of 1000mm. This is increased to up to 2.5m or greater where located close to, or above former, trees. However, in more stable sub-soils, such as gravel or firm sand, a shallower foundation can still be constructed, e.g. at least 450mm deep in firm sand, though the risk of frost attack must be recognised.

There have been great advances over the years in the development of new forms of construction, materials and components. Pressure caused by increased demands on land use, allied to our greater knowledge, has resulted in the development of a variety of alternative foundation techniques. Some have been introduced because they are more economic, others because they allow inherently more difficult sub-soils to be built over.

Traditional strip foundation

Trench-fill foundation

Drains and pipes should be above the foundation where possible and not too close.

Minimum thickness 150mm.

Depth must be suitable for the sub-soil type.

It is rarely less than 600mm; it may vary between 750mm and 1000mm (or more) depending on the date of construction.

Depth can be less in more stable sub-soils.

Width of foundation depends on bearing capacity of ground.
Strip foundations need additional width to provide working space.

Trench-fill foundation

This is a modern alternative to a strip foundation that is suitable for use down to depths of about 1000–1500mm below ground level. Below this level, it becomes uneconomic. Its advantages are its simplicity and relative cheapness because its use of mass concrete avoids the need for blockwork or brickwork below ground level. It can be found in buildings constructed from the 1970s.

Trench-fill foundations on a modern housing estate.

Raft foundation

A raft foundation is basically a slab of reinforced concrete at, or just below, ground level. It acts as both a ground-bearing slab over the floor of the building and a footing to the walls. It enables the loads of the building to be spread over a large sub-soil area and allows areas of either known or suspected weak sub-soil to be bridged over. Often used above former underground mining areas, it requires careful design by structural calculation. A twentieth-century approach to a variety of sub-soil problems.

Raft foundation

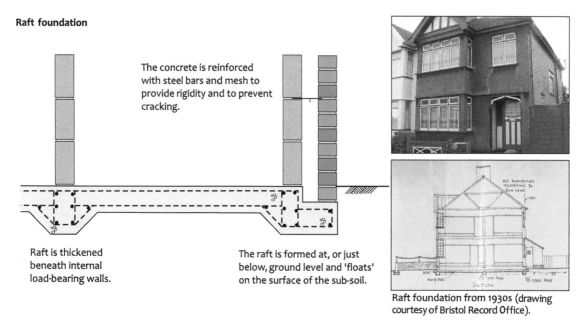

The concrete is reinforced with steel bars and mesh to provide rigidity and to prevent cracking.

Raft is thickened beneath internal load-bearing walls.

The raft is formed at, or just below, ground level and 'floats' on the surface of the sub-soil.

Raft foundation from 1930s (drawing courtesy of Bristol Record Office).

Piled foundation

Filled ground, such as former quarries, and ground with a weak sub-soil profile, such as alternating layers of soft clay and other similarly soft sub-soils, may require piling in order to properly carry the building load down to a sub-soil layer capable of supporting the load. The piles support ground beams which act as bridges beneath the walls above. In these circumstances, the ground floor might well be a reinforced concrete slab or pot and beam construction spanning the full distance between structural walls to avoid any problems of inadequate support. There is a variety of different types of concrete pile ranging from mini-piles, suitable for depths down to about 2m below ground level, to driven or bored piles which can be used down to much greater depths – 8m or more. This form of foundation is a late twentieth-century technique as far as low-rise domestic construction is concerned and is rarely found in buildings constructed before 1970, except as remedial underpinning.

Piled foundation

Modern steel-cased concrete piles

Ground level

Ground beams span between the piles. Beams may be at, or just below, ground level.

Layers of weaker sub-soil or made-up ground

Plan

The piles are taken down to a sub-soil capable of supporting the load of the building. They can be end-bearing and/or friction bearing.

In modern construction, piles are usually formed in concrete or steel. In Amsterdam, timber piles were common. Rot in the piles has caused extensive differential settlement in this building.

Conclusion

When examining any building for movement problems, reference must be made to the date of construction of each part. Consideration will need to be given to the implications of dealing with different ages, types and styles of construction within the same property. Initially, records should be examined to determine previous use(s) of the land, e.g. an infilled pit. Unfortunately, it may be difficult to ascertain these facts without some element of detailed and, possibly, destructive investigation, e.g. the removal of render/plaster or the excavation of trial holes. Foundations and their defects, by their very nature, are invariably difficult to precisely identify without some form of excavation. Table 3.1 summarises the common structural defects of foundations at the construction stage.

Table 3.1 Common structural defects of foundations at construction stage (new build)

POTENTIAL PROBLEM AREAS	IMPLICATIONS	TYPICAL DEFECTS
Insufficient depth of foundation (especially in shrinkable clay sub-soils)	Seasonal movement due to changes in water content of sub-soil Movement due to trees and shrubs	Cracks in walls especially at 'natural' lines of structural weakness, e.g. windows, doors, junctions with extensions and bays
Insufficient width of foundation	Load insufficiently spread over sub-soil Tendency for settlement to occur	Cracking in walls above ground level
Soft spots in sub-soils	Foundation is inadequately supported by the sub-soil leading to settlement.	Cracking of walls above the affected section of foundation

▶

POTENTIAL PROBLEM AREAS	IMPLICATIONS	TYPICAL DEFECTS
Insufficient steps in strip foundation on sloping ground	Tendency for building to slide or creep especially if there is soil erosion	Displacement movement of building / section of building. Cracking in walls
Trees and large shrubs in close proximity (in shrinkable clay sub-soils)	Sub-soil subject to changes in water content due to desiccation (lack of moisture) and heave (excess moisture)	Cracking in walls especially along natural lines of structural weakness
Recent removal of trees or large shrubs (especially in shrinkable clay sub-soils)	Sub-soil subject to a period of heave (over several years)	Cracking in walls especially along natural lines of structural weakness
No provision of compressible material to piles and ground beams in shrinkable clay sub-soils (to allow for heave)	Any expansion (or heave) of the sub-soil will lead to structural movement	Upward movement with cracks along lines of weakness in structure
Clay sub-soil with high sulfate content	Sulfates attack concrete foundations and below-ground cement mortar jointing	Movement in above-ground walls. Expansion and deterioration of below-ground cement mortar and concrete
Inadequacy of structural support from landfill (made ground)	Excessive settlement over a prolonged (often indeterminate) period of time	Cracking of walls. Complete collapse of building
Close proximity of drain runs which are deeper than foundations and/or lacking concrete cover	Inadequate support to the foundations	Cracking of walls above ground level Walls out of plumb due to lateral displacement

Foundation defects: the causes

Introduction

There is a wide range of reasons for foundation failure, but often it is because of inadequate design and/or poor construction. Historically, this was due to lack of technical knowledge of construction and poor understanding of how sub-soils perform when under load. With more recently constructed houses, it has to be accepted that such failure is often due to incompetence, e.g. lack of proper site investigation, poor design and/or bad site practice.

Left: Not all defects are as dramatic as these. The Temple Church in Bristol was started in 1398. The tower soon leant because the alluvial soil was too weak to support the heavy load of the tower. The upper stage (constructed at a different angle in an attempt to straighten it) was built in 1460. According to local legend, the Americans, stationed in Bristol at the end of the Second World War, intended to demolish the tower, incorrectly assuming the lean to be the result of bomb damage. Right: The 'defect' was caused by an explosion at Bridgenorth Castle during the English Civil War in the seventeenth century (right-hand photo courtesy of Peter Smith).

The investigation of any suspected foundation failure will generally involve assessment of movement that is initially apparent in other elements of the structure, e.g. walls, floors, roof. It is only after evidence of above-ground movement has been analysed and its wider implications considered that physical investigation of the underground parts of a building's structure will be undertaken (probably by means of trial excavations). The examination of the foundation, or lack of it, will include investigation of the sub-soil beneath the building. At the same time, reference will need to be made to topographical factors such as hills, valleys, water courses, trees and vegetation, in order that the effects of such external influences upon the structural performance of a particular building are fully recognised. Similarly, other factors of a man-made nature will need to be considered with regard to their effect upon the building, e.g. proximity of underground drains and services, proximity of recent construction, traffic movement, and industrial activities. (See Chapter 5 for a more detailed discussion of site investigation.)

The extent of foundation problems can be determined from the number of insurance claims made annually for subsidence. In a normal year, they are between 30,000 and 40,000, but after dry periods they can rise, e.g. in 2003, after a drought year, they amounted to a total of over 55,000. The total cost of insurance claims peaked at over £500 million in 1991, while the average cost in recent years is £400 million per annum.

Initial settlement

Certain sub-soils, such as clays and similar materials, may be subject to varying amounts of compression due to the pressure applied by a new building. Often, this is due to water contained within the sub-soil being squeezed out as the sub-soil consolidates under the newly applied load. Such settlement will be of a variable degree depending upon the type of sub-soil. A shrinkable clay sub-soil with a high water content will consolidate further and over a longer period of time than a firmer clay with less water. Movement of this nature will be at its greatest in the first few years after construction and will normally cease within 10–20 years of erection. Most buildings will suffer from it to a greater or lesser degree.

Normally, the effects of such movement are within acceptable levels of damage to the structure and finishes of a building, with cracks up to 5mm across. These can be filled as required for decorative purposes. The cracks can occur anywhere within floors, walls and ceilings, but are commonly experienced at the junction between two or more elements, i.e. at the weakest points in the construction. Sometimes initial settlement is also linked to seasonal movement (see later section of this chapter) and the cracks can open and close accordingly.

Originally constructed in the seventeenth century on a shrinkable clay sub-soil, this building experienced extensive initial settlement.

Occasionally, the damage caused by initial settlement can be greater than 5mm or be spread over a longer period of time. If the movement is significant and proves to be progressive, then remedial under-pinning and crack repair may be required.

It also needs to be recognised that even apparently firm sub-soils, such as gravels and softer rock formations, may compact slightly under the initial load of a building. In these circumstances, any cracking of the structure and finishes will probably be of minor nature.

Subsidence due to sub-soils containing organic matter

The construction process requires the removal of all topsoil incorporating organic or vegetable matter. This is usually to a depth of 150–225mm across the whole area of the building. Normally, such removal will have disposed of all soil containing vegetable matter. However, certain deeper sub-soils, in particular peat that contains high levels of organic matter and water, will also be unacceptable.

The effects of compressible sub-soils

Cracking pattern will be unpredictable. Cracks tend to be wider at the bottom.

Ground level

Settlement due to a compressible sub-soil can be extensive and uneven.

Shallow foundation on organic material (peat).

Any applied load will compress such a material and result in movement of the building above. These types of sub-soil should be completely removed or, if of sufficient depth, they will require the use of deep foundations, such as piles, bearing through the weak layer onto a firmer stratum beneath.

The effect of building on an easily compressible sub-soil can be cracking within the walls and other elements of the structure as the inadequately supported foundations settle. Such settlement can be extensively and unevenly distributed around a building, resulting in significant cracking with no discernible pattern. The pressure from the weight of the building will continue until the organic sub-soil is fully squeezed and compressed. If the sub-soil is deep, catastrophic failure of the building will occur. Remedial under-pinning may be carried out, but often with great difficulty. Frequently, the only choice when dealing with this problem is the demolition of the building.

Subsidence due to seasonal movement

A number of sub-soils are affected by seasonal change. This is due to seasonal variation between wet/cold and hot/dry weather. The extent to which a sub-soil is affected depends mainly on its water content. When a great amount of water is present, normally in winter, the sub-soil will expand and, conversely, it will contract when there is a reduction in water content, normally in summer. The presence of trees and vegetation, or their removal, can accelerate or exacerbate the effects of such changes.

Frost heave

This is a problem that can occur when the ground has a high water table or after a period of high rainfall. Water in the soil expands as it turns to ice in cold weather. The effect upon the sub-soil is known as frost heave and it results in the ground expanding, primarily upwards. Sub-soils that can be adversely affected in this way include silts, fine sands and chalk.

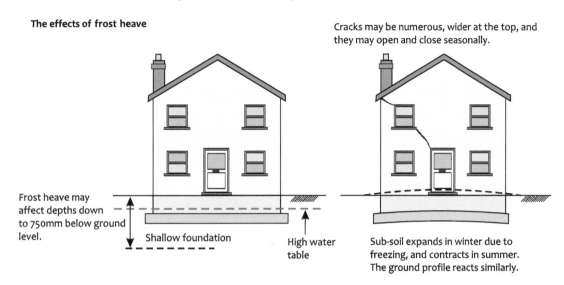

The effects of frost heave

Cracks may be numerous, wider at the top, and they may open and close seasonally.

Frost heave may affect depths down to 750mm below ground level.

Shallow foundation

High water table

Sub-soil expands in winter due to freezing, and contracts in summer. The ground profile reacts similarly.

In the UK, it is unusual for frost to affect a greater depth than 700mm below ground level. When frost heave does occur, it will have a very noticeable seasonal effect with cracks in a building opening and closing between winter and summer. It can sometimes be severe enough to require under-pinning down to a non-frost-susceptible level below 700mm to avoid its effects.

The effect of frost heave can be seen in this boundary wall. The top of the wall is bowed and the crack is wider at the top than the bottom.

Shrinkable clays

Clay soils are capable of absorbing and releasing large amounts of water, but not all clays do so. They range from firm clays with relatively low water content to shrinkable clays that contain high levels of water. The higher the water content of the clay, the more susceptible it is to climatic change and seasonal movement. Shrinkable clays can expand or contract by between as much as 50mm and 75mm seasonally. They are also subject to greater initial settlement as water being squeezed out under a new building's load will cause the sub-soil to consolidate. A look at the geological map of England and Wales reveals that the clays that are subject to shrinkage and swelling are to be found mostly south and east of a line drawn from Gloucester to Hull. Those areas north and west of this line tend to contain a higher proportion of sand and are less prone to shrinkage.

Distribution of shrinkable clays in England

In summer, shrinkable clay sub-soils will normally tend to contract due to reduction of their water content. This reduction of the soil moisture content is known as desiccation. It can be made worse by the presence of trees and shrubs that take up moisture through their roots. The resultant contraction of the sub-soil can lead to subsidence – the building above will sag/move downwards. In such circumstances, cracking will tend to be wider at the bottom rather than the top. However, the presence and effect of trees may confuse the damage pattern.

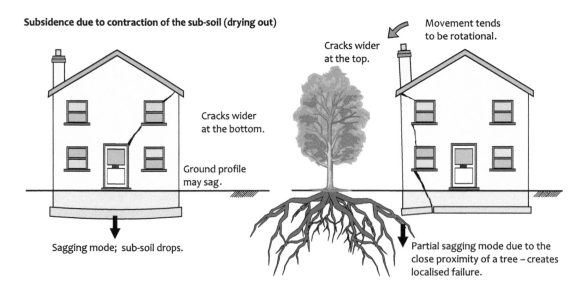

Subsidence due to contraction of the sub-soil (drying out)

Cracks wider at the bottom.

Ground profile may sag.

Sagging mode; sub-soil drops.

Cracks wider at the top.

Movement tends to be rotational.

Partial sagging mode due to the close proximity of a tree – creates localised failure.

Conversely, in winter and in periods of excessive rainfall, shrinkable clays will tend to expand as water content increases. The increase in volume of the clay leads to an effect known as heave – the building above will rise upwards (known as hogging). The effect is rather similar to frost heave (as described previously) in which the cracks tend to be wider at the top and narrower at the bottom.

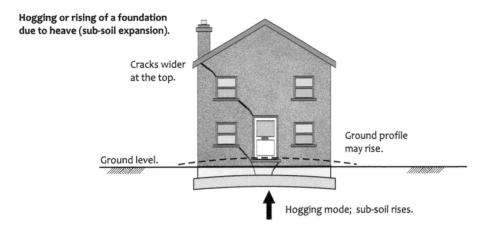

Hogging or rising of a foundation due to heave (sub-soil expansion).

Cracks wider at the top.

Ground profile may rise.

Ground level.

Hogging mode; sub-soil rises.

On investigation, this movement crack was found to be as a result of heave (a nearby tree had recently been removed). Note that the crack is wider at the top than the bottom.

The effects of seasonal changes in shrinkable clays can generally be avoided by ensuring that foundations are formed at a sufficient depth to avoid the problems of subsidence and heave. This was understood, although not well practised in the late nineteenth century. At this time, and for much of the twentieth century, a sufficient depth was thought to be 750mm, but research in the past 30 years indicates that a depth of at least 1200mm is needed (more if it is near trees – see p. 76). The research was originally instigated following the very long hot and dry period of 1975–1976 when a considerable number of buildings were affected by subsidence in shrinkable clays. The increased knowledge gained from that experience and subsequent similar hot, dry periods in the 1980s and early 1990s has led to a requirement for deeper foundations in these clays. The research has also indicated that in many cases the movement sustained by a building in such excessively dry spells can be matched by that movement closing back up as sub-soil conditions return to their norm in subsequent seasons. Unfortunately, this closing-up process does not always return a building to its original state of stability or condition. It can also be delayed by further dry weather conditions after the original hot spell.

This trench in a shrinkable clay sub-soil has been excavated to a depth of at least 1000mm to avoid the effects of seasonal change.

When subsidence due to a shrinkable clay sub-soil is suspected, the best approach to any investigation is to proceed slowly and methodically unless there is the likelihood of collapse. Initial investigation should indicate whether the movement is due to subsidence or heave, but it is usually more difficult to confirm that the movement is progressive, i.e. still continuing. Often, this can only be found out by monitoring the building over a period of time. This could be for weeks, months or even years. Monitoring may reveal a variety of conditions – the movement has stabilised and is not getting worse; it is progressing slowly/quickly and is worsening; it has changed direction and the cracking is closing. Once monitoring has confirmed what is happening, decisions can be made on any remedial work needed. (Monitoring is discussed in more detail in Chapter 5.)

These cracks Left: appeared after a long hot summer. The house is built on shrinkable clay which is susceptible to changes in moisture content. Right: Investigation of this building revealed a shallow foundation under the single-storey extension resulting in differential settlement. The cracking is both vertical and horizontal – running along the line of the DPC to behind the water pipe and along the stone sill.

Subsidence due to trees and shrubs: proximity to buildings

Movement due to the close proximity of trees to a building is a common problem. This is due to the effect of the tree on the water content of any sub-soil that it is growing in. Trees require water to survive and grow. Much of this is taken up through the root system. A mature poplar can take up to 50,000 litres of water from the sub-soil each year. Other trees, including oak, can have a similar effect and, the more mature a tree, the greater the amount of water it requires.

The root radius of a tree is often equal to or greater than its height above ground level. In some cases, e.g. willow, poplar, elm, the radius can be up to twice the height. The older a tree is, the greater the radius, and the greater its effect upon the sub-soil (Table 3.2).

Table 3.2 Popular species of trees known to cause sub-soil problems

SPECIES	MATURE HEIGHT (METRES)	LIKELIHOOD OF DAMAGE TO CLOSE BUILDINGS
Oak	25–30	1 (most likely)
Poplar	30–35	2
Ash	25–30	3
Plane	25–30	4
Willow	25–40	5
Elm	25–30	6 (least likely of the six shown)

Note: It should not be assumed that the above list of trees is comprehensive. There are a great number of other species that are known to cause problems, including many so-called small ornamental species such as hawthorn. Even shrubs and climbers, such as pyracantha, wisteria and rambler roses, have been identified as the causes of building movement. Also some leading authorities differ with regard to likely position.

The cycle track in the foreground has subsided by approximately 75mm. The cause is the oak tree behind which is also close to an adjoining garage in the background. Hopefully, the foundations of the garage were constructed to take account of any desiccation of the sub-soil by the oak tree.

Left: Local authorities are often liable for subsidence in adjoining properties caused by trees planted in streets. Right: Even a supposedly safe species, such as silver birch, can pose problems, in this case, affecting the adjoining pavement.

Sub-soils containing trees are affected to a varying degree by the type, age and size of the trees and, in some cases, by the number of trees growing. Gravels, firm clays and, to some extent, sandy sub-soils are less affected than shrinkable clays and other water-sensitive sub-soils. The effect of any tree growing in one of the latter types of sub-soil is to reduce the volume of the sub-soil as it takes up ground water. This is on top of any drying out due to seasonal change (see p. 82) and leads to contraction of the clay and the downward movement of any foundation supported. The result can be cracking of the walls and other structural elements of any building above due to subsidence. At its worst, it can affect the complete building.

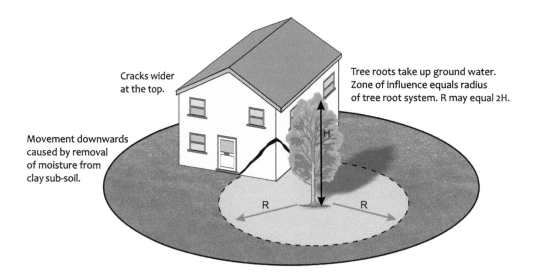

Cracks wider at the top.

Tree roots take up ground water. Zone of influence equals radius of tree root system. R may equal 2H.

Movement downwards caused by removal of moisture from clay sub-soil.

Seasonal changes can influence the effect of trees upon sub-soils. In particular, prolonged warmer and dryer spells can accelerate sub-soil shrinkage because moisture taken up by the tree root system is not replaced quickly enough to prevent drying up and desiccation of the sub-soil. In such periods, the root system may expand in its search for moisture and thus increase the ground area adversely affected by subsidence.

Generally, the further north and west a property is situated in the British Isles, the higher the rainfall that it will experience. Any moisture deficit caused by trees and shrubs will be more readily replaced, the greater the distance that the building is located from London and the south-east of England. The National House Building Council (NHBC) recognises this by allowing a progressive reduction in the depth of a foundation of a building that is close to trees, the further north and west from London that the building is situated.

Further problems can be caused by tree roots due to their physical presence and strength. As the tree grows, the radius of the root system increases and the individual roots grow in size. This, in time, can lead to the displacement of the surrounding sub-soil, upwards and outwards pressure on foundations and walls in the vicinity of the root and even growth through underground walls of softer/weaker materials.

The Building Research Establishment (BRE) has undertaken considerable research into the effect of trees, subsidence and heave in clay sub-soils. This is wide-ranging and continuing. Analysis of its database indicates that oak trees are by far the most damaging species of tree. However, the recommendation by the NHBC of up to 2.5 metres (or more where appropriate) deep foundations for all new housing near trees should be adequate to prevent damage by seasonal movements. Unfortunately, there are a great number of houses and other buildings with much shallower foundations than 2.5 metres (some of quite recent construction) that were built before stricter guidelines were issued for buildings erected close to trees. Furthermore, building owners and occupiers generally do not understand the problems caused by the close proximity of trees to buildings and frequently exacerbate the potential problems by new planting, often of totally inappropriate species. Similarly, trees are often removed by owners, occupiers and developers, sometimes leading to problems associated with heave.

Another recent contributory factor is the modern practice of householders paving over front gardens or having extensive patios. These impervious surfaces increase the zone of influence of any nearby trees, leading to a greater potential for the drying out of the sub-soil and the effects of subsidence, and also contribute to flash flooding.

Trees: the effect of their removal

Left: The close proximity of this protected oak meant that deep foundations were required to the adjoining house. Right: This pyracantha is not a problem – yet!

The removal of trees often causes a process of water re-absorption in a susceptible sub-soil such as shrinkable clay. This is known as ground heave or ground swell because the sub-soil has a tendency to expand as it takes up groundwater that would previously have been absorbed by the roots of the removed tree. The affected area will extend to the limit of the root system of the former tree and the process of ground heave may occur over a period of several winter seasons until the sub-soil has a water content approaching that normally found when unaffected by trees.

The effect upon any building within the radius of the former root system is an upward movement of the foundation or footing. Cracking in walls above will tend to be wider at the bottom than the top. The movement may affect part or all of a building and may continue over several years.

An example of a pollarded tree in a street in Bristol.

If a tree is causing concern, the advice of an experienced arboriculturalist should be sought to consider effective tree management procedures, e.g. alternative pruning remedies such as crown thinning, crown reduction, pollarding or root severance or a root barrier can all be considered.

In recent years, the NHBC has identified a problem where piles and ground beams have been installed as a foundation to avoid heave from tree removal affecting new houses. Void formers or compressible material are provided alongside and under the beams to protect against the effects of heave. In some cases, poor fitting of the sacrificial expansion elements has led to wet concrete leaking into the voids or compressible material. This has led to the heave forces being transmitted to the building and has resulted in extensive damage.

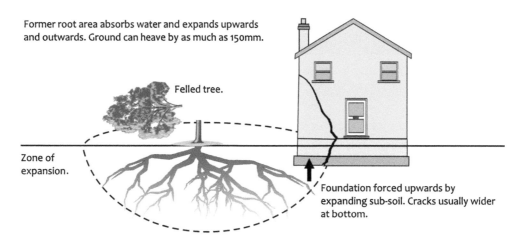

Former root area absorbs water and expands upwards and outwards. Ground can heave by as much as 150mm.

Felled tree.

Zone of expansion.

Foundation forced upwards by expanding sub-soil. Cracks usually wider at bottom.

Section   **Elevation**

Reinforced concrete ground beam

Reinforcement cage

Steel reinforcement in beam tied to reinforcement in piles

Depth to suit local conditions

If the clay swells it will compress the collapsible layer, but will not damage the building. The collapsible layer can also be fixed at the sides of the beam if there is a danger of lateral expansion (see right-hand picture). The compressible material is at danger from leakage of wet concrete which hardens within its voids and defeats the objective of a collapsible layer.

Shrubs, hedges, etc.

The close proximity of shrubs to a building, especially if mature or in large numbers, can have the same effect as trees, i.e. sub-soil shrinkage and building movement. Conversely, the removal of plants, particularly dense hedges and mature species, can lead to ground heave problems.

Greenfield developments on shrinkable sub-soils can be adversely affected by ground heave because they often involve the removal of hedgerows as well as associated ditches.

The Joint Mitigation Protocol

This is a method of subsidence claims management in the Greater London Area to deal with subsidence of adjoining property that is suspected of being caused by street trees. Published in 2008 by the London Tree Officers Association, nine of the 32 London Boroughs as well as three insurance companies had signed up to it by the end of 2012.

Differential settlement

This is the term used to describe the circumstances in which separate parts or elements of a building are subject to quite independent movement, e.g. a bay window or rear annexe of a house settles downwards while the main part of the building remains stable. This is often due to the fabric of the separate areas being quite distinct in form and/or construction, or the foundations being formed at different depths. Any part of a building formed in lighter

On investigation, the cause of the movement at the junction of this terraced house and its rear annexe was found to be extremely shallow Victorian foundations beneath the annexe, leading to differential movement.

construction will impose lower loads on its foundations and settlement may be less than for the main structure. Alternatively, shallower foundations will be more susceptible to climatic change, sub-soil shrinkage and the effect of trees or their removal.

The effect is that the two parts of the building settle or move at different rates or even in opposite directions. Cracking will tend to occur at the interface between the parts and will often indicate that they are tending to slip or slide past each other. Sometimes, the effect of differential movement will be movement of one part away from the next, possibly linked with settlement as well. Differential settlement or movement can occur in buildings of all styles, ages and periods. It is commonly found in older buildings where attention to foundation depth and detailed construction was less strictly controlled than in more recent years.

Left: This lean-to extension was built on inadequate foundations. The rotational movement is most evident at the top. Right: This extension also has a shallow foundation – in this case, the movement has been exacerbated by a fractured drainpipe.

Typical late Victorian terrace

Shallower foundations to bay

Shallower foundations to annexe

Differential movement may be vertical (up/down). It may also be rotational.

Cracking may occur in walls, ceilings and floors along movement line.

Plan

Dotted line indicates movement plane, i.e. interface of different forms/areas of construction where cracking can occur.

Subsidence caused by drains

A surprising amount of building movement is the result of underground drainage problems, which themselves may have been caused by settlement of the building. Such a circumstance is where a drain has been cracked or crushed due to the effect of building load. This can occur where the drain has been laid too close to the building and at a depth lower than the foundation or footing. The broken drain will tend to leak water into the surrounding sub-soil and can affect it in a number of ways that result in movement problems. It may, for example, wash a sandy sub-soil away from under a foundation (the cause of up to 20 per cent of insurance claims for subsidence) or it may increase the water content of a shrinkable clay, making it more sensitive to building load and thus lead to settlement or subsidence. Alternatively, the drainpipe may not fail, but the trench in which it lies may cause problems. Generally, any back-filled trench will not be as capable of providing as good support as the original sub-soil. Any pressure on the back-fill will tend to consolidate it, leading to settlement of any close building.

Drainage problems

Corners of buildings are often subject to the effects of drain problems as builders take the shortest route to save time and cost.

IC

Plan

Corner of building subject to settlement

Inspection chamber

Section

GL

Angle of load from the foundation

Softer/less compact soil in backfilled trench can cause instability. Drain cracks under building pressure. Displacement/crushing of drain alone may be sufficient to cause subsidence. If drain cracks or leaks, water adversely affects sub-soil and causes subsidence.

Drain-related problems can sometimes be exacerbated by trees, particularly poplars and willows, which can be extremely aggressive in their search for water. It is not uncommon for such tree roots to break into drain runs at weeping or defective joints to obtain moisture, especially in dry climatic conditions. Tree roots may also exploit the situation where a drain has already been damaged by building settlement. The leaking drain results in increased root growth of nearby trees and this can lead to more extensive drain problems and soil, especially if of a sandy nature, being washed away, creating increased building movement.

These exposed rigid mortar joints have cracked and are leaking into the soil.

Other causes of drain failure include damage by traffic, ground movement, e.g. landslip, rigid pipe joints or inadequate movement joints where pipes pass under buildings or join inspection chambers.

Flooded sub-soil caused by physical damage from tree roots

Drain-run within radius of root system broken into – possibly at joint.

Water leaking out of broken drain will affect sub-soil. Sandy sub-soil may be washed away. This may cause part of the foundation to drop.

In shrinkable clays increased water content may make sub-soil more susceptible to movement. It may swell as it absorbs moisture.

Swelling clay cracks foundation.

Subsidence due to landfill

The NHBC has stated that construction over poor landfill is the largest single cause of foundation failure for new houses – over 50 per cent of all failures. It must be recognised that any landfill site will be inherently unstable over a long period of time and that, if built over, appropriate foundation design must be undertaken. This means the use of piled foundations bearing on the sub-soil beneath the fill. This avoids the effect of the filled ground consolidating. Alternatively, a ground improvement process, such as vibro-compaction, may be undertaken in conjunction with reinforced concrete raft or beam foundations.

Recent pressures upon land use have resulted in the increasing development of poorer-quality land, including reclaimed land and former fill sites. However, this is not a new approach to land use. The Victorians would often raise the ground level on river flood plains to overcome flooding and subsequent generations over the last 100 years or so have constructed on these filled areas as towns have expanded.

Filled ground, however well formed, will need time for the rubble, waste and other matter to consolidate. In normal circumstances, a minimum period of at least 25 years, and perhaps

Initial ground profile of landfill site, e.g., old quarry or railway cutting.

Ground profile after 20–25 years. Landfill can still be unpredictable for many more years.

Consolidated fill

Any new building will be subject to settlement unless suitable piled foundations are used.

The quarry on the left was closed down some 50 years ago. Since then, it has been filled in and built on. Look at the two men in the bottom right to get some idea of scale.

longer, will be needed before the ground is sufficiently stable to build on. Unfortunately, the quality control of the fill material has, in the past, tended to be very poor or lacking altogether. Instead of inert building and waste materials being deposited, large amounts of organic and other inappropriate materials have been allowed to be used for fill. Apart from the adverse effects of obnoxious gases and combustibility, the rotting of organic matter and rusting of metals can create voids in the fill that may take decades to consolidate properly. This means that even apparently stable filled ground may still be subject to unacceptable levels of settlement.

Most landfill is associated with former quarries, gravel and clay pits, open-cast mines and railway cuttings. However, the unusual must also be considered if a building is suffering from settlement, e.g. raised flood plains, back-filled former streams and deep culverts in town centres.

Generally, the problem caused by landfill is settlement of ground and building. Occasionally, the opposite may occur – in both Teeside and Glasgow there have been instances of expansion of the filled site caused by the swelling of pyritic shales used as fill material. Pyritic shale contains iron sulfide that oxidises and expands.

It is wise to undertake at least a desk study when developing land or buildings where there is the possibility of fill having occurred. However, only a physical survey involving trial holes and/or boreholes will provide the fullest evidence of sub-soil conditions (see Chapter 5 for further information on investigations).

Top left: This end-of-terrace house is on a sloping site and partly built on filled ground (behind the retaining wall). Bottom left: Viewed from the rear the movement is self-evident. Top right: There is extensive cracking in the end wall. Bottom right: The adjacent property seems to be crack-free.

This 1950s house was built on a reclaimed industrial site, the fill included shale (a mining waste). This expands when wet and has led to heave affecting the floors, but not, fortunately, the foundations.

Subsidence due to mining-related problems

Many areas in England and Wales have been extensively mined for coal, lead, copper and other ores and minerals as well as stone for building. Some areas are well known for mining, e.g. Yorkshire, Nottinghamshire and South Wales (coal), Cornwall (lead and copper), Cheshire (salt). Others may not be so well known because mining ceased many years ago, e.g. Bristol, North Somerset and South Gloucestershire (coal mining), Woolwich and Norwich (chalk), Dudley and Bath (limestone), Droitwich (rock salt).

The effect of mining upon any buildings above depends upon the type of sub-soil over the mineral-bearing stratum. Strong, dense rock such as granite tends to result in little, if any, adverse effect upon any building. However, where the sub-soil above is of a weaker or softer nature, there is a tendency for the land above the mined area (known as the overburden) to be subject to subsidence. Often, the miners will have left pillars of solid material to support the ground above the mine, the intention being that there would be a bridge effect between each pillar. More often than not, this does not work as, over time, the bridged areas collapse causing the ground above to subside. This results in low spots above the mined areas and a lunar-like landscape. Sometimes, the thickness between the mine roof and the buildings above is very shallow, e.g. only 2 metres, in some cases, above the limestone mines of Bath.

Map showing main coal mining
areas in England, Scotland and Wales.
Most pits in these areas closed during the latter
half of the 20th century.

This map of mainland Britain gives an indication of how extensively one type of mining – coal mining – has affected many areas, while the photograph shows a street in Droitwich, a town that has suffered subsidence because of brine extraction in the past. This street used to be level.

Pillar and chamber mines can be found throughout the UK (photo courtesy of mine-explorer.co.uk).

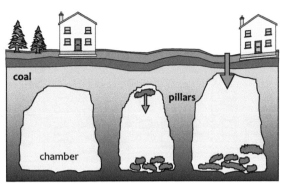

As the roof of the chamber collapses, the surface of the ground develops a 'lunar' landscape.

Left: Limestone mines below Bath were the subject of a government-sponsored engineering scheme to stabilise the weak rock overburden beneath the city. This was completed in 2009 and cost £165m (photo courtesy of Bath & North East Somerset Council). Right: This Victorian terrace was affected by mining subsidence some 100 years ago. The buildings were constructed with flexible lime mortar, so cracking was avoided.

Such subsidence might take place rapidly or over a longer period of time. This means that any location where mining has taken place in the past must be viewed with caution even if there is no evidence as yet of subsidence occurring and a thorough investigation should be carried out (see Chapter 5).

Older buildings would almost certainly not have been designed or constructed to take account of movement of this nature. Similarly, modern buildings would not have taken mining problems into account unless direct need had been proved or anticipated. All of these buildings would probably be affected by ground movement that is the result of mining having taken place beneath. Such movement will normally tend to be stealthy rather than suddenly catastrophic. Many older buildings so affected have a greater tolerance of movement because they are constructed with lime mortar. This results in a more flexible structure. On the other hand, modern cement-jointed cavity walls, while being stronger than comparable monolithic construction, are also more brittle, with the result that they are less tolerant of movement. With both older and modern buildings, massive subsidence may result in eventual collapse.

Original ridge line

Terrace of houses subject to mining settlement. Cracking will relate to degree of settlement. Doors, window heads, etc. will lean, tilt and stick. Subsidence may lead to collapse.

Uneven settlement of original foundations (where they existed)

The modern approach is to build on raft foundations and provide movement joints between units to allow settlement without damage.

Shallow mine workings should also be stabilised.

Modern buildings in mining areas that are subject to subsidence should be constructed with raft foundations so that they can 'float' on the surface of the ground. Longer buildings and terraces of houses should have movement joints incorporated to allow differential settlement to occur without causing structural problems.

Left: The catastrophic failure of this garage was due to unknown deposits of gypsum being washed away (photo courtesy of Azadour Guzelain). Right: The failure in this terraced Victorian house was caused by the collapse of a coal mining tunnel. The houses on either side were unaffected as they were constructed with bricks set in lime mortar and its tolerance of movement can be seen here.

A further problem can be that caused by the covering, if any, of shafts of former mines, where no record has been kept of their position. Often, the cover consisted of timber baulks or sleepers with a metre or two of overburden placed on top. It is not unusual for these to have been subsequently built over and the first indication of their presence may occur when the timber sleepers rot. The overburden gives way and the structure above is suddenly left unsupported.

The Coal Authority was established by the UK Government in 1994 and keeps records of all known coal mineshafts and tunnels. However, many are unrecorded, especially before 1872 and, prior to 1961, and even the accuracy of available records is questionable. The Coal Authority will provide information (where it has records) on any environmental and stability risk from any past, current or proposed underground and surface coal-mining activity that affects property. It will advise on a mineshaft's recorded position (i.e. not necessarily its actual position) if it is within 20 metres of the main building on a property or within 20 metres of its boundary. In a typical year, there are approximately 100 events linked to mine workings, although very few involve structural damage. The Coal Authority also keeps records of brine extraction for salt in England and Wales.

Sometimes, a shaft can open up that is not caused by mining. Many old wells were capped by similar means and occasionally one of these collapses. There will almost certainly be no record available, but any property constructed over the site of a former building may be subject to this problem.

Subsidence due to coal mining affects many countries. In Pennsylvania, USA, the mines are quite shallow – subsidence is a common problem. As a consequence, it is quite often excluded from insurance policies (photo courtesy of the Department of Environmental Protection, Pennsylvania, USA).

Landslip

Landslips, or landslides, are mass gravitational movements of soil or rock. The landslip may be slow, quick, intermittent or continuous. Many sections of the British coast suffer from landslips which cause coastal erosion and instability, e.g. Devon, Dorset, Kent, Suffolk, Yorkshire and the Isle of Wight.

This example of coastal landslip, although small in scale, would have a devastating effect on any buildings in close proximity.

However, landslip is not just a coastal problem. Inland landslips have occurred in recent years in the Wye Valley in Gloucestershire, at Daventry in Northamptonshire and at Halifax in Yorkshire. Railway and highway cuttings and embankments throughout the country are frequently affected by landslips and adjoining properties can be adversely affected. Waste tips are notoriously prone, the worst disaster being that at Aberfan in South Wales in 1966, when the school and adjoining houses were engulfed, killing 145 people. In the winter of 2000–2001, there were more than 500 instances of landslip reported after the wettest winter since records began in 1727. Further landslips occurred in the extremely wet weather of 2012; landslips in Whitby, Yorkshire, threatened a medieval church and also resulted in the demolition of a row of houses.

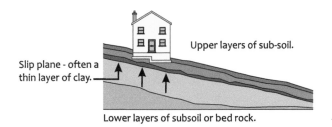

Upper layers of sub-soil.

Slip plane - often a thin layer of clay.

Lower layers of subsoil or bed rock.

Terraced ground at base of slip.

The red arrow indicates direction of land slip (often a result of a rise in ground water pressure after intense rain).

Landslips occur for a variety of reasons, e.g. on sloping ground with clay soils, there is a danger that the upper layers can slowly move downhill when they slide over lower layers. In many cases, susceptible soils or waste heaps can slip as a result of a prolonged period of rain, e.g. over the autumn and early winter of late 2012. It can often be recognised at the site

Left: This aerial photograph shows a massive coastal landslip near Lyme Regis (photo courtesy of Sealand Aerial Photography). Right: Building damage following a landslip at Lyme Regis in the mid-1960s (photo courtesy of Lyme Regis Museum).

investigation stage by the terraced appearance of the ground. Soil creep can occur in gradients as shallow as 1:10 and buildings with shallow foundations can obviously be damaged.

General guidance on new development in a coastal landslip area can be found in the National Planning Policy Framework, published in March 2012.

Landslip prevention can be an expensive business. Here, a new retaining wall is being constructed to protect the buildings in the background and the road below from landslip.

Sulfate attack

This is a problem that can occur where there is a high water table linked to moving ground water and a sub-soil (often ancient sedimentary clay) that contains calcium, magnesium, and/ or sodium sulfate. Such clays can be found in England anywhere below a line drawn between Lancaster and Middlesborough. Sulfates in the presence of moisture will attack any cement-based material (e.g. mortar in brickwork/blockwork, concrete in foundations) due to a chemical reaction between the sulfate and one of the constituents of cement – either calcium hydroxide or calcium aluminium hydrate. The result is a rapid expansion of the compound formed by the reaction.

The chemical process is discussed in more detail in Chapter 6 as the sulfates found in the clays used for the manufacture of some types of bricks can lead to similar problems in external walls. Sulfate attack also often occurs in burnt colliery shales which were sometimes used for hardcore for solid ground floor slabs in the first half of the twentieth century (see Chapter 7).

The effect on that part of the building below ground level is to expand and crack concrete foundations. The mortar joints in brick or concrete block walls beneath damp-proof course level will suffer similarly. The expansion resulting from the sulfate attack may lead to upward movement of the wall and may also be seen as cracked and broken down mortar joints below DPC level. Foundations subject to such an attack may disintegrate if left long enough. The process will be accelerated if the concrete, or cement mortar, is also subject to frost attack. In any case, as the foundation deteriorates progressively, it will become less capable of supporting the imposed building loads, leading to settlement. Where such sub-soils are identified, any construction below DPC level should incorporate sulfate-resisting cement in floors, walls and foundations. If the problem is discovered in an existing building, it may result in the need to completely remove and reconstruct the affected elements.

This photograph of a rendered brick retaining wall gives some idea of the effects of sulfate attack on brickwork and cement-based material. The impact of sulfate attack on below-ground brickwork, blockwork and concrete foundations can be imagined.

A separate type of sulfate attack has been recognised since the late 1980s – thaumasite sulfate attack (TSA). This attacks concrete rather than mortar, requires low temperatures (generally lower than 15°C) and the presence of carbonate, e.g. limestone, which is provided by the coarse and fine aggregates in the concrete. It does not depend on the presence of calcium aluminium hydrate and is becoming an increasing problem, in particular where there is an increased presence of water, e.g. in basement retaining walls. Unfortunately, the use of sulfate-resisting cement does not provide a defence against this problem.

Conclusion

This chapter has examined the major causes of foundation failure in domestic buildings. It has also described typical patterns of cracking associated with problems of settlement, heave and subsidence. In the future, the implications of climate change, with increasing temperatures, wetter winters and dryer summers, will have to be taken into account when assessing building performance and failure.

In practice, the pattern of cracking often provides the first clues in what can be a lengthy diagnosis. However, be prepared for the unusual. The cracking in the single-storey extension in this photo was caused by impact damage. Brake failure, not foundation failure, accounts for this apparent defect.

Finally, it should be noted that buildings affected by foundation failure may not require drastic or expensive remedy. Remedial action involving tree removal or pruning or relining drains is more often the solution than underpinning.

Building movement: walls 4

Introduction

A building will suffer some degree of movement throughout its life. Often, this may not affect its performance in any significant manner. On the other hand, the movement may be so great, or happen in such a way, that performance is adversely affected.

The walls of newly constructed buildings commonly suffer cracking as a result of the initial drying out of the materials used in the building process. This is usually caused by the evaporation of the water used in the process and can take up to 12–18 months after construction has finished before it ceases. Normally, the severity of the cracking is within acceptable limits and is insufficient to require more than minimal repair. Usually, filling in of the cracks alone will suffice.

Similarly, a building's structure will be subject to thermal and moisture movement caused by changes in climate and/or internal living conditions affecting building materials. This will result in expansion and, conversely, contraction of both the materials and the elements of structure that they form. Mostly, the amount of movement will be insufficient to adversely affect the overall performance of either the material involved or the element concerned or the building as a whole, especially if expansion in one season is offset by contraction in the next. Modern design and construction should incorporate expansion details to accommodate such movement, but in practice both older and modern buildings can suffer structural failure from these problems, the latter because, often, insufficient allowance was made for movement.

These four photographs show buildings that have suffered extensive wall movement. Older buildings often can sustain considerable movement because they incorporate timber framing or lime mortar. However, the integrity of the arched opening depends wholly on the head of the timber doorframe. The right-hand detail of one of the buildings in the middle photograph emphasises how much movement brickwork in lime mortar can accommodate.

Construction summary

Function of the wall

There are two primary functions of the wall. They are:

1 **To support floor and roof loads and transfer them to the foundations.** *This may be undertaken by either external or internal walls.*
2 **To give environmental protection to the internal areas of the building.** *This is performed by external walls.*

In addition, a wall may be required to perform a further function:

3 **To define separate areas of use.** *This is normally undertaken by internal walls or partitions.*

Walls are divided into **load-bearing** and **non load-bearing**, depending upon whether or not they perform function 1 above. The following diagrams indicate the principles of each.

One Brick thick ext. walls.   Half Brick thick int. walls.

A typical mid-Victorian terrace. Butterfly roof bearing onto load-bearing walls.

Alternative eaves detail.

One Brick thick ext. walls.   Half Brick thick int. walls.

Traditional double-pitched purlin roof. Internal wall is load-bearing.

Modern construction

Trussed rafter roof spans full width. First-floor partitions carry no load from roof structure.

Ground-floor walls support first floor. They may have foundations or be supported on thickening in slab.

Whichever functions a wall is undertaking, it must have suitable strength, stability and durability to perform properly. These factors must be carefully assessed at the design stage and each will influence the final choice of form and construction. Table 4.1 (see page 107) presents the common defects found in walls.

Walls: a brief history

The first walls were probably constructed from branches and other vegetable matter before heavier and larger pieces of timber were used. Bricks of sun-baked clay have been used to form buildings for at least 4000 years and stone for probably longer. Generally, people used whatever material was readily available locally and this can be seen in the localisation of construction forms and types of materials used. This state of affairs continued until the nineteenth century and the Industrial Revolution, which transformed transportation and industrial production. During the nineteenth century, canals and then railways enabled materials to be transported more or less anywhere, although local style and appearance still tended to be followed.

Left: An early timber-framed building showing its wattle and daub construction. Middle: An example of Georgian freestone cladding over a rubble and brick wall behind. Right: A Victorian sandstone garden wall showing how through stones tie together the two skins of a solid wall.

In England and Wales, the oldest domestic buildings still standing were originally constructed of stone, brick, or timber frame with infill panels – probably of wattle and daub, often replaced with brick. Few are older than the fourteenth century and from the eighteenth century onwards they are generally of brick or stone monolithic (solid wall) construction. Brick walls were initially constructed using a simple bond, e.g. a garden wall bond incorporating alternate courses of stretchers and headers, but more sophisticated bonds were developed for thicker walls to enable greater loads and heights to be sustained, e.g. English and Flemish bonds. (See Chapter 6 for details.)

Typical solid wall construction – nineteenth century

Monolithic (or solid) wall strength and stability depend on thickness. The taller the wall the greater the thickness at the base. Wall reduces in thickness as it gets higher. Each step is normally hidden as shown.

One brick

One and a half brick

Two brick

One brick

One and a half brick

One brick

One brick 9" (225mm)

One brick 'equivalent' 13" (325mm)

Foundations vary depending on date, location and size of building.

By-laws adopted by many towns required walls built in rubble, or rubble and brick or stone, to be 50% thicker.

External cavity walls started to appear in the 1830s, but were very uncommon until the twentieth century. From the 1930s, such walls, formed with two skins or leafs of brick, were the normal form of construction for housing and it is extremely unusual to find solid external walls in buildings constructed since that date. Concrete blockwork became the norm in the second half of the twentieth century, although many modern cavity walls have a brick external leaf. Thermal requirements for cavity walls mean that current construction requires a lightweight concrete block internal leaf plus a high level of added insulation as a minimum.

A typical cavity wall from the early 1970s

100mm concrete block
50mm cavity
100mm facing brick
(actually 102.5).

Wall ties at every 6th course of brickwork tie the two skins together.

Ties may be at every 4th course in early walls where both skins are brick.

Upper-floor joists can either be built into the inner skin or supported on joist hangers.

Inner skin is load-bearing and supports roof structure, upper floor and, in some cases, ground floor.

Separate DPCs in each skin minimum 150mm above ground level.

Ground level

Cavity is filled below ground level to prevent blockwork leaves being squeezed together.

Four common options for modern cavity walls

100mm facing brick
75mm cavity batts
125mm lightweight block

100mm facing brick
75mm cavity
25mm cavity insulation
125mm lightweight block

100mm facing brick
50mm cavity
125mm lightweight block
insulated dry-lining

18mm render
100mm lightweight blocks
50mm cavity batts
100mm lightweight block

Material and labour shortages after each of the World Wars led to alternative construction approaches being sought, most involving the use of innovative framed construction, mostly concrete or steel and, less frequently, timber. The numbers of these houses built in the inter-war period were relatively small (approximately 50,000), but between 1945 and 1960 approximately 500,000 were erected. Subsequently, since the late 1970s, timber-frame houses have become a significant percentage of new construction totals. (See Chapter 17 for a summary of their defects.)

Table 4.1 Common structural defects of walls at construction stage (new build)

POTENTIAL PROBLEM AREAS	IMPLICATIONS	TYPICAL DEFECTS
Below-ground brickwork or blockwork in sulfate-bearing clay sub-soil	Sulfates in sub-soil will attack cement mortar.	Expansion and cracking of mortar below ground plus movement in wall below and above ground
Lack of adequate movement joints	Excessive thermal expansion in long lengths of wall	Vertical cracking at regular intervals in long sections of wall and at short returns Horizontal movement above the line of the DPC
Inadequate restraint provided to walls from floors and roof	Walls will tend to move horizontally.	Bulging and bowing of walls
Lack of lintels over the openings – small and large (both above and below ground level)	Unsupported masonry	Movement cracking above opening Dropped head of opening
Lintels with insufficient end bearing	Crushing of supporting brickwork or blockwork due to excessive point-loading	Lintel drops at one end Vertical cracking above end of lintel
Insufficient wall ties	Lack of restraint on individual leaves of a cavity wall	Bulging of wall externally and/or internally
Incorrect mortar mix	Inherent weakness in the structural capability of the wall	Disintegration of mortar jointing Bulging and bowing and/or buckling of walls.

Wall movement: the causes

Reference has already been made in Chapter 3 to movement as a result of failure of, or defects in, the foundations or the sub-soil beneath a building. There is also the need to assess whether the apparent movement under investigation is as a result of a failure or problem in the part of the building above ground level. Normally the investigation will be triggered by evidence of movement in the latter and the assessment will involve finding whether the cause(s) are above and/or below ground related.

As with foundation failure, there can be a wide range of reasons for wall failure, many due to either inadequate design and/or poor construction. The underlying factors are also similar – historically, lack of technical knowledge; more recently, incompetence leading to poor design and/or bad site practice.

Distortion of walls (bowing, bulging and buckling)

Walls distort for a number of reasons. Some are poorly built which results in distortion from the start; some acquire distortion due to failure of the wall and/or adjoining elements; others have distortion thrust upon them when unforeseen loads are imposed. It is a problem that can affect any solid masonry or cavity wall and may occur for a number of reasons. In approximate terms, it is only when a solid wall is more than one-sixth of its thickness out of plumb in any storey height, or its overall height, that distortion needs serious consideration. For a cavity wall, the relevant thickness is that of the thinner leaf.

The causes of distortion leading to bowing, bulging and buckling include the following.

Poor building practice

Walls are sometimes constructed so poorly that they are inherently distorted. This can often be observed in older buildings and is, unfortunately, found in some of recent construction where walls can be seen to be out of plumb. Even though the wall may be distorted, it can still be stable and, unless there is evidence of current progressive movement, it is often acceptable to take no further action apart from monitoring at regular intervals.

Distortion

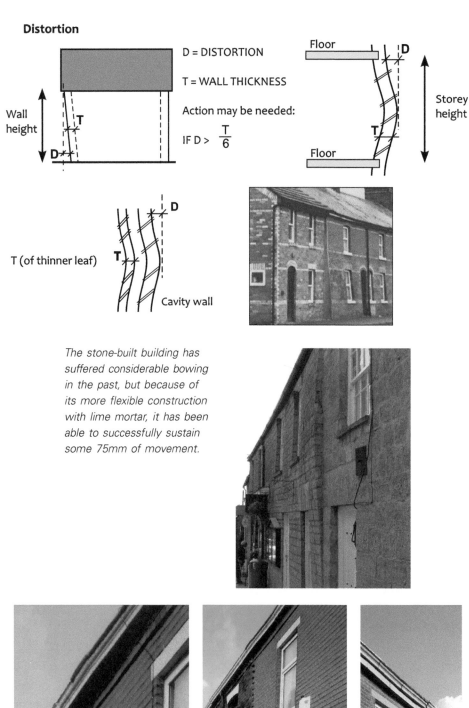

D = DISTORTION

T = WALL THICKNESS

Action may be needed:

IF D > $\dfrac{T}{6}$

Wall height

Storey height

Floor

Floor

T (of thinner leaf)

Cavity wall

The stone-built building has suffered considerable bowing in the past, but because of its more flexible construction with lime mortar, it has been able to successfully sustain some 75mm of movement.

Examples of bowed walls. The bowing on the front wall in the left-hand photograph was so great as to require its reconstruction. At the party line, the junction of the new and adjacent walls shows that the bowing is obviously a problem along the line of the terrace.

Subsequent distortion

This can be caused by a number of factors including the following:

Inadequate thickness of the wall. The relationship between a wall's height and its thickness is known as the 'slenderness ratio'. If a wall is tall and thin, it will have a greater tendency to buckle or bow under applied load than a short and thick wall. Historically, this has resulted in walls that are very thick at ground level and are stepped in thickness as they get taller. Modern knowledge of how walls perform, and the use of structural calculations in the last 50 years or so, have meant that in recent years taller, thinner walls have been successfully constructed.

External walls in solid construction should be at least one brick thick (225mm, 9"). In some houses. usually 3 to 4 storeys or more, the wall thickens towards its base. Problems of buckling should not, therefore, occur as a result of an indaequate slenderness ratio. Floor joists and roof timbers built into a wall also offer some restraint. The house above has an 18" front wall in stone (450mm), 4" internal walls (100mm), and a rear wall 9" (225mm) at the top and 14" (350mm) at the base (drawing courtesy of Bristol Record Office).

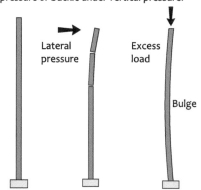

Tall thin walls will tend to fall over under lateral pressure or buckle under vertical pressure.

Lateral pressure

Excess load

Bulge

If wall is too slender (ratio of width to height), or if there is inadequate restraint, or loading is too high, buckling and cracking may occur.

Lack of lateral restraint or support. A large unrestrained or unsupported area of wall will tend to lean or fall over especially if it is subject to external load or pressure (thrust). Walls are normally restrained by being tied to adjoining walls, floors and roofs. Alternatively, walls can be supported by the provision of buttresses. Either approach can provide the necessary support. Current statutory requirements and good building practice should result in all elements of construction being properly and fully tied together.

Ground level

Buttress with level or raking brickwork.

Pier

Buttress

Effect of lack of restraint.

Possibility of cracks externally

External wall inadequately tied to internal walls and upper floor.

Tie rods added to provide restraint.

Lower standards were applied formerly. The older the building, the more likely it is to suffer from problems associated with a lack of restraint, e.g. walls were often butted up to each other, bonding was either non-existent or minimal. Often, only one or two courses of bricks per storey were toothed into each other; floor joists running parallel to a wall would have no lateral ties to support that wall. Where it is possible to do so, restraint can be provided by the addition of tie bars or rods.

No restraint from floor or roof.

Ground and first floor plan

restraint

restraint

Section A–A

If a wall is not adequately restrained it may move. This cracking is slowly getting wider.

Subsequent thoughtless alterations to a building can result in restraining elements being removed without the provision of alternatives.

The effect of alterations

Chimney breast removed to increase floor area (removes buttress effect). May be full or single storey height.

Internal wall removed to change layout – restraint reduced. Even a new door opening in an existing wall can have the same effect.

Staircase installed – floor joists no longer restrain external wall.

Remedial work being undertaken to stabilise the movement of the flank wall previously shown. Metal ties are being installed at floor level along the length of the wall.

Thrust from other elements. This is often associated with the problem of roof spread where a roof construction which is inadequately tied together results in an eccentric thrusting force being applied at the top of a wall. The affected area of wall is pushed outwards. The appearance may be a bulge or bow, although often brickwork may become stepped as the individual courses slide across each other.

Thrust can also be exerted laterally by the thermal expansion of roofs and floors especially when they are of concrete construction, e.g. a concrete flat roof. (Refer to the later section in this chapter on thermal movement.)

Excessive loading. Buildings are not always used as originally intended and this may result in over-loading of the structure. An example of this is the greatly increased amount of furniture and personal possessions commonly to be found in the modern family home. Cheaper Victorian and earlier housing generally had thinner/lighter forms of construction for upper floors and walls than would be acceptable now. Apart from any over-stressing of the floors, the higher-than-designed-for loads may result in unforeseen pressures causing bulging/bowing of adjoining walls.

Buildings are not always what they seem. This stone building in Bath looks quite solid. However, in some Georgian buildings the top storey , or even the whole facade, is built in thin facing stone with no brick or rubble backing.

Loss of bonding between materials. The bulging, bowing or apparent buckling of a wall may not be due to structural failure. Older, solid masonry walls are often formed of several different materials bound together in a series of layers by a mortar (probably of lime). Walls of all ages

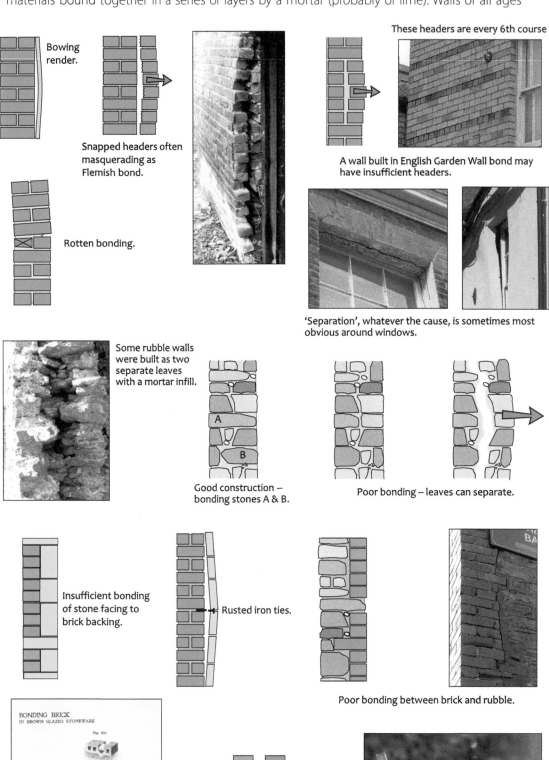

Bowing render.

Snapped headers often masquerading as Flemish bond.

Rotten bonding.

These headers are every 6th course

A wall built in English Garden Wall bond may have insufficient headers.

'Separation', whatever the cause, is sometimes most obvious around windows.

Some rubble walls were built as two separate leaves with a mortar infill.

Good construction – bonding stones A & B.

Poor bonding – leaves can separate.

Insufficient bonding of stone facing to brick backing.

Rusted iron ties.

Poor bonding between brick and rubble.

BONDING BRICK
IN BROWN GLAZED STONEWARE

Wall tie failure.

A corroded tie–installed in the 1920s.

Clay wall ties–1890s to early 1930s.

and types may be finished with a thin skin of facing material bonded to the structural element behind. Historically, this was frequently achieved by the use of iron ties, the modern equivalent being metal cavity ties. The bond between the surface material and the underlying material(s) may fail, e.g. iron ties may rust.

In some circumstances, the appearance of the wall suffering such a failure may give the impression of a more serious problem involving building movement. A check should be made on whether movement identified on one face of a wall is repeated on the other face – lack of such evidence may indicate a failure of bonding rather than a loading problem.

The bulging between the windows in the central photographs is caused by the thin external facing of ashlar losing its bond with the brick backing. This is often caused by rusting of the tie irons. Traditionally, these were made from wrought iron which, although less brittle than cast iron and therefore more tolerant of minor movement, is prone to rust.

Thermal movement

The external walls of a building are, by their nature, exposed to the elements and are, therefore, subject to the effect of temperature changes. This often results in expansion and contraction of the wall.

The exact amount of expansion experienced will depend upon the materials used in the construction of the wall. In some cases, the excessive length of the wall or its detailing may influence the amount of movement. Usually, any expansion due to the effect of solar heat gain will be followed by contraction as the affected element cools down during subsequent climatic change. Unfortunately, the amount of contraction does not always equal the original expansion and the net result is permanent cracking. The cracking actually occurs during the cooling period of the cycle. It will also be found that the affected wall expands/contracts above any DPC – over which it tends to slide. This factor provides supporting evidence that the problem is caused by thermal movement.

The terrace expands as it gets hot. If there is a DPC, it may slide over it and over-sail the substructure. As the terrace contracts during a cooler period, it is unable to pull the building back to its original shape; cracks therefore occur. Over the years, the movement can get worse as dust and debris fill the cracks. If there is no DPC, lateral movement is sometimes less prevalent at ground level – this can lead to a slight outwards lean higher up the building.

Cracking may be less obvious in buildings constructed with a lime mortar.

In each of these two photographs, one in the centre of a brick wall, the other at the end, thermal expansion of brickwork has caused the movement to take place above the DPC.

If shop windows are at the end of a terrace, long-term thermal movement can lead to distortion. The photo (below right) shows the flank wall of the terrace. At some point in the past, the wall has been anchored back to the first party wall to try and limit further movement.

Modern design should incorporate the use of expansion joints to accommodate this type of movement, although they are sometimes omitted because of poor design or construction. Older buildings are unlikely to have any such provision. However, those built with a lime-based mortar will be more likely to avoid any serious problems as, being a more flexible material, it can accommodate thermal movement more successfully than the cement-based mortar used in buildings since the early twentieth century. (An important point to note is that the materials used in movement joints suffer deterioration due to age and fatigue and may themselves, unless periodically checked and renewed on a regular basis, become a defect.)

Walls can also be affected by thermal movement in adjoining elements or at weak details in the wall. Thermal movement of other elements of construction can also cause problems due to the lateral pressures placed on adjoining walls as a result of the expansion of roofs and floors. (Refer to the previous sub-section on 'Thrust from other elements'.) The two buildings that follow are suffering from two quite separate problems. Expansion of a concrete roof has caused the horizontal displacement beneath the parapet wall, while expansion due to moisture movement in the long length of the flank wall has led to the vertical cracking (see next section).

Moisture movement

Expansion and contraction of the fabric of a building can also occur because it is affected by moisture or subsequent drying out. The cause of the moisture may be weather-induced (e.g. rain, snow), a defect (e.g. a leak), condensation, or as the result of the building process.

Contraction of a material due to initial drying

Brief reference has already been made in this chapter to the effect of the cracking caused by initial drying out of a building once the construction process has been completed. Shrinkage cracking of this nature is quite normal. It is the most common form of cracking found in newly constructed buildings. Portland cement products, lightweight concrete, sand-lime bricks, calcium silicate bricks, some plasters and timber are all subject to initial drying out and cracking. With cement-based products, the greater the cement content of the mix, the more marked the drying shrinkage. While such cracking can be extensive and may occur over a long period of time (up to 20/25 years), it rarely results in serious problems, but it is irreversible.

The photographs show two examples of cracking caused by the initial drying out of sand-lime bricks. These bricks were frequently used for construction in the 1960s and 1970s.

Expansion of a material due to moisture absorption

On the other hand, clay products, such as bricks, undergo initial expansion as moisture is re-absorbed after firing. This is only slight, between 0.1 and 1.0mm/m in unrestrained brickwork, but the effect is increased in long/large areas, where the overall movement can be considerable. It results in vertical cracking that is similar in appearance to that of thermal movement. Brick panels in concrete-framed buildings can also be affected – the brickwork is constrained by the frame/floors/roof and expansion results in bowing and horizontal/vertical cracking of the panel (see last photograph in previous Thermal Movement sub-section).

The majority of moisture-induced expansion occurs in the first few months, but it can continue for up to 20 years after construction and is irreversible.

Not all moisture-induced movement will be of such great extent that it adversely affects a building's performance. Failure may more often lead to weather ingress problems rather than serious structural failure. However, the modern designer should have anticipated moisture (and thermal) movement by incorporating movement joints with appropriate filler material capable of expanding or contracting as necessary.

Failure of arches and lintels

Often, movement cracks in a wall will be observed above window and door openings. They may be indicative of failure of the wall itself or its foundations. However, the problem may be as a result of the failure of the arch or lintel above the opening.

A range of lintel problems can be seen. Left: The window head is collapsing as a result of the inner timber lintel deflecting because it is over-loaded. Middle: The timber lintel had only render as weather protection. Right: The stone lintel has been affected by foundation settlement.

In traditional masonry walls, openings were normally formed with flat brick or segmental rough arches above. If the mortar in the joints of the arch weathers excessively, or the wall around the opening is subject to some movement, the result can be slumping or movement within the arch itself. Ultimately, this may lead to its collapse, together with any area of supported wall above.

Area of brickwork supported by lintel is quite small due to brick bonding. In rubble walls, the area (and load) supported can be much higher.

Gauged arch.

Axed segmental arch.

Timber lintel 100mm wide (half brick).

Arch too near quoin.

A rough arch formed in half-brick rings.

Folding wedges

Arch failure can be a consequence of mortar failure, rot in the backing lintel or building movement. In many cases, the timber lintel, if sound, will prevent total collapse. Many rendered buildings hide simple arches.

In the mid-twentieth century, the concrete boot lintel was popular for a time. The problem with this particular type of lintel is that it is not built into the external skin of the cavity wall into which it is incorporated and this often results in rotation of the lintel. Even simple reinforced concrete beam or plank lintels can fail if the ends are inadequately supported where built in.

Victorian and older buildings often had window and door openings formed with one or more timber lintels. In external walls, the timber lintel was commonly positioned behind the external facing material of brick or stone so that it did not show on the elevation. However, the lintel was, frequently, insufficiently far enough behind the face of the wall to be unaffected by damp penetration. This may result in the lintel being affected by wet rot or possibly dry rot.

Any lintel so affected will eventually lose structural capability and this can lead to collapse of the timber itself and loss of support to the wall above the opening. Timber lintels are also subject to the threat of insect infestation and a bad attack can result in similar failure of the lintel and wall above (see Chapter 15).

Lintel section, plan and elevation

Lintel rotation

Concrete boot lintels (photo–below left) were popular from the 1930s to the 1950s. Some of them were only supported by the internal leaf. The weight of the external leaf could cause rotation in the lintel, leading to cracking in the brickwork.

The lintel in the middle is suffering from carbontation. This is explained in more detail in Chapter 17. The cracking is caused by rusting of the reinforcement. The lintel on the right has insufficient bearing – it should be at least 150mm either side. To make matters worse the lintel is bearing on a brick off-cut (Left photo courtesy of Jeff Ayling).

Left: The timber lintel is behind a sandstone window head. Right: Three timber lintels (traditionally known as bressummers) support the full thickness of the wall above; they are protected externally with only a render finish. In each case, penetrating damp has affected the lintels.

The cause of the movement in each of these photographs was found to be the failure of the timber lintel behind the stone brick window head.

Corrosion of steel and iron components

Unprotected steel and iron will rust when in the presence of water. The formation of the rust (or hydrated iron oxide) results in a four-fold increase in the volume of the affected metal. The only way to prevent this is to coat the metal to provide a barrier between it and any water. This is often achieved by the use of hot or cold galvanising. This process involves applying a coat of zinc over the complete surface of the metal. It will offer protection for between 20 and 60 years, depending upon the thickness and type of galvanising applied.

The expansion of the rusting metal can cause movement, spalling or cracking of the surrounding fabric of the building. It is a common problem in many houses with earlier, inadequately galvanised wall ties or concrete reinforced with steel and either contaminated with chlorides or carbonated by atmospheric carbon dioxide.

The rusting of the steel reinforcement in the window lintel and in the door jamb has led to the spalling of the concrete cover of each component.

Many buildings have unprotected steel and iron components, such as steel joists supporting floors and as lintels above openings, which are built into the structure. The problem of rust can be very serious where the metalwork is in an external solid masonry wall which, unless of some thickness, may be incapable of offering adequate protection from water penetration. Similarly, steelwork within the building may be affected by water leaks from defective services. In modern houses of cavity wall construction, any unprotected steel components should only be built into the internal skin of the external wall and should, in normal circumstances, not be affected by water penetration.

Left: A rusting rolled steel joist (RSJ) embedded in a solid brick wall is the cause of the problem. Right: A slightly rusty iron fixing cramp in a limestone wall. Extensive rusting and resultant expansion of such fixings are a common cause of loose stone facings on older buildings.

At its worst, rust-affected steelwork can completely lose its strength and may lead to total structural failure of any supported part of the building. The actual process of expansion as the rust progresses should be evident in its effect upon any adjoining structure. This may show in the forcing apart of the fabric vertically and/or horizontally with increasing strength as the defect progresses.

Typical wrought-iron or rolled-steel joist.

Inadequate depth of cover to steel allows it to be affected by moisture.

Moisture penetration depth

Rusting steel/iron expands up to four times original volume, applying pressure to adjoining structure.

Pressure from rusting (expanding) steelwork causes movement in wall.

Steel lintels can also rust and cause cracking.

Retaining walls and garden walls

Retaining walls

Retaining walls are subjected to a mixture of lateral pressures and they must always be designed and constructed to withstand them. The principal pressure is normally formed by the force exerted by the volume of soil supported, but this is often increased by further applied forces. These may include groundwater pressure, loads from adjoining buildings and loads from traffic on nearby roads. Common problems are set out below.

Slenderness of the wall

A retaining wall that is too slender to withstand the pressures being applied against it will tend to bow or lean outwards. This is a defect that is commonly observed in garden walls that are retaining soil. It is often as a result of too little thought being given to the design implications of any external landscaping. The higher the ground being retained, the thicker the wall, though it can be of stepped width – retained heights up to 900mm should have a 225mm-thick brick wall. This is required to become progressively thicker at the base as the height of the wall increases. The addition of piers or buttresses may allow a thinner wall to be formed, but never less than 225mm.

Drainage problems

All retaining walls require the provision of drainage points at frequent intervals in their length and height. Their provision is a common omission and a frequent cause of movement due to the pressure of the groundwater behind the retaining wall. The drain points should be backward sloping to prevent any weeping and the subsequent staining of the wall. Over time, the holes can become clogged, reducing their effectiveness. The use of a French drain behind the retaining wall is also highly effective in relieving groundwater pressure. Fine soil can clog up the drainage system, but this can be prevented by the installation of a geotextile woven mat behind the wall or behind the French drain (where installed).

Wall must be thick enough to withstand any applied loads/pressures.

GL

Building load
Ground pressure
Water pressure

Drain

Water table

Simple drainage details are often omitted leading to build up of water pressure.

Progressive thickening of brick retaining wall.

1B

900

Weepholes

A French drain improves drainage and relieves water pressure.

Weepholes should fall backwards slightly to prevent dripping.

Proximity of trees

A further cause can be the close proximity of trees and shrubs. The growth of the root system over a number of years can lead to increasing lateral pressure on any adjoining retaining wall.

Bowing

Leaning

Leaning

Ivy stem pushing against wall

As the small tree grows, the retaining wall is pushed out by the root growth.

Plan

Lateral pressure from root system causes bulge in wall. Bulge increases each growing season.

Sulfate attack

Brick retaining walls can suffer from sulfate attack. This can be avoided by the use of sulfate-resistant bricks and mortar or the encasing of the retaining face with water-proof material. (See Chapter 6 for more information on sulfate attack.)

Brick walls should also have movement joints at approximately 10 metre intervals. These are frequently omitted and can lead to movement due to moisture ingress or thermal expansion.

Moisture and frost problems

Moisture penetration leading to frost damage can be a problem where brick or stonework is susceptible to groundwater penetration. Suitable frost-resistant materials should be used in any retaining wall construction.

This old retaining wall is suffering badly from sulfate attack.

Garden walls

These may act as boundary markers or be dividing space within a garden. Older garden walls were constructed in clay bricks and/or stonework – both stone rubble and freestone were used. Modern walls may be of the same materials, but are often constructed with concrete blocks, sometimes rendered. The top should be protected from the weather and a variety of capping materials have been used including stone copings, concrete copings, clay tiles and bricks-on-edge, either laid flat or castellated (which is also known as 'cock and hen').

Because they are commonly regarded as of no great aesthetic or structural importance, garden walls are frequently ill-considered, leading to them being poorly designed, constructed, maintained and/or repaired. They are susceptible to a wide range of problems of which the following are the most common.

Shallow foundations

Victorian and older walls were constructed on the shallowest of foundations; frequently they had no foundation (see Chapter 3). Modern garden walls should have a foundation that is at least 500mm deep. This is particularly important where a wall is 2 metres high or taller, but often this is not the case. A lack of depth can result in instability, especially if the wall is of any height and/or in an exposed location.

Older stone walls

Older stone walls were commonly constructed with two skins infilled with rubble and lime mortar which bonded them together. Sometimes through-stones were incorporated. The top may be finished with a cock-and-hen capping or a coping. Over a period of time, one or both skins can detach from the backing, leading to partial or total collapse of the wall.

Proximity of trees

The close proximity of a tree and its roots can lead to the cracking and/or lifting of a wall as the tree grows.

Tall, thin walls

Half-brick walls constructed in long lengths without stiffening piers at regular intervals and tall, thin walls will both have a tendency to be overturned by high (in some cases, relatively low) winds. The exposure of the location will be an important element. These types of walls can be strengthened at initial erection by the inclusion of piers or by staggering or curving their construction, which permits increased height in exposed locations.

Left: This stone boundary wall was partially demolished to provide a vehicle access. It is still generally intact, but will deteriorate if left in this state. Right: This old stone boundary wall has suffered in recent years as the tree behind it grows in size.

Even with the piers and being sheltered behind the hedge, this 100mm-thick concrete block wall is too high.

Movement joints

Lack of adequate movement joints, especially in brick walls, can result in bowing and cracking due to moisture or thermal movement. Such joints should be provided at intervals of between 8 and 12 metres depending on the type of brick, e.g. 8 metres for engineering bricks. A frequent error is to provide the joint, but not continue it through the capping.

The expansion joint needs to continue through the capping bricks.

Frost damage

The omission of a proper capping or a defective capping can lead to excessive moisture penetration and frost damage to the face of brickwork or stonework. Similarly, the lack of a DPC can lead to frost damage at the base of a wall. Another cause of frost damage is the use of inappropriately soft bricks or stones. (See Chapter 6 for more information on frost damage.)

Left: These soft London Brick Company facing bricks were totally unsuitable for use in a garden wall. Right: This Cotswold limestone wall has extensive frost damage.

Mortar joints

In older walls, lime mortar joints can decay over time leading to unstable brickwork. Repair work, such as pointing up the face of a joint with a strong and inappropriate cement mortar, can result in the repair taking away the face of the brickwork or stonework. (See Chapter 6 for more information.)

Chimneys

Construction

Until the 1960s, most new houses were built with chimneys. After this time, few new houses were built with chimneys, although since the 1990s many new houses have chimney stacks above flues for gas fires rather than a traditional fireplace. Chimney stacks were usually built in brick (sometimes even in stone buildings) because brick is better able to withstand heat, and the complex flue shapes sometimes required were easier to form in small units (i.e. bricks). Before the introduction of the national Building Regulations for England and Wales in 1965, the construction of fireplaces and chimneys was controlled by generally accepted good practice and local bye-laws. Inevitably, construction details varied slightly from area to area.

With regard to chimney stacks, by-laws usually required that:
- *The stack had to be taken at least 3 feet (900mm) above the ridge line.*
- *The flues had to be surrounded by at least half-a-brick (102mm).*
- *The stacks could not be more than six times the height of the stack's narrowest dimension.*

The Building Regulations 1965 introduced a requirement that all flues should be lined (terracotta linings are the most common) and specified more demanding requirements regarding the position and height of stacks.

Left: A typical Victorian terrace in Stratford-on-Avon, each house having its own chimney stack. Right: An example of a stone cottage in Bristol with a brick chimney and flue.

This example (an original drawing from about 1905) shows chimneys front and back, each one serving two flues (drawing courtesy of Bristol Record Office).

Odd course and even course

In very good-quality work, the outer chimney wall is one brick thick. Half brick thick is much more common.

Odd course only shown

This example shows a gathered stack. In a 'handed' terraced property, the chimney is sometimes shared; hence the eight flues.

Defects

In general

It is easy to ignore chimney stacks when inspecting a building. Chimney stacks which look sound from the ground may look much worse on closer inspection. Sometimes, a look from an adjoining house window can provide a useful chimney view. A thorough inspection may require ladders; at the very least, it requires a good pair of binoculars.

The following section considers only structural problems associated with the flues and stack. Other problems due to damp penetration or operational defects are not discussed.

Tall, thin stacks

Chimney stacks are obviously in a very exposed position. In high winds, it is the taller stack, particularly if it is narrow, which is most at risk. Tall stacks are most likely to be found on houses with hipped roofs. Although by-laws adopted by most towns permitted a height six times the narrowest width, modern Building Regulations only permit a multiplier of 4.5. The Building Regulations are not retrospective, although the change reflects health and safety concerns. A stack which has been in position for several years may not be about to topple over, but it is still potentially dangerous. If there are any signs of movement, cracking or open mortar joints, a more detailed inspection may be necessary. Any stack that is leaning by more than 1mm in 100mm is unsafe and consideration will have to be given to (1) repairing it or (2) demolishing and rebuilding it.

Tall, thin stacks require careful examination – they are potentially very dangerous, particularly in high winds. Demolition, strapping or bracing may be required. (Right-hand photo courtesy of Ian Hancock.)

Sulfate attack

Cracked mortar joints or a pronounced and permanent lean in a chimney are often the result of sulfate attack. Sulfate attack can occur where clay brickwork, bedded in cement mortar, remains saturated with moisture for long periods. The reaction between sulfates, the cement in the mortar and water forms a compound known as tri-calcium sulfoaluminate which causes expansion and cracking of the mortar joints as it crystallises. Sometimes the face of the bricks can spall, most commonly around their edges. The source of the sulfates is often the bricks themselves, but they can also be introduced from air pollution and from the exhaust gases of slow-burning fuel appliances. This means that sulfate attack can occur in stacks built from any type of masonry.

Sulfate attack results in expansion and cracking of cement-based mortars. Cement renders can also be affected. In some locations, the sulfate attack is more pronounced on one side of the stack than the other; hence the leaning chimney in the right-hand photograph.

A common consequence of sulfate attack in chimneys is a pronounced lean caused by the different wetting and drying cycles between elevations, with that elevation facing the prevailing weather more readily suffering the problem. (See Chapter 6 for a more detailed discussion of sulfate attack.) As with leaning tall, thin stacks, demolition and rebuilding should be considered.

Chimney stacks constructed in limestone are also susceptible to sulfate attack. This is an example in the Cotswolds.

Failure of the withes

The thin half-brick walls (withes) which separate the flues can fail. Failure is usually caused by deterioration of the bedding mortar, i.e. the brick becomes loose, or cracking in the bricks themselves. Acid gases can also accelerate deterioration of flue brickwork, especially if it is of poor-quality bricks. It is not unknown for loose bricks to drop onto the grate. Loose render dropping onto the grate is more likely to be caused by failure of the parging (or rendered lining) of the flue.

The thin walls between the flues are known as withes. Where they are built as raking brickwork (where flues are gathered), they are prone to failure. Smoke escaping to another flue can cause 'smoking' and downdraft problems. Escaping flue gases from a gas fire (without a gas liner) can lead to carbon monoxide fumes leaking into adjoining rooms.

(Photo courtesy of Chimney Care.)

Some 150+ years use of the flue of this cottage has resulted in the disintegration of the sandstone external wall.

Failure due to lack of support

Some structural defects may only be visible from inside the roof space. Where chimney breasts at lower levels have been removed, the brickwork in the roof void (assuming it has not also been removed) should be properly, i.e. structurally, supported.

The stack on the left looks quite normal, but just below the roof line, and in the room below, the breast has been removed. Corbelled brickwork is all that supports the stack.

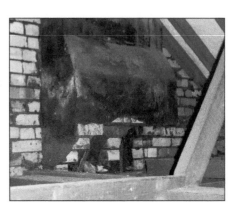

Other common defects in stacks include:

- *general deterioration of the mortar joints, particularly common where the brickwork has been pointed rather than jointed*
- *vertical or horizontal splits caused by the heat expansion of linings*
- *frost attack in the bricks or stonework*
- *potential damage caused by the weight (lever action) of aerials*
- *broken or loose pots; cracked flaunchings.*

Timbers should not normally be built into a chimney stack. Sometimes it is unavoidable. There is no risk as long as the timber is at least 225mm clear of any flue.

Splitting or cracking may affect stack stability.

Heavy aerials can affect the stability of tall stacks.

This freestone stack is in such poor condition due to age and inappropriate repair with cement mortar that it requires total reconstruction in new stone. Note the hole just above the roofline.

No amount of cement mortar can prevent the deterioration of this brick stack.

Over the past 20 years or so, chimneys have become more common in new construction. It is not unusual, for example, to find a stack containing a single flue from, say, a 'living' gas fire. Modern flues are usually built in pre-formed concrete or clay blocks hidden inside traditional stacks. However, some modern houses have artificial stacks made from glass fibre or stainless steel with brick-slip facings. These are not normally designed to contain flues – they are for appearance only.

An artificial chimney stack ready for lifting into place. These tend to be of lightweight construction, so a secure fixing is vital to protect against wind damage. Weatherproofing can also be a problem where flashings are not incorporated into the chimney in the factory (as is the case here).

Conclusion

It would be impossible in any textbook to set out a truly comprehensive compendium of all of the potential structural problems that may affect a building. There have always been infinite variations of construction technique, some good and others not so. Similarly, this applies to the many materials used in construction. Many traditional construction techniques would be unacceptable in modern buildings but, even with the advent of regulated building standards in the late nineteenth and twentieth centuries, the range of alternative construction solutions available still gives rise to failure.

Sometimes, this means that the problem being experienced may not be truly understood until a full (and destructive) investigation can take place. The reason for the movement in the building below was only identified when the external render was removed to expose the softwood timber plates, known as levelling plates. These had been installed by the builder partly to level up the poor-quality brickwork and also to provide a wall plate to the upper floor. This was a common form of construction in the nineteenth century. Years of water penetration, resulting in wet rot had led to disintegration of parts of the plates and movement in both the wall and the upper floor.

Alternatively, the process of destruction may have already taken place. The reason for the spectacular failure of this external skin of brickwork was the force of nature – lightning!

(Photo courtesy of Knightstone Housing Association.)

Finally, some further examples of structural failure and/or poor remedial treatment.

Top left: The movement in the flank wall has sheared straight through the concrete blocks of this modern garage resulting in a loss of bond between the two walls. Top right: This Victorian terraced house has been 'modernised' recently. Unfortunately, the artificial stone cladding at ground-floor level is masking structural movement – the crack can just be seen to the right of the roof of the bay window. Bottom: Some quite obtrusive and extensive metal strapping and ties have been used to stabilise movement in this flank wall.

Left: Various repairs to the chimney including rebuilding the top in brickwork and incorrectly applying a cement-based render over limestone rubble have failed to prevent deterioration of the stonework. Middle: The recently constructed brick retaining wall has a basic flaw – no consideration has been given to drainage! Right: Rusting of the reinforcement has caused the concrete of this boot lintel to disintegrate.

The solution to movement of an external wall is often the installation of restraint ties or stitches. Here are three examples, although whether they will all be successful in halting any movement is questionable. Left: The steel ties are much too short to be really effective – in fact, they may make the situation worse and it may be better to treat the crack as a sealed movement joint. Middle: A series of very short concrete stitches have been installed. Right: The surface-fixing of the angled tie to the external wall on the left is almost certainly going to fail – the tie needs to be built in, not merely bolted to the surface.

The investigation of structural defects 5

Both of these houses are suffering from structural movement. However, not all houses being investigated will be in such a self-evident poor condition.

Introduction

As stated in Chapter 3, it is important to recognise that evidence of movement in one element of the structure may be as a result of failure in a different element, e.g. cracks in a wall may be caused by failure of the foundation or movement of the roof structure. This means that any investigation should involve a full examination of all structural elements and components that might be causing, or be affected by, the problem under inspection. It is, therefore, both difficult and unwise to consider an affected element in isolation. Due consideration needs to be paid to the various structural elements – roof, walls, floors, foundations – and their effect upon each other. In particular, the close relationship between walls and foundations needs to be recognised.

One of the key principles of defect investigation is that of 'following the trail'. This principle has been stated in numerous legal judgments and should always be borne in mind when investigating any building defect. It is especially important when examining structural defects and involves reviewing all of the available explicit and implicit evidence in terms of both cause and effect. A simple example is to examine the outside face of a wall if a crack is found internally. Only when all available evidence has been considered can a judgement be formed as to the cause of the problem and its remedy. The courts have stated that it is essential to 'follow the trail' from the defect originally noted in order to draw the correct conclusions and undertake the proper corrective action. Failure to do so might be considered negligent, especially if one is acting in a professional capacity.

This Victorian house is suffering from the movement identified in the enlarged photograph. 'Following the trail' will involve a full examination of the external and internal areas of the building as well as the surrounding garden area. A walk-over survey and desk survey (see later section of this chapter) are also likely to be undertaken.

Further general factors to consider will be the age of the building and its current and past use. Its age will indicate the original construction in terms of general form of design and materials used so it is important to have knowledge of alternative types of design and materials used in any particular period of construction. The age will also have an important bearing on the defects encountered, e.g. those found in a Victorian house of monolithic construction being quite different from a 1930s house with cavity walls (see Chapter 2 for a more detailed discussion). Another important factor to consider is the use of the building since its construction. Many buildings, whether old or modern, will originally have been constructed for quite a different use from that currently noted. Some buildings experience several changes of use over their life. Ill-thought out and executed alterations, inappropriate changes of use, poor maintenance and repair, all need to be considered when undertaking investigation of structural defects as they can act as indicators of the likely extent and degree of the problem being experienced.

This chapter is divided into a number of sections. However, building failures and their investigation cannot be compartmentalised. A thorough reading of the complete chapter is recommended in order to be able to make the correct decisions as to which procedures will apply for any particular investigation.

Principles of investigation

Movement due to foundation failure

The failure of the foundations of a building almost inevitably leads to cracking in the walls. However, there are many other causes of cracking in walls, such as thermal movement or stress induced within the wall from floors and roofs. Only careful investigation and consideration of all the facts and implications will lead to the correct diagnosis.

Often, the only physical evidence of building movement due to foundation failure will be cracking or distortion of the walls above ground level. Initial assessment of the defects in the wall will be required, followed by further investigations as the causes are explored and a foundation-related problem becomes suspected.

The investigation may include one or, more usually, several of the following procedures.

This 1930s semi-detached house is suffering from movement in the front and flank walls. The causes are unknown and a full investigation is required before any judgement can be made and remedial works commenced.

'Walk over' survey

This is a physical review of the site and the surrounding area and is undertaken in order to assess local features having implications for the suspected problem, e.g. topography such as hills and dales; natural and man-made watercourses; excavations; filled areas of ground; ground erosion or instability; ground contamination: the proximity and condition of surrounding buildings; active/ recent building works; position and type of trees and shrubs; local place names (e.g. Fishponds, Clay Hill, Bog Lane, Quarry Way).

This village is a few miles north of Bristol. The sign indicates potential problems linked to coal mining that need to be investigated in any property to be surveyed. Further consideration needs to be given to the type of sub-soil; heath-land is often found to be formed over a sandy sub-soil.

The walk-over survey also needs to consider the proximity and types of trees, the effects of the newly constructed tower blocks behind the houses and the effect of the sloping ground.

Desk study

Reading the 1903 Ordnance Survey map of this area of Walsall reveals that it has a highly significant industrial and natural history that is not apparent on the modern aerial survey photograph, e.g. a colliery, old shafts, a glue works, clay pits, brick works, a stream, The Bog. The implications of the findings will have to be considered as part of the survey (photo courtesy of Get Mapping).

(Reproduced from the 1903 Ordnance Survey map.)

A desk study involves inspection of any of a number of sources of documented and electronic information sources including maps (e.g. current and old Ordnance Survey maps, geological maps), photographs from aerial surveys, plans deposited with local authority planning and building control departments; and plans from utility companies. It could also include maps and reports from the British Geological Survey, the Coal Authority, the Cheshire Brine Subsidence Compensation Board, and articles from local historical societies. Comparison between older and more recent editions of OS maps, etc. will reveal changes of land use, see example above.

Below-ground investigation

This can involve one or more of a number of approaches to identify the geotechnical and geo-environmental characteristics of the site. These include the excavation of trial pits by hand or machine to a maximum depth of 3 metres to examine the footings of the wall and the sub-soil immediately beneath it, and/or the sinking of small bore holes to obtain deeper sub-soil samples.

A trial pit can provide much basic information about the sub-soil, the depth and size of foundation, the ground just below the foundation, tree roots, water levels, broken drains, etc. Initial testing of the sub-soil may be carried out on a simple basis, e.g. squeezing a small sample in the hand to determine water content or driving a metal bar into the sub-soil to indicate its firmness/density. However, such tests can only provide crude/simple information and, generally, more sophisticated testing needs to be undertaken in a laboratory. Laboratory testing can determine soil type, bearing capacity, water content, organic content, chemical content, plasticity and other relevant information.

A trial pit is being excavated beneath an existing foundation in order to determine the type and condition of the sub-soil. On this site, cracking was caused by tree roots drying the clay soil. This type of site investigation is often an expensive, lengthy and potentially dangerous operation.

Hand-augered or mechanically bored holes, which can be drilled to greater depths, can provide small diameter (up to about 100mm) core samples for laboratory testing. The number of pits/bore holes will depend on the pattern of cracking and the results of any monitoring (see later section of this chapter).

For deeper investigation, a window sampler can be used. This is a steel tube with a series of windows along its length that can be driven-down mechanically to a depth of approximately 10 metres, depending upon sub-soil type. Each window provides a view of the soil type and samples can be extracted for laboratory testing. Alternatively, deeper boreholes can be driven using light percussion drilling (shell and auger) though special methods are needed to obtain undisturbed soil samples for testing.

Left: A soil sample has just been extracted from a borehole. Right: The soil sample is boxed ready to go off to a laboratory for testing.

Above-ground investigation

This involves one or more of a number of activities including thorough internal and external inspection of the building to identify where cracking and/or distortion is taking place, recording and assessment of the actual evidence (refer to next sections), the opening up of adjoining walls, floors, ceilings and roof voids to determine whether the element(s) involved are influencing the structural movement and/or the monitoring of suspected movement over a period of time (see later section in this chapter). Horizontal and vertical dimensions can be taken by means of tape measures, levelling instruments, plumb-lines and electronic equipment. A visual record needs to be kept; normally this will be by means of dimensioned record sketches. These can be supported with photographs; nowadays, digital cameras make this task relatively easy.

High-level inspection requires appropriate inspection apparatus. Left: A crane together with a cradle is being used (photo courtesy of Knightstone Housing Association). Right: An elevated platform (or cherry picker) is being used. In many investigations, a ladder may be sufficient.

In any case, it may take some time before a final opinion can be reached. Very occasionally the movement discovered is so great, or progressing so rapidly, that a decision is needed straight away. In such circumstances, human safety or the prevention of further damage elsewhere may necessitate immediate emergency works to prevent further movement. The building may even require demolition.

This is a rather extreme example of a house with structural problems. It appears to have already suffered partial collapse and the chimney is in a dangerous state.

The defect discovered may be due to movement or failure only in the element obviously affected. Alternatively, it may be part of a more complex problem involving structural failure elsewhere in the building, e.g. overloading of a foundation due to additional structural loads, or it may be found to be due to a cause other than movement, e.g. failure of a material because of impact damage. In any case, the investigation of both cause and effect should be as full as possible.

Movement associated with wall failure

Visible cracking in a wall

In the previous section, it was noted that wall movement may be indicative of any of a number of structural problems including failure of the wall itself or of another building element such as the foundation, a floor or the roof. If movement is discerned in a wall, a key part of any investigating procedure will be to examine the wall itself. The investigation should commence with the recording of basic information on each crack (Table 5.1).

Table 5.1 Cracks: key information that needs to be recorded and considered

FACTOR	POINTS TO CONSIDER
Element	• Which elements are affected?
Location	• Where in each element affected is the crack? • Does it run through window/door openings?
DPC	• Is the crack above/below DPC level? • Does it pass through it?
Length	• Identify the start and finish points in each element
Width	• Check width of each crack including any variation(s) along its length
Depth	• Is the crack through partial or full thickness of the element, e.g. wall?
Pattern	• Is the crack vertical, horizontal, diagonal or stepped? • Is there more than one crack in an element?
Effect	• Is there evidence of tension, compression, expansion, rotation or shear?
Alignment	• Are surfaces on either side of the crack and/or each face of an element level or is one side proud? • Is any affected wall out of plumb? • Is any affected floor or ceiling out of level?
Age	• What amount of dirt and/or dust is in the crack? Note: new/recent cracks tend to be clean, older cracks dirtier/darker
Relationship	• Is there other evidence of the crack extending into adjoining elements? • Of what construction are the adjoining elements? • Are apparently unrelated cracks linked in terms of cause?

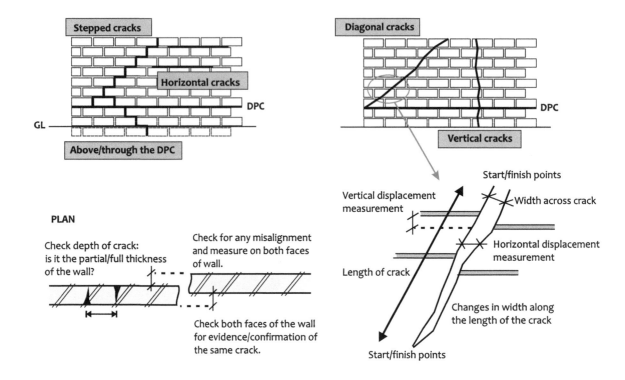

The result is often a fully dimensioned and annotated sketch or drawing. This may detail a single crack or be more comprehensive – perhaps covering a series of adjoining elements and the cracks that they contain.

There are a number of conventions for recording cracks, some of which are shown on the diagram. Determining when the evidence of movement should cause concern will depend upon the particular factors involved, e.g. type of building, form of construction, material(s) involved, geological information, topographical features. The amount of movement itself may be the trigger that sets in motion the need for investigation and repair. Considerable research has been undertaken by the Building Research Establishment (BRE) into the overall implications of crack size or width and they have produced a guidance list (Table 5.2).

Table 5.2 Classification of visible damage to walls with particular reference to ease of repair of plaster and brickwork or masonry (reproduced from BRE Digest 251 [Table 1])

CATEGORY OF DAMAGE	DESCRIPTION OF TYPICAL DAMAGE	APPROX. CRACK WIDTH (MM)
0	Hairline cracks of less than about 0.1mm width are classified as negligible. No action required.	Up to 0.1
1	Fine cracks which can easily be treated during normal decoration. Perhaps isolated slight fracturing in the building. Cracks rarely visible in external brickwork.	Up to 1.0
2	Cracks easily filled. Redecoration probably required. Recurrent cracks can be masked by suitable linings. Cracks not necessarily visible externally. Some external repointing may be required to ensure weather-tightness. Doors and windows may stick slightly.	Up to 5.0
3	The cracks require some opening up and can be patched by a mason. Repointing of external brickwork and possibly a small amount of brickwork to be replaced. Doors and windows sticking. Service pipes may fracture. Weather-tightness often impaired.	5–15 (or several of, say, 3.0)
4	Extensive repair work which requires breaking out and replacing sections of walls, especially over doors and windows. Windows and door frames distorted, floor sloping noticeably*. Walls leaning or bulging noticeably*. Some loss of bearing in beams. Service pipes disrupted.	15–25, but also depends on the number of cracks.
5	This requires a major repair job involving partial or complete rebuilding. Beams lose bearing, walls lean badly and require shoring. Windows broken with distortion. Danger of instability.	Usually greater than 25, but depends on the number of cracks

Note: * Local deviation of slope from the horizontal or vertical of more than 1:100 will normally be clearly visible. Overall deviations in excess of 1:150 are undesirable.

(IHS BRE Press, reproduced by permission.)

Left: This crack is about 5–6mm at its widest point. Right: The cracks in the building range from 5 to 50mm and there is an obvious danger of instability.

It should be pointed out that most buildings have cracks in them. The great majority are cosmetic and are usually caused by thermal or moisture movement. Cracking of a structural nature, particularly cracking caused by foundation movement, is not very common. Distinguishing the exact cause of different types of crack is not always easy and careful thought is needed before coming to a judgement. Use of Table 5.2 will assist in deciding just how critical is the amount of movement that has been identified. Note: BRE Digest 251 (Table 1) states that the crack width is only one factor in assessing the category of damage and should not be used on its own as a direct measure of it.

The cracks in this building are of a significant width. Shoring has been put in place because the suspicion is that this movement is progressive. Monitoring (see later in the chapter) is being carried out to determine the state of the movement.

Investigation of distorted walls

Apart from cracking, a wall may be subject to distortion, e.g. bowing or leaning over. When such a defect is identified, it will be necessary to measure a wall's thickness, the amount of distortion – both horizontally (with a spirit level or optical level) and vertically (by means of a plumb-line), and the overall extent of the wall area affected in order to determine the exact amount of movement.

Measuring a bulge

Area affected may be part or whole of wall/storey.

Horizontal and vertical measurements required internally and externally.
M = measurement required.

The construction and materials of the wall also need to be ascertained. The investigation should reveal whether the whole thickness of the wall is distorted or only one face is affected. The former would appear to indicate failure of the wall as a whole; the latter only one part – perhaps a loss of bonding between backing and facing materials. The examination should include the following information (Table 5.3).

Table 5.3 Information that needs to be recorded and considered

FACTOR	POINTS TO CONSIDER
Distortion	• What is the amount or depth of distortion? • What is the extent of the distortion (height/width, whole or partial thickness of the wall, etc.)?
Construction	• What is the form of a wall? • What are the materials used? • What are the thickness of the wall, the height of the wall, the type and size of openings, etc.?
Thickness	• What is the thickness of the wall or skin?
Loading	• Which type(s) and amounts of designed and unforeseen loads are imposed on the wall (including lateral and/or eccentric thrusts from other walls, roofs and floors)?
Performance of materials	• Are materials decaying? • Is adhesion/bonding being lost?
Evidence of lateral restraint	• Is the wall properly tied to adjoining elements, i.e. walls, roofs and floors?
Continuing movement	• Is the movement static or progressive?

The bowing in the wall of this terraced house in North Bristol became significant enough to warrant supporting scaffold until its cause and remedy could be determined.

The investigation may require an internal examination of the wall by means of a boroscope (a type of optical probe that is inserted into a pre-drilled hole in the wall) or by the removal of selected bricks/stones from the affected area. Further examination of adjoining elements (walls, floors and roofs) may also be necessary to identify discernible crack patterns and evaluate the implications for the distorted wall.

Investigation of movement linked to roofs or floors

The problems associated with floors and roofs are discussed in detail in Chapters 7 and 8 (floors) and Chapters 9 and 10 (roofs). However, as stated in the Introduction to this chapter, an investigation of structural movement needs to consider all potential causes and this must include the likelihood of floor and/or roof movement being involved.

The following factors need to be investigated when considering whether a floor and roof are linked to building movement.

The effects of vertical and horizontal forces from floors or roofs. These may be due to the self-load of the floor or roof, but may also be caused by superimposed or applied loads, e.g. furniture and/or stored goods on floors; wind, snow and/or rain loads on roofs. Problems caused by floor loadings, especially in older buildings, are often linked to a change of use of a building and also to the increased amount of furniture and personal possessions to be found in the modern household. All of these factors need to be taken into consideration when the investigation is undertaken.

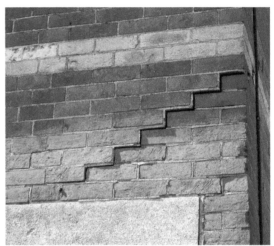

The cause of this movement is not apparent from an external inspection. The investigation will need to consider possible causes relating to the foundation, the wall, the adjoining floor and the roof.

Lack of restraint from adjoining elements. A properly constructed floor or roof should be linked to a wall, especially an external wall, in such a way as to act as a tie and thus provide restraint. Modern Building Regulations require mechanical connection between the floors, roof and external walls. However, this was not the case in Victorian and earlier buildings. Any investigation of building movement will need to include close examination of any adjoining floor and the roof, if only to positively eliminate the other element as a cause of the problem.

General examination of the building

Once the nature and position of any cracks, bulges, etc. have been recorded, a more detailed inspection of the building and not just the affected area can begin. The purpose of the inspection is to provide further evidence to support an accurate analysis. This process will include an assessment of the crack's significance (see Table 5.2) as well as its cause.

The nature and scope of the inspection will be dictated, to some extent, by the nature of the client brief. Whatever the nature of the brief, the following guidance should be regarded as a minimum procedure for any inspection (Tables 5.4 and 5.5).

Table 5.4 Examination from the inside of the building

ACTION	POINTS TO CONSIDER
Look in the roof space	• There may be signs of cracking in the walls or at junctions of walls • There may be obvious roof defects such as lack of gable restraint or lack of effective ties preventing roof spread
Check decorations	• Review each room and try and assess the age of decorations (see left-hand photograph below)
Look for cracks in corners of rooms and at vertical junctions with partitions	• Do not forget the junctions of walls and ceilings • Pay particular attention to the junction of any annexe to the building
Check ceilings for level	• Sagging ceilings may indicate undersized joists
Inspect inside built-in cupboards	• Cracks, etc. tend to get overlooked inside cupboards and old decorations are often not renewed
Check the operation of windows and doors	• Look for openings out of square, binding windows, doors that have been planed down and cracks at architrave mitres
At floor level, look for any gap below the skirting	• This can suggest that the floor has dropped or that the wall behind the skirting has moved
Check whether a floor has a slope	• Sometimes difficult to determine. A marble can be used.
Try and assess joist depth	• A 'heel-drop' test may confirm undersized joists or lack of effective strutting • Jumping up and down also works, but may cause a great deal of vibration with older floors

Table 5.5 Examination from the outside of the building

ACTION	POINTS TO CONSIDER
Inspect roof for problems	• Look for signs of dishing, etc. in the roof (dishing may suggest roof spread)
Check whether the building is vertical	• Use a plumb bob on external walls, e.g. from a bedroom window (see right-hand photograph above)
Check for movement cracks	• Check for gaps between window frames and the surrounding wall
Review openings on different floors	• Look for vertical movement between openings on different floors
Check whether the building is level	• Check whether masonry joints are horizontal using a spirit level, water level or laser level
Check for recent building work	• Note any recent structural alterations or additions such as patio doors, extensions, roof conversions
Identify recent repairs	• Check whether cracks been repaired – if so, how long ago?
Review proximity of trees and shrubs	• Identify tree and shrub species, age, position and height, etc. • Consider whether trees and shrubs are still growing or not
Review lack of trees and shrubs	• Consider whether trees and shrubs may have been removed
Consider type of sub-soil	• Identify, if possible, sub-soil type and consider need for trial pits
Inspect drainage	• Identify provision for drainage and positions of drains, soakaways, etc. • Lift inspection chamber covers to make sure there are no blockages • Flush WC to see how water runs away • Consider drains test to check for leakages
Review garden layout	• Note any significant changes to gardens, paths and patios (they may affect drainage) • Similarly, consider similar changes to adjoining property – buildings and land
Review evidence from adjoining property	• Consider the condition and proximity of nearby buildings and land as well as current and past use

All internal surfaces of a building must be carefully inspected. Left: The wallpaper is split along the line of movement. Right: Similarly, this shows the importance of full inspection externally – the wall is over 100mm out of plumb.

Monitoring of movement

It is current practice to avoid costly remedial works, such as underpinning, whenever possible. This is, in part, because research into failures related to sub-soil since the severe drought of the mid-1970s has revealed that, in the long term, movement in the majority of severely affected buildings will return (more or less) to a 'safe' or acceptable level of movement distress as subsequent weather patterns return to normal. Therefore, if it is suspected that the movement has not already ceased, consideration should be given to monitoring its progress prior to making any decision on remedial works. However, there is often pressure from householders and some insurance companies to undertake underpinning, in particular, as it is seen as a complete and final solution to any movement problems, whereas no work other than minor repair may actually be necessary in the long term.

Sometimes the amount of movement requires temporary shoring or scaffolding to arrest progressive movement. In such circumstances, further monitoring may not be necessary if the period of movement can be already determined. Note: the shoring on the right appears to be of a more permanent nature.

Monitoring involves the regular (usually monthly) inspection and recording of the amount of movement. This is commonly achieved by means of fixed devices, such as tell-tales or crack markers (studs), on the face of a wall. It can also be helpful to keep a photographic record as well, but only in support of a system of measurement as photographs are not normally accurate enough to calculate the precise amount of movement and its direction.

Normally, monitoring would not be undertaken for a period exceeding 12 months, though occasionally it is carried out for up to 24 months. For movement where a clay sub-soil is involved, the period should cover a full seasonal cycle of summer–winter–summer with readings every 4–6 weeks. At the end of the selected period, there should be evidence available to indicate: (1) whether the movement is static or progressive; (2), if the latter, its direction; and (3) whether it is seasonal. This will enable an appropriate decision to be taken on the amount and type of remedial work required.

There are a number of different methods of monitoring. All involve the regular periodic reading of a measuring device. They include the following methods.

Visual devices

The simplest method is to measure the crack with a ruler or to measure across a couple of pen or pencil lines either side of the crack. Alternatively, a glass tell-tale can be attached either side of a crack by means of mortar dabs, but it provides only limited information. If it falls off, it will confirm that movement is still taking place and if it remains in position that the movement has ceased. Neither simple measurement nor glass tell-tale indicates the direction of any movement. Glass tell-tales are also, unfortunately, commonly subjected to vandalism. Both of these methods are of no use for long-term monitoring.

A better method, externally, is to fix two screws either side of the crack and measure the distance between the screws with an appropriate gauge such as a calliper with a digital read-out. A more sophisticated and accurate version is to use three screws so that both horizontal and vertical movement can be measured. Internally, a system of three pencil dots can be used.

Left and middle: The remains of a mortar tell-tale. Two marked strips of tape are shown in the centre; their limitations can be clearly recognised. Right: A three-screw rather than a two-screw monitoring system.

Left: The plastic tell-tale is manufactured in two halves; each half is fixed either side of the crack. If the crack opens, or closes, the red cross-hairs show the amount of horizontal and vertical movement relative to the graded background. Right: This tell-tale measures movement across a crack and also 'out-of-plane' movement.

A number of companies produce plastic tell-tales and these are considered to be the most effective of the simpler visual devices. They are manufactured in a range of types, the simplest being fixed straight across a crack and measuring in two dimensions (horizontally and vertically), the more sophisticated being able to measure in three dimensions and/or cracks at the junctions of adjoining walls. The problem with these is that they only really measure to an accuracy of 1mm when an accuracy of ±0.1mm is often required. Some surveyors consider the use of three screws and a calliper to be a more accurate method.

Optical levelling

Differential movement or distortion can be measured with an optical level and staff. Variations in level can be recorded and plotted as shown in the graphic. This is difficult in buildings that are rendered or which are built in stonework because commonly a level line, e.g. a brick joint, is needed as a starting point. This obviously assumes that the mortar bed is level in the first place. However, any building can be level-monitored by fixing markers (often stainless steel pins) into the face of the wall. Level and verticality monitoring is normally undertaken every 6–8 weeks over a period of 6 months. This period has been shown to be sufficient to provide clear evidence of the pattern and the extent of any movement.

The photograph shows a levelling team at work. Plotting the results as a graph or as an isometric model clearly shows the nature and extent of any movement.

Electronic monitoring

Remote monitoring by means of electronic equipment is rarely used for investigating movement in smaller domestic properties because of expense, but it can be of benefit in long-term monitoring of historic/important buildings or those parts of a larger building that are difficult to access. A mixture of sensors such as linear variable differential transducers (LVDT), which measure linear displacement/strain, electro-level gauges, which measure angular displacement, and piezoelectric and strain gauge accelerometers, which monitor vibration, can be connected to a data logger. The information is collected over a period of time via transmission to an office-based computer, either by a modem or wirelessly. Alternatively, it is possible to connect the data logger to a manually held portable instrument. The data collected can be stored and viewed at any time by the monitoring team.

Electronic monitoring equipment is also used where non-destructive investigation methods of structural problems are required, in particular to determine what is happening within the structure. In each case, the use of such equipment requires a high level of expertise to interpret the findings, which are often based on continuous readings taken over a period of time. The devices include:

- **Ultrasonic measurement.** *This makes use of ultrasound to detect objects and measure distances. It can be used to locate areas of decay and movement within the structure of a building.*
- **Impulse radar.** *This senses and measures distances to objects. It is used to locate and measure cracks and other internal discontinuities, such as flues and voids, within walls and floors.*
- **Infra-red thermography.** *This measures minute variations in the infra-red radiation emitted by all structures. Although normally associated with locating and measuring heat loss, it can also be used in suitable weather conditions to locate and assess movement in a wide range of building structures and surfaces.*

Analysing the findings

The findings of the investigation and associated monitoring should enable the surveyor or other construction professional to analyse and then make appropriate judgements about the following over-arching factors:

1. *the cause(s) of the movement;*
2. *identification of whether the condition is passive or progressive;*
3. *advice on how to remediate its condition and effect(s);*
4. *advice on financial implications.*

The key point about any investigation and associated analysis is that they should both be as comprehensive as possible to ensure that the decision made (e.g. to remediate or not) is the most suitable for dealing with the problems being experienced by the building.

The exact method of analysis will depend upon a number of factors including: public safety; the type of construction; the use of the building. One simple method is set out in Table 5.6.

Table 5.6 Five key criteria to consider when reviewing a building suffering structural movement

CRITERIA	COMMENTS
1. What is the scale of the damage?	• See Table 5.2
2. How does it affect the building now?	• Is it purely cosmetic, i.e. just affecting the building's appearance? • Does it affect the use or serviceability of the building, e.g. are the windows difficult to open, do cracks allow rain penetration? • Does it affect the stability of the building, i.e. is it safe? • See Table 5.2
3. Is it likely to get any worse?	• Are there signs that the movement has ceased or is it progressive? • Should it be monitored (see earlier section of this chapter)?
4. How might it affect the building in the future?	• Is there a future risk of collapse?
5. What is the cause?	• Is it caused by defective foundations and or poor ground conditions? • Is it caused by problems in the superstructure?

The cracks on the face of this 1930s house need investigation. They are considered to be of a structural nature and the visual evidence is that they indicate cavity wall tie failure.

Providing a detailed list of crack types and possible causes is an almost endless task because there are so many factors and permutations to consider. The brief summary of causes in Table 5.7 relates to traditional masonry buildings (solid or cavity construction). (A more detailed discussion of causes and effects can be found in Chapters 3 and 4.)

Table 5.7 Summary of cracks and possible causes in traditional masonry buildings

EVIDENCE OF MOVEMENT	MOST LIKELY CAUSE(S)
A crack which appears to be static (i.e. there are no signs of progressive movement) and the crack is no more than a few millimetres wide	• Shrinkage of new materials – this is mostly likely to be evident inside a building; usually at junctions of building elements • Settlement of new buildings – this will happen almost immediately on sandy sub-soils; it can take a few years on clay sub-soils • Cracking caused by loading – most common around lintels and other beams
A crack which opens and closes over the course of several months Note: cracks like this tend to 'ratchet' as they accumulate dirt and dust	• Thermal movement in the superstructure – cracks caused by thermal and moisture movement are normally between 1mm and 5mm wide • Seasonal expansion and contraction of clay sub-soil
A pattern of cracking that suggests some form of foundation or ground problem, e.g. Cracks which show on both faces of a solid wall Cracks which show on both faces of a cavity wall, i.e. on the outside of the outer leaf and the inside of the inner leaf Cracks which taper, i.e. either wide at the bottom and narrow at the top or vice versa Distortion in door and window openings Walls out of plumb and ground floors out of level Cracks which run across the DPC, i.e. they appear both above and below the DPC Cracks which are often greater than 5mm wide Broken drains or disrupted service ducts	• Subsidence – cracking is likely to be stepped and tapered (wider at the top) • Ground heave – the crack is likely to be wider at the bottom • Whether or not the crack is stepped depends on the nature of the masonry: if the masonry shears, the cracks may be almost vertical • If drains have cracked, the effect will depend on the nature of the sub-soil: sands may be washed away, clays may heave
A wall that is leaning or bulging This can occur with or without associated cracking	• Thrust from a badly tied roof • Lack of restraint at the gable (trussed rafters not strapped) • Lack of restraint at upper floors • Lack of restraint from chimney breasts, partitions and return walls • Lateral thrust from expanding solid ground floors • Loss of bonding between the material forming the skin of the wall and its backing
Cracks and/or distortion in a chimney stack	• Vertical cracks in chimney stacks are usually caused by cracking of clay linings • Horizontal cracks are more likely to be caused by sulfate attack • Pronounced leans are often caused by sulfate attack

It needs to be noted that in old buildings the use of flexible lime mortars may often limit cracking. Old cracks are also often hidden by 'post-event' renders or filling.

The bottom right-hand graphic above shows how heave in the central section of a building can be confused with subsidence on either side.

Conclusion

It should be understood from reading this chapter that an investigation into any suspected structural movement must be extremely thorough. The person carrying out the investigation should also be prepared for the unusual. Many would, correctly, suspect subsidence as the cause of the following defects (photos top left and bottom left p. 152). However, how many would think the chimney damage (top right p. 152) is caused by an earthquake. The key factor in any investigation, however methodical and detailed, is that the investigator must not ignore evidence, however trivial, and must not jump to conclusions.

(Photo courtesy of Department of Environmental Protection, Pennsylvania, USA.)

(Photo courtesy of Matthew Jackson Photography.)

(Photo courtesy of Cheshire Record Office.)

'Follow the trail': some further examples

Left: Not all ground surveys are as easy as this where the sandstone bedrock is clearly evident. Right: A clay sub-soil can be seen just over a spade's depth below ground level. The amount of roots in the top-soil may indicate the close proximity of a tree or a large shrub; this needs further checking.

Left: The building has experienced past movement that necessitated a metal tie being installed across its width. The wall is now suffering from lateral movement that requires investigation. Right: Street trees (whether pollarded or not) always require consideration of any potential effect on adjoining buildings.

Evidence of past movement always needs to be checked. Left: Have the concrete stitches stabilised the structural problem? Internal inspection of a building should not be over looked, either. Right: The split woodchip lining paper indicates fairly recent movement. Is it progressive or has it stabilised?

External walls: brickwork and stonework 6

Introduction

This chapter looks at defects in external walling materials. It concentrates on brickwork and stonework. It also briefly considers the historical development of these two materials and comments on aspects of construction which might affect their durability.

Historical background

Brick is perhaps the main material which helps define the appearance of a modern house in this country. However, brick has competed with stone in terms of popularity and status for many centuries. There is evidence of brickmaking going back to Roman times and after the Dark Ages it re-emerged in East Anglia – an area with little available building stone but with clay deposits. Generally, however, until the sixteenth century, timber had been the dominant structural material for most vernacular buildings. Brick and stone started to be adopted more commonly for a number of reasons, including timber supply shortages and, perhaps more significantly, the disastrous experience of fires which led James I, and later, after the Great Fire of London in 1666, Charles II, to issue proclamations insisting on the use of stone or brick.

The dominance of either brick or stone has waxed and waned over the centuries. Partially this was a factor of geography and, therefore, availability of materials. It was also affected by the development of trade skills and by advances in technology which allowed for more efficient quarrying of stone, or excavation of clay, as well as industrialised kilns for brick production which improved the ability to control both quantity and quality. The creation of canals, and later, the rail system had a dramatic effect, particularly on the ability to distribute bricks to areas with no clay deposits. Less pragmatic, but perhaps more significant, factors were the issues of style and status. For example, brick was popular in the Restoration period and the artistic skills practised by the bricklayers in both decorative and precision work during this era were, and still are, admired. The Georgian period is often characterised by the stone buildings of cities such as Bath, but brick was a fashionable material for developments in, for instance, Liverpool, London and Bristol. Because of the links between style and status, the aspiring classes looked to these materials to demonstrate social position. In the Regency and early Victorian periods the appearance achieved by ashlar stone was aspired to, but because stone was expensive, the look was often achieved by using cheaper bricks covered externally with stucco (a type of render) with lines drawn out to mimic the mathematically precise ashlar blocks. Unfortunately, builders often used substandard bricks in this type of work. This damaged the general reputation of bricks and the practice still causes technical problems in these buildings today. There were, however, surges in the status of brick, including that which arose during the 'Battle of the Styles', one example being the Queen

Anne revival in the 1870s. At other times in the Victorian period, brick was equal to stone in status, partly because of the increasing precision with which bricks could be made. This precision, allied to the wide range of colours, finishes and shapes that became available, appealed to the Victorians' sense of pride in technological achievement as well as their liking for colour and decoration. Clearly though, this appeal was not universal; many late Victorian houses in cities such as Bristol have stone walls at the front with brick being relegated to the side and rear elevation, where it was often rendered. There are, however, geographical areas where brick was used to the front elevation with stone being relegated to the side and rear.

Overview of brick and stone defects

Given the UK climate, it is inevitable that external walls may remain wet for long periods. Water penetration is, in fact, the common denominator in most of the defects referred to in this chapter. Although the durability of a wall depends to a large extent on the correct specification of stone/brick and the mortar, the overall design, practical detailing and regular maintenance are all key factors affecting durability. Many of the defects referred to in this chapter can be designed out in modern buildings by proper attention to such items as DPCs, flashings, roof overhangs, copings and projecting sills. Material selection and detailing should also take into account the potential effect on brick or stone of exposure to frost and the sun.

The exposure of a building will be related to:

* **The position of the building.** *Is it particularly exposed by being on the coast or on top of a hill? Is it relatively exposed because of its height or perhaps sheltered by surrounding buildings or trees?*

- **Geographical location and orientation.** *Some elevations will face the prevailing wind and rain, others will be warmed by the sun, while the north elevation will remain cold. Sheltered positions, with little exposure to sunlight or wind, may restrict evaporation, resulting in water remaining in the wall longer.*

Particular features which are vulnerable include: chimney stacks, parapets, aprons under windows, garden walls, and retaining walls, i.e. areas which receive a lot of rain, or are saturated because of contact with the ground. Plain vertical walls are not generally a high risk if they are adequately protected by roof overhangs, correct sill details, etc. However, horizontal projections are vulnerable, and may wet the adjacent wall if they shed water inadequately.

Brick

At least 90 per cent of the bricks produced in the UK are made from clay. Concrete bricks, which were introduced in the1950s, are also produced, as are calcium silicate (sand lime) bricks. Calcium silicate bricks were first patented in the 1860s in the USA and developed in a number of different countries over the following century. They are produced from hydrated lime and sand: sometimes crushed flint is added (flint lime bricks). Calcium silicate bricks are generally resistant to frost attack and are virtually free from soluble sulfates. Their popularity appears to go in and out of fashion, though they did gain a poor reputation in the 1960s and 1970s. This was to some extent a result of designers' failure to deal adequately with the tendency of the bricks to undergo moisture shrinkage.

Solid and cavity walls

In order to fulfil its structural function, a brick or stone wall has to be bonded. Modern walls are built with two single skins, usually with the outer protective skin separated from the internal load-bearing skin by a cavity. The internal skin is nowadays usually constructed in blockwork (unless it is a timber-framed house). The two elements are tied together by wall ties. Where the outer skin is in brick, this is usually bonded in stretcher bond. Some houses of course will be constructed with blockwork to the inner and outer skin and finished with an external render.

1960 to early 1990s

Galvanised tie

Stretcher bond

Cavity 50 to 75mm wide

1990s onwards

Stainless steel ties

Modern cavities are usually insulated – with either flexible batts or rigid boards.

Cavity 50 to 100mm wide

Before the cavity wall became common, solid walls were the norm. Although half-brick thick walls were sometimes used for small extensions on poorer-quality houses, solid walls were at least 225mm, or one brick thick. Solid brick walls were usually constructed in English or Flemish bond.

English bond is made up of a row of stretchers followed by a row of headers. Flemish bond is made up of alternate headers and stretchers in the same row. English bond produces a stronger wall, though Flemish bond tended to be considered the more aesthetically pleasing. In the Victorian period a number of other bonds were developed, including more economical, but less stable, garden wall versions of English and Flemish bond. An English garden wall bond, for example, might consist of three or four rows of stretchers followed by one row of headers, and a Flemish garden wall bond might have three or four stretchers followed by a header in the same row. Despite their name, garden wall bonds were often used in houses, though they were usually confined to side or rear walls and were often rendered.

English bond.

Flemish bond.

English garden wall bond.

Flemish garden wall bond to the gable end elevation of a house.

The four photographs above all show queen closers used at returns or openings.

Header bricks were used to tie solid walls together. Unfortunately in some speculative construction, headers were 'snapped' (i.e. cut in half) and therefore would not perform this function.

Timbers would sometimes be built into a wall in order to add stability. Although there is some debate as to their exact purpose, they may have been intended to provide lateral constraint and continuity while the lime mortar set. Another reason for burying timbers in the wall was to provide easier fixings for timber panelling and the ends of floor joists.

Snapped headers were often used where internal and external bricks were of different sizes. They were also used to save money. Facing bricks were much more expensive than commons. 'Snapping' a header created two facing bricks.

Note: these headers might be 'snapped'.

Snapped headers caused (and still do cause) defects because there is a lack of tying-in between the inner and outer face (see photograph on p. 112 as an example). Bonding/levelling timbers are vulnerable to rot and beetle attack; if they lose their strength, they will no longer resist the compressive loads in the wall and bulging and/or cracking may occur.

Other examples of Victorian bonding were created with cavities, for example, the rat trap bond and Dearne's bond.

'Rat trap' bond was formed by laying bricks on edge. The void in the centre could fill with rainwater.

Rat trap bond

Dearne's bond was similar but the headers are laid flat.

Both of these bonds, while economical, were short-lived because they allowed water penetration to the interior. They also lacked the stability of English or Flemish bond.

Although there are examples from earlier periods, cavity walls, as we would recognise them today, were essentially introduced in the Victorian period, when it became accepted that having two separate skins helped in preventing penetration of moisture to the interior and when it was technically easier to economically construct thinner individual skins. Cavity walls are referred to in documents from the beginning of the 1800s. However, their introduction was slow and developed at different rates in different parts of the country. They tended, for instance, to be adopted more readily in areas of high exposure where rainwater penetration was a problem.

Victorian

inner leaf

Late Victorian wall ties

Cast iron *Wrought iron* *Clay*

1930s - 1950s

1 Brick– 225mm
or 9 inches
(or thicker)

Some Victorian walls were in cavity construction. The outer leaf often
had snapped headers to give the appearance of Flemish Bond.

Modern walls are usually formed
with a lightweight block inner leaf.
For examples see earlier chapter.

By the 1930s, cavity walls were commonplace in new construction, though they were not necessarily dominant in all parts of the country.

The early devices used to tie together the two leaves of the cavity were made from a variety of materials including, bricks or, variously, cranked, glazed, perforated and hollow stoneware. However, the popular approach in the mid-Victorian era, and the norm towards the end of the century, was to use metal ties. Metal ties were either cast or wrought iron. Bitumen was sometimes added as a protective coating against rust. The ties were quite bulky, perhaps because of an assumption that rusting would occur. Because of their size and the type of metal used, the ties were relatively stiff and did not easily accommodate movement between the two skins. By the time that cavity walls were becoming the preferred method for external wall construction, mild steel ties were the norm. Again, protective treatments were rather hit and miss, though galvanising with zinc was used as an alternative to bitumen. Modern ties are made from galvanised steel, stainless steel or plastic. Copper ties have also been used.

Brick categories

Brickwork is generally a durable material with modern methods of manufacture ensuring good quality control. Problems experienced in the past, such as under-firing (leading to softness) or over-firing (leading to brittleness), deterioration through reactive materials, lamination, distortions, and differing sizes, are now rare.

Modern bricks are categorised by strength, salt content and frost resistance. The category of frost resistance considers whether or not a brick is suitable for use where there is repetitive subjection to both water saturation and freezing temperatures. At present, clay bricks are categorised as:

F0 – which means the bricks are not frost-resistant and should not be used externally.

F1 – which means that the bricks are moderately frost-resistant.

F2 – which means that the bricks are frost-resistant in all normal situations and degrees of exposure.

Soluble salt content is categorised as:

S0 – which means there is no restriction on the amount of soluble salts that they contain.

S1 – which means that there are limits set on the amount of soluble salts that they contain.

S2 – which means that there are lower limits set than for category S1.

By no means do all the soluble salts that may cause problems come from the bricks. They can be present in the sand used in the mortar, and they can enter through rainfall, or from the ground.

Mortar

Mortar helps to distribute the load through a wall and it seals against water ingress. It is made from a fine aggregate (usually sand) and a binding agent (nowadays, in new buildings at least, this is usually cement). When mixed with water, a chemical reaction called hydration occurs and the mortar sets. Before cement came into common usage, mortars were based on lime and sand. Modern mortars use cement as a binding agent though lime should be included in order to make the mix more plastic and workable. Lime also gives better adhesion to the brick and it improves the mortar's ability to cope with thermal and moisture movement. The properties of mortar can be changed by varying the proportions of the constituent materials. A weaker mortar mix (that is containing proportionately less cement) is more flexible and therefore more able to resist the stresses imposed by movement. With such a mix, if cracking does occur, it will tend to happen at the weaker mortar joints, which can be more easily and cheaply repaired than the bricks. Modern design should theoretically mean that there is no unforeseen movement but the adage, that the mortar should be weaker than the brick, still generally holds good. This is particularly so in older buildings, where repointing in a strong mortar can lead to damage in brick or stone where the mortar has not performed as a sacrificial item or has failed to allow water to evaporate adequately. However, if a mortar is too weak, it will not be sufficiently durable. Stronger mortars are durable because they will absorb less water and are more resistant to frost attack. Paradoxically, a stronger mortar may reduce durability because of its tendency to shrink and crack, thus allowing water into the wall.

Liquid plasticisers are sometimes used in place of lime. They usually work by introducing air into the mix in order to break down internal friction and therefore increase workability. They do not, however, impart the other qualities to the mix that lime does. In fact, they can reduce the strength of the bond between the brick and the mortar.

Defects in brickwork

Defects can arise in new brickwork because of poor design or specification, the use of sub-standard materials and poor standards of workmanship. These factors, which depend on quality assurance at the design and construction stage, are not specifically addressed in this chapter.

Frost attack

This is a problem that usually occurs in older bricks, and those that were under-burnt on firing. In newer construction, failure through frost attack tends to be confined to areas of severe exposure, or where the frost resistance of the brick was incorrectly specified.

The ability of bricks to resist frost attack is determined by their pore structure (in particular, the percentage of fine pores). Frost attack occurs through a combination of excessively wet brickwork and freezing temperatures. When water turns to ice, there is a 9 per cent increase in its volume. This expansion can produce stresses within the brick, which cause spalling, with the brick face flaking off and/or crumbling. Mortar is also subject to frost attack. In the deteriorating state, both elements readily absorb water which in turn increases the rate of frost attack. Although the risk of frost attack is increased where the saturated brick is subjected to particularly low temperatures, it is the rapidity of the freeze–thaw cycle that causes the damage. Because the process is progressive, frost attack can lead to total disintegration of the brick.

The images above show examples of the effect of frost attack on brickwork.

Frost attack is a potential problem where brick walls become saturated. Saturation can occur because of the failure of the design to protect brickwork or where an inappropriate brick type has been selected in an exposed position. It is also possible that individual bricks may be of

poor quality due to having been inadequately fired or because they contain impurities. The use of mortar which is too strong can also be a contributory factor in frost attack. If the mortar is too strong, then it tends to suffer from shrinkage cracks which allow water to penetrate. In addition, it will be difficult for water to evaporate through the joints which means that water stays within the brick longer.

The use of cement-rich mortars may contribute to frost damage in brickwork.

Boundary walls are particularly prone to frost attack. The sides and the top of the wall are exposed to the elements and they are therefore easily saturated, particularly if the wall has inadequate copings. Their exposure means that they are also subjected to extremes of temperatures, including freezing conditions. The tops of the walls often suffer the most because of radiation heat losses to the night sky.

In the photographs above it can be seen that the clay bricks of moderate frost resistance perform satisfactorily on the house where overhanging details at eaves, verge and at sills to window openings protect against excessive wetting. However, frost damage has occurred where the same brick has been used in areas where there is no such protection – such as the brick on the edge coping to the boundary wall and the unprotected brickwork below it. (Photos courtesy of Mike Hammet (Brick Development Association.)

The external walls of a house should be less vulnerable than boundary and retaining walls to frost attack. Roof overhangs and other key design details will offer protection from saturation. This protection will also guard against very low temperatures and in many cases the heat losses from inside the house will keep the temperature of the brickwork above freezing. There have been cases where frost attack has occurred in older brick cavity walls following the insertion of thermal insulation into the wall cavity. Cavity insulation, by its very nature, lowers the temperature of the external leaf and restricts evaporation into the cavity.

Wall treatments such as the use of silicone (as a remedy for rain penetration) can also cause problems because, when applied inappropriately, they may inhibit the drying out of the brickwork and therefore increase the likelihood of frost attack.

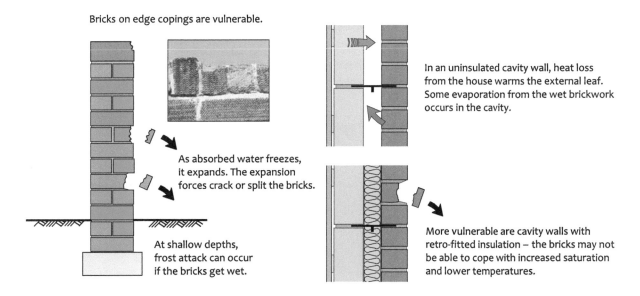

Bricks on edge copings are vulnerable.

As absorbed water freezes, it expands. The expansion forces crack or split the bricks.

At shallow depths, frost attack can occur if the bricks get wet.

In an uninsulated cavity wall, heat loss from the house warms the external leaf. Some evaporation from the wet brickwork occurs in the cavity.

More vulnerable are cavity walls with retro-fitted insulation – the bricks may not be able to cope with increased saturation and lower temperatures.

Efflorescence

Efflorescence is a common sight in new brickwork. It is caused by soluble salts in solution being brought to the surface as water in the wall dries out. It is usually a harmless, temporary problem, often occurring in spring following a wet winter. The main concern is the unsightly appearance caused by the white staining that it produces. Persistent efflorescence may indicate a design or construction fault which allows the brickwork to become, and to remain, saturated.

The three images above all show the white efflorescence stains caused by the deposition of white soluble salts on the brick surface.

Efflorescence is caused by a number of soluble salts including the sulfate or carbonate compounds of calcium, sodium, potassium or magnesium. The salts may originate in the bricks or they may be introduced through the mixing water, the cement or sand used for the mortar mix, or even from the ground on which the bricks were stacked. Additional sources may include sea air and unfortunate site practices such as the use of washing-up liquid as a mortar plasticiser (it usually contains sodium chloride – common salt). The salts may therefore emerge from the bricks and mortar. As the salts are water-soluble, they are often removed by rainfall, though they can usually be brushed off if their appearance is causing concern.

Although it is usually a harmless problem, there have been cases of damage caused by efflorescence. Crystallisation of salts just below the surface of the brick can cause spalling. This is known as cryptoflorescence. The problem is often associated with magnesium salts.

Left: Efflorescence on a newly completed building that was not given adequate protection against rainfall during construction. Right: The same building about six weeks later where the efflorescence has been washed away by the subsequent rainfall.

Cryptoflorescence is associated with a large build-up of salts and usually occurs where old, relatively weak, bricks are re-used inappropriately, particularly in areas of excessive dampness. It can also occur through salts deposited by the run-off from limestone or from air pollution. Damage may also occur if the brickwork has been covered by a surface treatment because the salts may crystallise behind the treated surface and force it off. The effect on the bricks is similar to that caused by frost attack.

The photographs above show damage to brickwork caused by the intensive crystallisation of salt, i.e. cryptoflorescence. In this case it may have been cause by the salt contained in sea water as the house is close to an exposed coastal cliff.

The soluble salts produced by the reaction between acid rain and the calcium carbonate in limestone (see later for fuller explanation) can cause problems in adjacent materials. When the soluble salts are formed, they may run off, say, a limestone coping, onto a brick element below. As these salts crystallise, they can cause decay in materials that would otherwise not be particularly vulnerable. The salts involved may also cause lime staining/lime run-off (see below).

Lime run-off

Where excess water flows through cementitious material, another type of staining can occur. Water can dissolve calcium hydroxide (free lime) which is then deposited on the brick face. The calcium hydroxide is a soluble form of lime which is created as Portland cement hydrates. The source of the lime may be the cement from mortar joints or it may come from concrete or cast stone elements, for example, a coping above a brick wall or a floor slab built into the brickwork.

The run-off is often seen 'dribbling' from weep holes. The calcium hydroxide reacts with carbon dioxide in the air, producing a hard crystalline formation of calcium carbonate. The initial staining can be removed with water and brushing before it carbonates but once the reaction has occurred, an acid solution will be necessary to remove it.

Two examples of lime run-off.

Lime run-off flowing from the floor slab behind this balcony wall (with a close-up of the 'stalactites' formed from the hardened calcium carbonate).

Lime run-off flowing from a weephole.

Lime run-off flowing from under the coping with efflorescence showing on the face of the brickwork.

Other stains

Brickwork can be stained by a variety of materials including:

- **Vanadium salts.** *These salts produce a yellow or green efflorescence on new brickwork. They are generally best left to weather away naturally. The salts occur naturally in certain clays (usually, but not exclusively those that are used to produce buff/lighter-coloured bricks).*
- **Iron staining.** *When it occurs, it usually appears as a stain to the mortar joint. The staining can come from metal embedded in the structure, or it can be derived from the bricks or the mortar sand. Iron staining can be removed mechanically if the mortar is still relatively weak; otherwise chemicals may have to be employed. If it appears on the brick, it should be allowed to weather away.*
- **Manganese staining.** *This manifests itself as a dark brown or black colour. The approach to treatment is similar to that used for iron staining.*

Lime blow in bricks

This is caused when clay bricks contain small amounts of lime. When the bricks are fired, the lime is converted to calcium oxide (quicklime). When the bricks become wet, the calcium oxide begins to slake. The process of slaking is vigorous and in this situation can cause an eruption on the face of the brick. Lime blow in plasters and renders is discussed in Chapter 11.

Sulfate attack

Sulfate attack is a serious problem as it can cause crumbling of the mortar joint and expansion and instability in the wall. It is caused by a reaction between sulfates in solution and a constituent of Ordinary Portland Cement known as tricalcium aluminate. Sulfate attack relies upon a number of conditions occurring simultaneously, that is, it requires water saturation over a relatively long period, a source of sulfates and reasonable amounts of tricalcium aluminate. Even where these factors exist simultaneously, the attack will take a relatively long time to develop. The rate of deterioration is affected by the quantity and the type of sulfate – the sulfates of magnesium and potassium are the most aggressive.

The reaction between the sulfates and the tricalcium aluminate forms a compound known as calcium sulfo-aluminate. This compound expands as it forms, leading to cracking in the mortar joints, followed by general deterioration and the loss of the mortar's 'bonding' function, as the face of the joint spalls and the mortar cracks and crumbles. The cracks can be at the edge of the mortar joint or through the middle. The expansion, which leads to leaning and bulging, exacerbates the instability in the brickwork caused by the deterioration of the mortar. Sometimes the face of the bricks spall, most commonly around their edges.

The source of the sulfates can be the bricks themselves but they can also be introduced from the ground, or from air pollution. An additional source is the exhaust gases from slow-burning fuel appliances.

Sulfate attack usually occurs in situations which are particularly exposed to relatively large amounts of water. Examples of vulnerable situations are illustrated as follows.

The two photographs above show sulfate attack on a roof-level parapet wall.

Sulfate attack in retaining walls   Saturation of the wall can be caused by high levels of ground water, defective drainage, ineffective damp proofing details and evaporation through the wall.

Sulfate attack has just started to crack the horizontal bed joints in this retaining wall.

The three photographs show sulfate attack on garden walls.

The cracks caused by sulfate attack to this garden wall have been made worse by the plant growth.

Sulfate attack to a retaining wall.

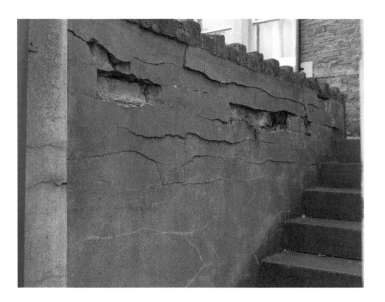

A common consequence of sulfate attack in chimneys is a pronounced lean caused by the different wetting and drying cycles between elevations. In chimneys, additional sulfates may be deposited by the combustion process and additional water may be introduced from condensation within the chimney itself.

Sulfate attack in chimneys

Rain

Additional sulfates are provided by exhaust gases from fires. Additional moisture is provided by exhaust gases condensing inside the cold upper parts of the chimney.

Sulfates in the bricks combine with cement by-products in the mortar to form calcium sulfo-aluminate. As this crystal forms, it expands.

Tall chimneys can develop a noticeable lean because the wetting and drying cycle is uneven due to nature of prevailing winds. The wetter side of the chimney (through rainfall or condensation) suffers the most expansion.

The effects of sulfate attack are often made worse by applying render to 'at risk' elements. Render which is too strong, i.e. has too much cement in it, may shrink and allow rainwater into the wall. Because of its density, a strong render restricts the rate of evaporation. A sulfate attack will also cause cracking in the render, thus more water can get in and this further exacerbates the situation.

The cracking showing through this render is typical of sulfate attack.

The horizontal cracking caused by sulfate attack can be distinguished from that caused by wall tie failure (discussed later) because it may occur in every joint. Also there is often a characteristic white colouring to the mortar as it deteriorates. Because sulfate attack involves water saturation, it is often accompanied by frost attack.

The cracking to the mortar beds shown in the two photographs is indicative of sulfate attack.

Cavity wall tie failure

The main cause of failure of cavity wall ties is the rusting of metal ties, though there can be other causes, such as a failure to properly bed the tie in the mortar joint, poor-quality mortar reducing the bond between tie and mortar, or not installing the requisite number of ties.

The obvious danger with rusting wall ties is the possible collapse of the outer leaf of the cavity wall. Other consequences of rusting wall ties are:
- *The rust will have a significantly greater volume than the original metal. This expansion of the tie may cause cracking and distortion of the structure, particularly where strip ties have been used.*
- *The rust-induced expansion in strip ties can lead to secondary damage, such as a redistribution of loads, buckling and bulging of walls, and damage to the roof as the external leaf increases in height.*

Strap ties taken from a house built in the 1920s.

- *The less bulky wire ties will not generally produce enough expansion to induce cracking unless the joint is abnormally thin or the mortar is very dense. Unfortunately, wire ties produced before 1981 (see following) had less rust protection than strip ties and therefore are likely to have a shorter life; a particular problem because failure can occur without the outwardly visible warning signs produced by the cracking.*
- *The cracking will also reduce the weather resistance of the wall, which in turn accelerates the rusting process.*

Expansion in the joints can lift the roof.

Ties like this (vertical twist ties) are likely to cause the widest cracks in the bed joints and the most vertical movement in the wall.

As ties rust and fail, the external leaf loses its restraint. In addition, the expanding joints can cause bowing and bulging in the external leaf.

Wire ties do not cause as much expansion as twist ties but, because they can rust without causing cracks, there may be no warning signs of structural instability.

Cracking is not always easy to spot – here it's more obvious because of the mastic 'repair'.

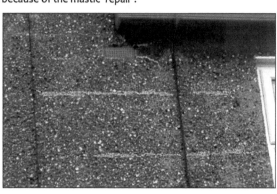

If there is extensive expansion in the mortar joints the loads from the roof can be transferred to the outer leaf. This can, in extreme cases, cause rotation in the foundation.

The term 'corrosion' refers to a partial or complete wearing away or a dissolving or softening of a substance by a chemical or electrochemical interaction between the metal and its environment. Rusting is a type of corrosion and it is caused by the interaction of water with iron or steel in the presence of oxygen. Rust is a hydrated iron oxide, which, though solid in form, is weak and brittle. Galvanising is a means of protecting iron or steel by coating it with zinc. The zinc protects the steel electrochemically. Of the two metals, zinc is the more chemically active and it will slowly corrode in preference to the steel. The thicker the original coating of zinc, the longer it will last. The zinc protection reduces in thickness but it does not undergo a chemical change. Eventually the zinc coating will disappear.

The process of rusting can be speeded up by:

- **Aggressive chemicals.** *Black ash is a product of coal mining which was sometimes added to mortar. The high sulfur content of black ash produces a weak sulfuric acid when wet for long periods. This sulfuric acid can attack a galvanised coating.*
- **Chloride salts**, *which may come from marine sands or may have been added to the mortar as accelerators, can, even in small amounts, speed up corrosion.*
- **Carbonation.** *Where ties are well bedded in mortar, protection against rust is provided by calcium hydroxide, which is formed as the cement hydrates. Unfortunately, as the mortar slowly carbonates, this protective alkaline layer is destroyed. Carbonation will occur more quickly in permeable mortars.*
- **High levels of moisture ingress** *in the building, or particular sections/elevations, because of their exposure.*

The portion of the tie in the outer leaf will usually be the most severely affected because of exposure to rain penetration. However, condensation may also produce enough moisture to allow rust failure to occur.

In 1945, a British Standard was introduced which set out, among other things, a minimum thickness for the galvanising layer. The standard identified the two basic shapes of tie: vertical twist and butterfly. The standard specified that the galvanising should be twice as thick on the vertical twist as it was on the butterfly. Unfortunately this minimum thickness was reduced in 1968, because it was thought that the standard was excessive. In 1981, the standard was increased and specified the same thickness for both types. The reason for this was the increasing evidence regarding performance in use and the realisation of the scale of the problem. The problem of rusting was thought at one time to be localised, as it was believed that it occurred mainly in areas where a catalyst, for example, black ash mortar, increased the likelihood of breakdown. It was later appreciated that the problem was more widespread and the causes more complex. The Building Research Establishment has suggested that all steel or iron wall ties inserted prior to 1981 are at risk of (premature) failure. (Premature relates to the notional 60 years' life of a building.) Because cavity walls tended to be first introduced in areas with high rain penetration, including marine environments with their associated salts, there can be a potent combination of age and aggressive environments.

It is worth noting that stainless steel wall ties are available. Stainless steel is a ferrous alloy which contains at least 10 per cent chromium. Since chromium is more reactive than iron, a 'self-healing' impermeable layer of

Unlike sulfate attack the cracks associated with wall tie failure correspond to the horizontal mortar courses containing the ties.

chromium oxide forms naturally on the surface of the steel. Chromium oxide is durable in a wide range of exposures and prevents the formation of rust. Where there is exposure to chlorides, for example in marine conditions, care must be taken in selecting the appropriate grade.

Wall tie failure can often be identified by the horizontal cracking that results from the expansion caused by the rusting process. Unlike sulfate attack the cracks will coincide with the position of the ties (and will obviously only occur in cavity walls). However, because butterfly (i.e. wire) ties will usually not cause cracking, it may be necessary to carry out investigations where the problem is suspected or anticipated. Of course, ideally, strip ties should be identified before the cracking occurs. This might involve a visual inspection of the suspect tie, either by using an optical probe or by removing bricks. It should be borne in mind, however, that a tie can be in good condition in the cavity but poor in the outer leaf and therefore exposure of the end embedded in the outer leaf is a sensible step.

Ties can be assessed visually for rusting. The presence of red rust, the product of the oxidation of iron, indicates severe rusting, while the presence of white deposits or a blackening of the galvanising indicates that the zinc is corroding. Measurements of zinc thickness will enable the projection of the remaining life. (The Building Research Establishment gives some useful guidance on this in Digest 401.) Metal detectors can determine the exact pattern of tie placements and can help in establishing if a defect is caused by wall tie failure.

Disguising wall tie replacement is difficult, particularly where houses are rendered. Painting will help but it is very difficult to match the texture of the render. In plain brickwork, extensive repointing will be necessary.

Weathering and disintegration of joints

Brickwork can be jointed as work proceeds or pointed after completion. The mortar can be finished with one of a number of different profiles.

(Photo courtesy of the Brick Development Association.)

The profile of a joint has an aesthetic effect – because of the way it casts shadows and reflects light – and it affects the durability of a wall. Some profiles are more effective than others in preventing water ingress. The nature of the joint can, in fact, be more significant than the porosity of the brick in terms of weather protection. A tooled joint using, say, a bucket handle is effective. The tooling compresses the mortar slightly, thus reducing its permeability. At the same time tooling ensures a good seal between mortar and brick. Struck joints are equally effective though they require slightly more skill.

Creating a bucket handle joint.

Recessed joints are problematical because they do not easily shed rainwater but allow it to sit on the edge of the brick and, though tooling may compress and seal the joint, the brick is at risk. Flush joints often leave small cracks between the brick and the mortar. Also, they are often 'buttered over' the joint, increasing the likelihood of water penetration.

Deterioration of mortar will reduce the wall's resistance to rain penetration and may, in an extreme situation, lead to structural instability. A deeply recessed bed joint could reduce both lateral and compressive strength. The deterioration may be due to inadequacies in the mortar mix or damage by frost attack or sulfate attack.

Repointing in the incorrect mortar can be an aesthetic problem in that it alters the way a wall will look. It can also be a technical problem because, for instance, a dense (i.e. too much cement) mortar will not allow the brickwork to 'breathe' adequately and/or it may shrink and crack and so allow water ingress. In the right-hand photograph the bricks of both houses are the same but the mortar colour and profile changes the aesthetic appearance dramatically.

Stonework

The existence of many stone buildings, both as single examples, or coherent settlements such as in the city of Bath, is a testament to the durability of stone, as well as a reminder of the aesthetic and symbolic appeal that made it such a high-status material.

There is evidence of stone buildings in this country dating back to the first century B.C. The Romans used stone and brick but, as mentioned, there was a decline in the general use of both materials after they left. Nevertheless, the Saxons used stone for significant buildings, as evidenced by the simple churches of the period. The Normans brought their own craftsmen and trained the Saxons to build monumental buildings such as cathedrals and castles, establishing stonemasonry as a recognised craft. However, stone was not used for poorer dwellings until much later, mainly because wood was readily available. Gradually, stone became available to a wider spectrum of society (or perhaps more accurately there was a broadening of strata in society and more people were able to afford stone), until by the 1500s it was in relatively common use.

Today stone remains a material of status – it is often still used for significant buildings, particularly those that are supposed to signify power, gravitas, or trustworthiness.

The nature of stone

Stone is a natural material which is derived from one of three basic rock formations: igneous, metamorphic or sedimentary:

Igneous. *These rocks were originally formed from the cooling and solidification of molten magma – the name is derived from ignis, the Latin for fire. Building stones derived from igneous rocks are generally durable. Granite is an example of an igneous building stone.*

Metamorphic. *As the name suggests, these rocks were produced by changes brought about in existing rocks. The changes mainly occurred through the influence of pressure or heat or both. Examples of building stones derived from metamorphic rocks include marble and slate.*

Sedimentary. *These were created by the laying down of sediment produced from the wearing away of existing rocks, or from the shells and skeletons of living organisms, or by chemical deposition. These particles were deposited in layers on the floors of lakes, rivers, seas and deserts. The sediment particles were cemented together by other minerals and compacted by self-weight and sometimes by earth movements.*

Limestone and sandstone are examples of sedimentary rocks. As they are the most widely used stones for building, this chapter will focus on these two materials. Although limestone is still being formed today, the stones that are most commonly used in buildings come from the Jurassic age and were created between 136 and 195 million years ago.

Limestone as a building material has been used extensively in the southern part of the UK for many centuries. It has helped to define the character of significant monuments such as the cathedrals of Wells, Ely, Exeter and Salisbury, the Palace of Westminster and the Tower of London, as well as cities such as Bath and geographical areas such as the Cotswolds. Limestone is found in the Jurassic belt, which covers an area from the coast of Yorkshire, through Lincolnshire and Northamptonshire and then through Oxfordshire, the Cotswolds and on down to Dorset. There are also older deposits of limestone in Wales and Cumbria.

Sandstones can be found in quite a wide geographical area from Sussex to the north of Scotland. The most important sandstones are those produced in the Carboniferous period

– between 280 and 345 million years ago. These sandstones can be found in large areas of the north of England and the Midlands, as well as Gloucestershire, Devon and Cornwall. Sandstones were used in, for example, the creation of Edinburgh New Town, important houses such as Chatsworth, and cathedrals such as Durham, Hereford, Liverpool and Coventry (new and old).

Sedimentary rocks were formed in parallel, mainly horizontal, layers or beds (diagonal bedding sometimes occurred due to changes in wind or water direction). Bedding planes are the defining points of an episode of sedimentation – they mark where sedimentation stopped before recommencing at a later time. Joints in rocks are formed by movements, tensions or shrinkage in the rocks. They will divide the horizontal bedding planes vertically, as well as providing further horizontal delineation. Sometimes the shrinkage forces will have been strong enough to pass through one bedding plane into the next. The blocks defined by the joints and bedding planes make quarrying easier but they may cause problems in later repair or extension work because they determine the size of stones that can be extracted.

Limestone

Limestones are made up mainly of calcium carbonate which accumulated at the bottom of seas and lakes. However, very few limestones consist entirely of calcium carbonate. Many also contain clay, sand, or organic material.

Limestones can be categorised according to their mode of formation:

* **Chemical.** *Produced by water passing through rock formations and picking up calcium carbonate. Water contains carbon dioxide and this reacted with the calcium carbonate to form calcium bicarbonate, which is soluble. The water containing the calcium bicarbonate then flowed into a larger body of water, such as the sea, where the precipitation of calcium carbonate took place.*
* **Organic.** *The main constituents of these stones are the fossilised shells of once living organisms.*
* **Detrital (clastic).** *These are derived from the decomposition and erosion of existing limestones.*

As mentioned above, limestones are all made of similar material – with the exception of magnesium (dolomitic) limestones which consist of calcium and magnesium carbonate – but their structure can vary considerably in ways that will affect both how easy they are to 'work' and their durability. Some are very coarse while others are very fine; some contain very significant and obvious amounts of original material which has not been ground down, such as shells. Unfortunately, some of the more durable limestones are not very easy to work or they come from narrow beds.

Limestone has been quarried at Doulting (Somerset) for nearly 2,000 years. In the thirteenth century it provided the stone for Wells Cathedral. Sometimes fissures do not become obvious until the stone is cut.

Sandstone

Sandstones were formed from components derived from the breakdown of other rocks. The deposits formed were transported by water, and to a lesser extent wind and ice, and deposited at the bottom of larger bodies of water such as lakes and seas. Sandstones are made up of a variety of materials but the essential material is quartz. Sandstones were produced when grains of quartz were cemented by various compounds. They are, in fact, categorised by the cementing medium or matrix. Because quartz is largely inert, the cement largely determines the weathering resistance of sandstones. The types of cement found include:

- **Siliceous.** *These are silicas, they are inert and therefore produce durable stones. Effectively, these are quartz grains cemented by a quartz matrix.*
- **Ferruginous.** *These are iron oxides and have poor durability.*
- **Calcareous.** *These consist of calcium carbonate and are reasonably durable except in urban environments where they are exposed to air pollution.*
- **Dolomitic.** *These are calcium and magnesium carbonates and as with calcareous matrices, they are vulnerable in polluted environments.*
- **Argillaceous.** *These are clay and have poor durability. They are particularly susceptible to frost action.*

Construction of stone walls

Stone walls are defined by the manner employed in dressing the stone. The two general categories are: ashlar and rubble. Ashlar is the description given to squared stone which is presented by being dressed to a smooth face. Ashlar consists of mathematically precise blocks separated by very thin joints to produce a 'classical' appearance.

Ashlar limestone blocks.

Rubble was a popular approach in the Victorian era. It was used because it was cheaper than ashlar but, in many instances, it was deliberately chosen because of its appearance. In part, the preference for the natural rustic look of rubble was a generalised reaction against the plainness of classically derived Georgian architecture. There are a number of variations, often defined by the shape of the blocks (e.g. whether they were squared or not) and how well defined the horizontal coursing was. Three examples are shown as follows.

Often, rubble and ashlar were used together, with rubble being framed by ashlar work at the corners of elevations, or ashlar bay windows being inserted into rubble walls. Brick was also commonly used as the dressing around the windows and doors of rubble-faced houses.

Defects in stonework

Stone is a natural but complex material. Under the general headings of sandstone and limestone, there are many stone types, with different durability and weathering characteristics. The discussion of defects is therefore inevitably generalised.

Pore structure

There is still much debate about the key factors affecting durability in stone and empirical evidence is somewhat contradictory. However, it would appear that while natural durability can be affected by a number of factors, probably the most significant is pore structure. Pore structure is significant mainly because it affects the amount of water entering and moving through the stone. Also, salts that may damage the stone can be transported by water and accumulate in the pores. The critical factor is not the total amount of space created by the pores but how they are structured. Stones with low porosity will not allow much water penetration and are therefore less likely to suffer from salt and/or frost attack. Stones with a high porosity will allow more water in but, if the pores are large, the water will tend to be able to evaporate reasonably quickly. However, if there is a large network of fine pores, capillary action will be high – but evaporation will be relatively low. In addition, a stone with large pores is less likely to suffer salt damage than one with small pores. This is because the larger space is more likely to be able to accommodate the expansion pressures of salt crystallisation.

Incorrect bedding

As mentioned previously, sedimentary rocks were laid down in beds. Defects can arise in use if a stone is incorrectly placed in a building in relation to its bedding plane.

When it is placed in a wall, the stone should generally lie in its natural bedding position. That is, the layers should run horizontally, in the manner in which the stone was originally formed. The stone is stronger in this position and is also less vulnerable to defects. If the stone is face-bedded (the layers are vertical), it is more vulnerable to damage through crystallisation of salts and/or frost action. This is because the mechanical actions involved find it relatively easy to push off the bedding layers as there is no restraint from the adjoining stones.

Bedding sedimentary stone

Sedimentary stone should normally be laid in its natural bed, i.e., with the layers horizontally, the way it was formed over thousands of years. However, copings and sills are usually edge bedded because they are exposed on their top side.

The damaged stone above has been incorrectly laid. It has been face bedded and has started to delaminate. Failure occurs over time through the wetting and drying cycle, possibly exacerbated by frost attack.

The laying of stone in a face-bedded manner is not uncommon. It may have happened through ignorance of the consequences, or through difficulty in determining the layering pattern – it is easy to distinguish with some stones but is less so with others. For some features of a building, such as the cornice or coping, laying the stone in its natural bed will be incorrect because it will be exposed on its top side. In these situations, the stone should be edge-bedded. With a string course, the corners of a coping or carved stone, no solution will be entirely satisfactory. In these situations, attention to protective detailing will be particularly important.

Stone delaminating because of face bedding.

A coping stone delaminating.

Salt crystallisation

This involves salts in solution passing into the stonework and entering the pores of the limestone or sandstone. There are a number of salts which can cause problems and they may come from a variety of sources. Sources may include building materials such as concrete, bricks or mortar, as well as soil and the air. As the stone dries out, salt is deposited at the surface (efflorescence) or within the stone.

Salt crystallisation damage to limestone.

Crystallisation within the pores will exert pressure, often resulting in damage. The possibility, and extent, of the damage will be related to the type of stone, the type of salt, and the characteristics of the pores – mainly their size and their arrangement.

The potential for damage is increased because of the repetitive nature of the mechanical action caused by the cycle of (re)dissolving and (re)crystallising. The damage usually shows as a powdering of the surface. In some cases, it can look more dramatic by causing splits in the stone as the mechanical expansion works against an existing weakness, such as incorrect bedding or a fissure.

Because some of the salts will absorb water from the air (i.e. they are hygroscopic – see Chapter 16), they will cause damage even if there is no direct wetting from the more obvious sources such as rain, rising damp or plumbing and drainage faults. The relative humidity at which they will absorb water will vary with the salt, but, for some, it will be below the normal level for an occupied building. In a coastal area there will be a potent combination of salts from the sea and a high relative humidity.

Road salt has crypto-fluoresced within the pores of the stone shattering it.

Acid rain

A significant cause of damage is the interaction between limestone and the products of pollution in the form of acid rain. Acid rain is produced by industrial and domestic activities and, increasingly, by road traffic. 'Acid rain' is a generic term which includes other weather conditions, e.g. fog, mist, etc.

There is a popular impression that damage to limestone through acid rain is rather like the process involved in dissolving an antacid tablet in water. In reality, though acid rain can directly dissolve stonework, this action is relatively insignificant compared to the reaction between the acidic rainwater and the calcium carbonate in limestone.

The major problem is the production of soluble gypsum. The main way in which this occurs is when sulfur gases go into solution with rainwater and form sulfurous acid (H_2SO_3). Sulfurous acid then combines with oxygen from the air. The resultant sulfuric acid reacts with the calcium carbonate ($CaCO_3$), which is the main constituent of limestone, and forms calcium sulfate ($CaSO_4$). The calcium sulfate then takes up water and crystallises as gypsum ($CaSO_4 2H_2O$). The resulting pattern of decay of the limestone is determined, essentially, by its exposure:

- *On exposed areas of the building, rainwater will wash away the gypsum. Gypsum is slightly soluble in water but more soluble in acidic water or rain that contains common salt – as might be the case in a marine environment. What occurs is a gradual weathering away of the stone, though the visual effect can, from a distance, look like cleaning or renewal.*
- *In sheltered areas of the building, a relatively impermeable skin will start to form. The skin can be up to 12mm thick and in polluted urban environments it will contain dirt from the atmosphere. The skin may stay in place for a long period (this might be affected by the characteristics of*

the particular stone), or it may form blisters that burst to reveal a powdery decayed interior. Crystallisation and recrystallisation may also be going on underneath this skin, creating large areas of damage. In sheltered areas, damage to magnesium limestones can be particularly severe because both calcium and magnesium sulfate are produced. The calcium sulfate again creates a skin but below it the more soluble magnesium sulfate will often be found forming crystals. This crystallisation often results in relatively deep caverns of decay in the stone, which are revealed when the skin breaks down. This is sometimes referred to as cavitation.

Although the durable sandstones that are cemented by a quartz matrix are inert, and therefore resistant to sulfur-based gases, those sandstones which are cemented together by a calcium carbonate matrix (calcareous sandstones) can be severely damaged by the chemical reaction with acid rain. The reaction between the calcium carbonate and the acidic gases is the same as described above, that is, sulfurous gases in solution react with calcium carbonate to produce relatively soluble calcium sulfate. However, as in this case the cementing matrix is being attacked, the result is the release of large amounts of otherwise stable quartz crystals. The damage is often worse than in limestones placed in a similar environment.

Dirt on stone buildings may cause problems because it can contain soluble salts. The dirt comes from particulate air pollutants which are cemented onto a limestone wall by the gypsum produced from chemical attack. The particulates may be derived from various animal, mineral or vegetable sources but the most common are tar, diesel residue, oil, and carbon deposits. As stated previously, the gypsum, and therefore the dirt, will be removed by rain on exposed surfaces of limestone, but it will remain on sheltered surfaces. However, with sandstone the particles become 'bonded' to the stone's surface and will not be washed away by rain.

The dirt on this limestone has been washed away in places by rainfall.

Frost attack

Frost will tend to be a problem in those areas which get both excessively wet and are subject to freezing. The vulnerable areas have been identified at the start of this chapter, but it is worth reiterating that frost damage is not very common in sheltered plain walls, except, perhaps at below damp-proof course level. As with brickwork, pore structure is a significant factor in determining susceptibility to frost attack and the process of attack is the same.

Where it does occur, frost damage can dislodge quite big pieces of stone either in vulnerable areas, such as parapets or copings, or where the stone has been damaged by other agents.

The two photographs above show the effect of frost attack on a stone coping.

Contour scaling

This is another problem associated with salts. Contour scaling occurs in sandstones and is thought to be the result of the pores of the stone being blocked with calcium sulfate. This seems to happen even when it is not a calcareous sandstone. The effect is a breaking away of a rather thick crust from the face of the stone. The effect produced by contour scaling resembles that caused by incorrect bedding, but it occurs irrespective of the way the bed lies. It is thought that the damage is caused by the different reaction of that part of the surface material with the blocked pores compared to the underlying body of stone when subjected to thermal and moisture movement stresses.

Contour scaling to sandstone.

Soft beds

In some stones a relatively softer and less durable layer of material may be present due to the different materials that were laid down millions of years ago. This can lead to a defined line of weakness as the soft bed will deteriorate at a faster rate than the rest of the stone.

Expansion of metals

Iron and steel cramps have traditionally been used as fixing devices in stonework for many centuries. When these metal fixings rust, they expand and can fracture the stone. Additionally, stone cavity walls may suffer from the cavity wall tie problems discussed earlier in the chapter.

Damage to ashlar stone caused by the expansion of rusting metal cramps. The photo shows where mortar has been removed to expose the cramps. (Photos courtesy of Stephen Bond.)

The rusting metal fixing cramps which caused the cracking shown to the previous building. (Photo courtesy of Stephen Bond.)

Expansion of metal fixing cramps from rust has caused the damage to this stonework. (Photo courtesy of Stephen Bond.)

Damage to stonework caused by the rusting expansion of these drainage pipes.

Damage to stonework caused by the expansion of metal railings.

Extraction and dressing

Stone can be damaged in the quarry if it is extracted by the use of explosives which may cause internal fractures. Damage can also be caused by excessive tooling of the surface of the stone.

Organic growths

Plants and lesser organic growths on stones are common. Sometimes they are seen as adding to the charm of stone buildings, sometimes they are regarded as unsightly. The different perspectives depend to some extent on context – random growths may be thought of as detracting from a classically designed building, for example, or, in other situations, valued as part of the patina of age.

- **Lichens.** *Lichens are a symbiotic association of fungi and algae. They are hardy in the sense that they are able to derive sustenance from a number of sources: the fungi element seeks out water and salts and the algae element manufactures food by photosynthesis. Lichens can produce acids which can etch into the surface of soft stones such as limestones or calcareous sandstones. Finely carved stonework may be particularly vulnerable. The fine rootlets of some lichens may cause minor mechanical damage. The potential for damage is reduced by the fact that lichens are extremely sensitive to air pollution.*

- **Algae.** *Algae are a common form of vegetation on buildings. They prefer conditions where damp, warmth and light are available as they manufacture food by photosynthesis. They are usually green in colour, but can be red, brown and blue. Like lichens, their acidic secretions can damage stone. They also collect both water and dirt and can affect the transpiration rate of walls.*
- **Mosses.** *Mosses can act as 'sponges' and retain moisture; perhaps enough in a concentrated area to cause frost attack. Like algae, their presence will tend to indicate excessive dampness. They can secrete acids which may cause etching in some stones. They have also been known to develop enough to cause root damage to very soft stones.*
- **Fungi.** *Fungi can grow on stone if they can find a food source (they cannot manufacture their own, unlike plants). Fungi can produce acids and therefore possibly contribute to the decay of limestones.*

Higher plants

The biological growths mentioned above could create the conditions for other plants, which may be more damaging to buildings, to grow. Biological growths do this by retaining moisture and by providing humus when they die.

Clearly any plant or tree could potentially seed and grow in stonework. Some may cause damage, some may not. For example:

- **Ivy.** *The roots of ivy can cause mechanical damage. Usually the damage is to mortar joints that they have lodged in, but the fine rootlets can split stone if they penetrate through crevices.*
- **Virginia creeper.** *This is usually harmless as the plant merely attaches itself to the wall by suckers rather than through a root system.*
- **Bacteria** *are thought to be able to affect decay but from a practical viewpoint the damage caused is generally insignificant. Bacteria cause damage by either starting or adding to the production of chemicals that then attack stone. An example is by oxidising sulfurous acids in polluted atmospheres, thus forming sulfuric acid.*

 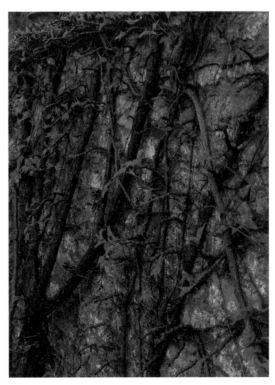

Virginia creeper.

Birds and bees

Birds can damage stone (or brick) in two ways: they can cause mechanical damage by pecking for salt or grit. They can also cause decay in limestone and calcareous sandstone because their droppings release acids. Starlings and pigeons are the main culprits.

Masonry bees can cause a small amount of damage as they bore into soft stone and brick to make a nest.

Damage caused by masonry bees in stonework and brickwork. (Photo courtesy of S. Gilbert.)

Poor repair/conservation techniques

Although repairs are not discussed in this book, there are instances where poor repairs can cause further defects:

Incorrect pointing. As with brickwork, an inappropriate mortar mix can cause problems in both the mortar and the stone itself. The main problem is usually related to having too strong a mortar. Because a strong (dense) mortar allows less evaporation, any moisture in the wall will have to evaporate through the stone. This may increase the likelihood and severity of salt-related defects. Additionally, as mentioned before, stronger mortars are relatively brittle and may be more susceptible to shrinkage cracking. This may increase the likelihood of rain penetration, which in turn may increase the likelihood of salt-related defects and frost attack.

Making the jointing stand proud of the face of rubble stonework is another common mistake. This is sometimes known as ribbon pointing. Ribbon pointing is usually formed with a very strong mortar. Aesthetically, it is wrong because it is so dominant – the impression is often of a mortar wall with stone inserts. Technically, the ribbon pointing performs poorly because it shrinks and allows water ingress, it acts as a ledge for retaining water, and it is brittle and so easily cracks and falls away.

Elevation

Shrinkage of mortar allows water penetration

Rainwater cannot be shed efficiently and dense pointing impedes evaporation

Section through traditional rubble wall with ribbon pointing

In freezing conditions ice forming in joints can force pointing away from wall

Ribbon pointing to rubble stonework. Note how the mortar is cracking and falling away in the bottom two images. This is a common sight in such dense mortars which are susceptible to shrinkage cracking.

Failure of protective coatings

Coatings are often applied to stone in an attempt to protect it. They are usually applied in order to do the following:

- *prevent water penetration;*
- *prevent stone decay.*

Additionally, given the number of bay windows where the stonework is painted, some coatings are presumably added because they are judged to improve the look of the stone. While there may theoretically be good reasons for applying a silicone-based water treatment to stone (i.e. where penetrating damp is occurring), they should not be used to try to arrest the effects of

stone decay. Where they have been used, there is evidence that the rate of decay is actually accelerated. This is probably because the water takes longer to evaporate when a coating is applied. Also, because salt crystals are often trapped behind the treatment, they can cause greater damage to the stone. Eventually, the forces exerted by the salts will probably break down the coating.

Defects in this limestone bay have been exacerbated by the application of paint.

Extensive damage can occur where bitumen-based solutions are applied. These are a particular problem as they are very effective in preventing the stone from 'breathing'.

Damage to this stonework has been accelerated by the application of a bitumen-based coating.

Conclusion

The common element in brick and stone defects is water. When excessive water combines with freezing temperatures or soluble salts, then defects may result. In practice, these situations often occur where insufficient consideration is given to the relationship between design, detailing and materials selection.

Some of the defects discussed may occur, irrespective of whether the wall is stone or brick. Others may only occur in one material or the other. There is potential for confusion because the naming of defects uses similar phrases to describe different problems. For instance, the term 'sulfate attack', used to describe the reaction between sulfates and Ordinary Portland Cement, should not be confused with the problem caused (in limestone) by the conversion of calcium carbonate to calcium sulfate.

Ground floors 7

Types of floor

Most Victorian and Edwardian houses of the middle and upper classes had timber ground floors. These same houses often featured small areas (usually circulation areas) of decorative tiling laid on early forms of concrete. However, in housing built for the working classes, many dwellings had ground floors made in whole, or in part, from stone flags, or large heavy clay tiles, laid directly on compacted earth or a thin layer of broken stone.

During the first half of the twentieth century most ground floors were built in timber and are usually referred to as raised timber, or suspended, ground floors. Nowadays, most ground floors are formed from insitu (i.e. cast in place) concrete or pre-cast concrete.

Clay tiles such as these were popular from the 1850s until the 1930s. They were bedded in mortar and laid on a screed or direct on a concrete slab. Their use was often reserved for those areas visitors would see (and admire), i.e. halls and lobbies. This example is from the Metal Agency Company Catalogue 1932.

Floor Tiles

No. 500. No. 501. No. 502.

No. 503. No. 504. No. 505.

Scale, ¾ inch to a foot.

Prices per square yard.

No. 500	..	14/-	No. 501	..	20/3	No. 502	..	14/-
No. 503	..	17/3	No. 504	..	26/6	No. 505	..	16/6
No. 506	..	19/9	No. 507	..	19/3	No. 508	..	19/3

When ordering Tiles from this page a correct plan of the floor should be given with figured dimensions and Design Number stated.

For **Red and Buff Floor Tiles**, see page 62.

Delivery, ex Works, Carriage Paid on 3 sq. yards.

Raised timber floors: construction summary

Timber floors, if correctly built, will last virtually indefinitely as long as insect attack and rot can be kept at bay. Unfortunately, many early timber floor structures were poorly designed and poorly constructed. If they become wet, and remain so, decay can be swift.

At its simplest, a timber ground floor comprises a series of beams or joists covered with timber boarding or, nowadays, a manufactured particle board such as chipboard.

The drawing above dates from the very end of the nineteenth century. It is typical of many artisan-class terraced houses up and down the country. In these houses the floor joists can be found running front to back or side to side (party wall to party wall). It was common practice to lay a concrete floor or flags in the scullery because this area, by its very nature, was often fairly wet.

In a typical late Victorian floor, the floor joists are supported by the external walls, the internal load-bearing walls between rooms, and intermediate sleeper walls. Timber floors require ventilation below the floor to keep the underfloor space dry. This was achieved by providing vents in the external walls. These early floors vary in quality, particularly in their resistance to the problems of dampness. Late Victorian houses, for example, were usually built with damp-proof courses (DPCs) in the external walls, internal walls and intermediate sleeper walls. However, many of these DPCs are no longer effective and the timbers are susceptible to rising damp. In addition, penetrating damp through the solid external walls, together with inadequate sub-floor ventilation, increases the risk of floor failure. The example shown has bare earth below the floor.

Several towns and cities adopted by-laws during the 1880s or 1890s which required a concrete slab or layer of asphalt below a ground floor. On this drawing (1899) the concrete slab ('oversite') has been added: whether it was added by the builder or by the Sanitary Authority we shall never know. Either way, many houses were built with nothing but bare earth below the timber floor joists.

Joists often built in to external wall.

DPCs not common until 1890s.

AIR

Vent

A void separates the joists from the earth.

DPC

The foundations (if they existed) could be of various types; see Chapter 3.

Intermediate walls (sleeper walls) provided support for the joists. This avoided the need for expensive, deep section, timbers.

Air flow limited by internal wall, although the odd missing brick sometimes helped to improve air flow.

The two photographs above show the under-floor void of a suspended timber ground floor in a Victorian house. The void is about 1 metre deep. Note that in the right-hand photograph the top of the stepped brick foundation is visible.

In the 1920s and 1930s, timber floors were constructed to higher standards. There were a number of improvements which included provision of better-quality DPCs in all walls, keeping the joists clear of the external walls, and the provision of honeycombed sleeper walls to aid ventilation. In honeycombed sleeper walls a number of bricks are omitted to encourage the through-flow of air. In addition, provision of the concrete 'oversite', sometimes found in earlier floors, became standard practice. The oversite provides a base for the sleeper walls and reduced the incidence of rising damp; it was normally laid just above external ground level. These early twentieth-century floors usually contained additional vents (compared to their Victorian counterparts) to ensure adequate ventilation of the underfloor void.

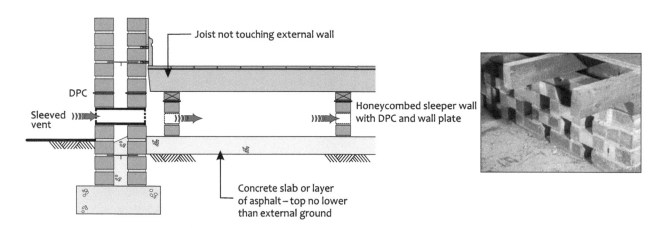

This photograph shows the under-floor void from a house built in the early 1930s. The sleeper wall, built on a concrete oversite (covered with later builders' debris) does not have a wall plate on top but there is a DPC. A few missing bricks at the base of the wall help ventilation.

Modern floors can still be formed in timber and a typical modern example is shown below. Nowadays, sleeper walls are still found though floor joists are usually supported by hangers built into the external load-bearing walls. In many small houses the use of deeper timber joists precludes the need for intermediate sleeper walls. Periscopic vents keep the outside vents well clear of ground level.

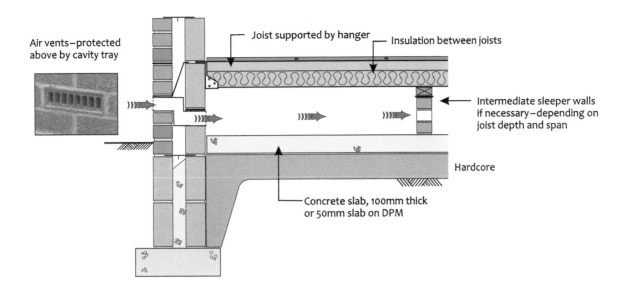

Air vents–protected above by cavity tray

Joist supported by hanger

Insulation between joists

Intermediate sleeper walls if necessary–depending on joist depth and span

Hardcore

Concrete slab, 100mm thick or 50mm slab on DPM

Concrete ground-bearing floors: construction summary

Although there are examples of concrete ground-bearing floors from the 1930s and even before, it was not until after the Second World War that they became the most common method of forming ground floors. This was, in the main, due to a shortage of timber caused by government restrictions on imports (eventually relaxed in the mid-1950s). Nowadays, although timber floors can be found, concrete ground-bearing slabs, and more recently, suspended concrete slabs, are much more common.

At its simplest, the floor comprises a slab, or bed, of in-situ concrete supported by a hardcore sub-base. The floor is independent of the external walls and load-bearing internal walls to allow for slight differential movement.

If the hardcore is over 600mm, a ground-bearing slab should be avoided. Long-term settlement in the hardcore is likely to cause cracking or movement in the slab. Suspended concrete floors, usually pre-cast, can be used in this situation.

Hardcore

The hardcore below the slab fulfils a number of functions. It provides a cheap bulk filler to make up levels and prevents the concrete from being contaminated by the sub-soil below. It helps spread the load evenly across the ground and should reduce the incidence of rising damp (a good-quality hardcore has an open, porous structure which limits capillary action). A good hardcore should be chemically inert, easy to compact and should not be affected by water. It should be compacted in layers to prevent subsequent settlement. The most effective hardcores are usually by-products from aggregate quarrying. Incorrect choice of hardcore, or poor site practice, are common causes of floor failure.

Hardcore in the process of being compacted.

Concrete slab

The concrete slab is usually about 100–150mm thick. Nowadays the required strength of the concrete is set out in the Building Regulations. Prior to this, it depended on local by-laws and generally accepted good practice. In some cases the slab is reinforced with a steel mesh, for example, where soft spots exist in the ground or in garage floors. The mesh prevents the slab from cracking due to uneven loading or minor settlement.

Steel reinforcement mesh laid over a damp-proof membrane and awaiting the pouring of the concrete.

The surface of the slab can be power-floated to produce a very smooth surface, usually smooth enough for vinyl tiles of sheet materials. However, power-floating has only recently become common and most ground-bearing slabs have a separate surface finish added towards completion of the building (see later section).

The ground floor shown right was laid in 2007. The slab is 100mm thick and laid on a 200mm layer of limestone hardcore (75mm 'down' – all the hardcore must pass through a 75mm sieve – the 'fines' aid compaction). The garage slab (reinforced) is in the front of the picture.

Damp-proof membranes

Although some early ground-bearing floors incorporated waterproofing agents in the slab or screed (the most common finish), most ground-bearing slabs contain a separate damp-proof membrane. They can be formed from a variety of materials, some of which are no longer readily available. They include the following elements (Table 7.1).

Hot-applied liquid membranes	**Cold-applied liquid membranes**	**Sheet materials**
• *Bitumen (distilled from oil)*	• *Bitumen solutions*	• *Polythene sheet*
• *Asphalt (bitumen and aggregate)*	• *Bitumen/rubber emulsions*	*(post-1960s)*
• *Pitch (distilled from coal tar)*		• *Bitumen sheet*

To be effective, these materials need to be continuous and need to form a good joint with the DPC in the external walls. The sheet materials can be found above or below the slab; liquid materials can only be used above the slab. DPMs are designed to prevent capillary action; they are not effective against direct water pressure. The diagram below shows a number of common DPM options.

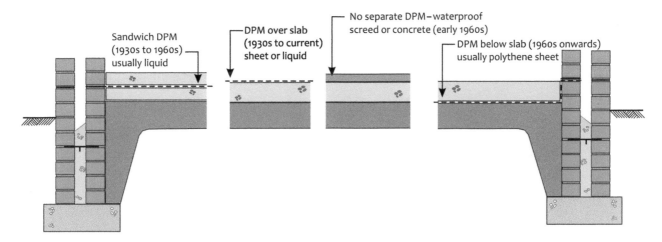

Membranes below the slab are relatively cheap and will protect the concrete from aggressive chemicals emanating from below. They should be laid on a sand blinding layer to protect them from any sharp stones in the hardcore. When the membrane is below the slab, thinner floor screeds can be used because a screed laid directly onto a slab will partially bond to it. However, in this situation, the floor slab will take a long time to dry, delaying the fixing of moisture-sensitive floor coverings. Some sheet membranes are not completely vapour-proof, either because of inadequate joints or because the material is not completely impervious, and are therefore not suitable under moisture-sensitive floor coverings, e.g. chipboard sheeting and cork tiles.

Membranes above the floor do not protect the slab from potentially harmful chemicals in the hardcore or ground but can offer the very best in terms of damp-proofing, particularly if hot-applied liquid DPMs are used. If sheet materials such as polythene are used, the screed cannot bond with the slab and will therefore have to be thicker to prevent it from cracking. With both materials, achieving continuity with the DPC can be difficult.

Early floor finishes

In the 1940s and 1950s, timber floor finishes were popular. These often took the form of floorboards fixed to dove-tailed battens secured by a screed, or rectangular battens fixed by clips or nails. This gave the floor the appearance of a traditional raised timber floor and was suitable for a period when few people had fitted carpets. The membrane below the screed was most likely to be formed from hot liquid materials such as pitch or bitumen though cold proprietary materials, applied in two or three coats, were also used.

Boards nailed to battens

Battens nailed or clipped to screed

Splayed battens secured in screed

Bitumen or pitch DPM

Block floors were also popular. These could be laid in a basket weave, stretcher bond or herringbone pattern. The design of the blocks could incorporate a dovetail to provide a good bond with the DPM. Other blocks were tongued and grooved. The blocks were usually laid in hot mastic (bitumen with added chemicals to improve flexibility) which also acted as the DPM.

Brick pattern edge

Basket weave

Wooden blocks set in bitumen or pitch

Tongued and grooved

Dovetail groove

Herringbone

Mastic bedding (hot bitumen) on screed

More recent finishes

From the late 1950s to the early 1990s, screeds were the most common floor finish. Screeds are formed from a mix of coarse sand and cement (usually in the ratio of about 4:1). Screeds could be carpeted direct though, at first, fitted carpets were rare and the screed was normally covered with asbestos tiles, or later, vinyl tiles. During the early 1960s, polythene membranes became more readily available. Although, as indicated earlier, these do not offer the best protection from capillary action, they are the most common method, nowadays, of forming the membrane.

Unlike liquid materials, polythene membranes can be used above or below the slab. However, liquid materials are more likely to be found where moisture-sensitive floor coverings were specified. The thickness of the screed depends on the position of the membrane. Where the membrane is below the slab, thinner screeds are acceptable as observed earlier, because there is some bond with the slab.

Modern floor slabs are insulated; the insulation, depending on its type, can be laid above or below the slab. A number of DPM, screed and insulation combinations are shown below.

DPM and Insulation below slab
38–40mm screed

38–40mm screed

DPM below slab, insulation above
65–75mm screed

DPM above slab, insulation below
50–60mm screed

DPM and insulation above slab
65–75mm screed

A modern alternative to the screed is particle board: chipboard and strandboard are the two most common products. They are laid on a resilient layer of insulation boards laid directly onto a clean concrete slab. They offer fast construction and good insulation. Where a membrane is positioned below a slab, an additional membrane is required above it to prevent trapped construction water from affecting the particle board. Problems of durability are discussed in a later section.

Tongued and grooved (18mm) chipboard laid on insulation boards

Ideally a polythene membrane should be sandwiched between the insulation and the chipboard. This prevents construction water trapped in the slab from affecting the chipboard. It also acts as a vapour control in the reverse direction, i.e. reducing the risk of condensation on the cold slab. If a DPM is laid over the slab several manufacturers suggest that the additional membraine is not required – although the risk of condensation will be increased.

In kitchens and bathrooms, moisture-resistant chipboard should be used.

Suspended pre-cast concrete floors

In many cases the use of a concrete ground-bearing slab is not suitable. These include:
* *where there are aggressive chemicals in the ground;*
* *where water tables are high;*
* *on sloping ground where uneven depth of hardcore may lead to differential settlement on ground likely to heave;*
* *where the hardcore is likely to be more than 600mm thick.*

In these situations, a pre-cast floor is more appropriate. They usually comprise a series of pre-stressed beams with a block infill. The floors are supported by the external and internal load-bearing walls and are not in contact with the ground. DPMs are not required as long as minimum recommended gaps between floor soffits and sub-soil are maintained.

Screed or particle board floor finishes can be specified as outlined earlier in the chapter.

The suspended pre-cast concrete floors shown in the two photographs can be finished with a board insulation material covered with a cement:sand screed or particle boarding. In early examples, i.e. when insulation requirements were less onerous than today, the screed was laid directly on the floor structure.

The floor in the photograph is a reasonably recent development; the polystyrene infill blocks preclude the need for additional insulation. The floor is covered with a concrete topping and power-floated. Due to their ease, and speed, of construction, suspended concrete floors are very common these days and are often used in situations where ground-bearing floors would be quite acceptable.

Common site errors during the construction process

PROCESS/STAGE	TYPICAL SITE ERRORS	FUTURE IMPLICATIONS
Slab substructure.	Some industrial wastes expand if wet. Hardcore not compacted in layers, or hardcore too deep. Trees felled or growing in clay soils.	Settlement or lifting of slab.
Concrete strength, etc.	Cracking caused by shrinkage. Concrete not laid in bays. Concrete too wet when laid. No surface protection on hot days. Frost attack.	Loss of strength
Uneven finish to slab.	Concrete too dry when laid. Concrete finished with lightweight tamp (arches in middle).	Problems with partition walls. Floating chipboard floors on insulation cannot provide level finish.
Screed.	Screed too dry or too wet. Screeds can crack if they dry too quickly or are laid too thin. Screed on wrong grade of insulation, joints not taped.	Cracking screeds, surface dusting. Damage to floor finishes.
Chipboard.	No perimeter gaps. Incorrect grade of chipboard. No vapour check below chipboard.	Lifting. Breakdown of chipboard.
Integrity of DPC and DPM.	Poor bond between DPM and DPC. Inadequate laps in DPM. No blinding on slab.	Rising dampness.
Choice of membrane not appropriate to floor finish.	Polythene DPMs not always suitable for moisture sensitive floor finishes.	Rising dampness and damage to floor finish.

PROCESS/STAGE	TYPICAL SITE ERRORS	FUTURE IMPLICATIONS
Concrete oversite (suspended timber floors).	Risk of ponding in wet spells and potential problems with high water table if oversite too low.	Dampness and mould growth, rot.
Timber joists (suspended floors).	Joists built into cavity. Build up of mortar in cavity. Joists undersized or at wrong centres.	Dampness. Deflection
Sub floor ventilation.	Blocked or insufficient number of vents.	Dampness, build-up of methane gas or leaked piped gas.
Insulation.	No edge insulation around slab. Gaps.	Cold bridging and condensation.
Radon gas protection.	Substructure not sealed.	Health risks.
Pre-cast floors.	Pre-cast floor not grouted. Beams too long. No sub floor void.	Cracking to screed. Damp penetration.

Defects in timber floors

Dampness

There are a number of defects which occur in raised timber floors. It is the early floors which are particularly at risk. Most of these defects are caused by dampness. Rising damp and penetrating damp can occur in a number of circumstances and these are shown in the diagram below.

If the cause of the dampness is not addressed promptly, dry or wet rot may occur.

In addition to a DPC, the adequate ventilation of the sub-floor void is vital if problems of rot are to be avoided. If moist air lingers in the sub-floor void, the moisture content of the floor timbers will increase until it is in equilibrium with the moist air. This problem will be exacerbated if there is no concrete oversite and if the ground is naturally damp, for example, where the water table is high. If the moisture content of the timber exceeds 20 per cent for long periods, it is at risk. It is, unfortunately, quite common to find builders' rubble and other rubbish tipped below the floor. These can block effective ventilation and, in some cases, provide a direct path for rising damp to reach the floor joists.

In extreme cases condensation may occur on the floor timbers. Warmth from the house will normally ensure that the timbers remain above dew point, in other words, the temperature at which condensation occurs. However, the use of an impermeable floor covering such as sheet vinyl will restrict evaporation into the house and condensation may occur on its underside. Any trapped water will slowly soak into the timber floor boarding.

The three photographs show the exposed under-floor void below a suspended timber floor in a Victorian house. Left: The floor joists, and the timber wall plate that they sit on, are in reasonable condition. This is mainly because this is an internal party wall with a working (chemical) DPC. Middle and right: The condition of the joists and wall plate on the opposite (external) wall in the same room. The end of the joists have rotted and so has the wall plate – which has completely crumbled away (see right hand photograph with the end joist also removed prior to repair). Note the grey debris is mainly bits of render that have fallen from the wall. The main source of the dampness that caused this problem is that the chemical damp-proof course had been bridged by a raised external ground level.

Many terraced houses built during the Victorian and Edwardian periods do not contain adequate provision for ventilation. The diagram below shows a typical early example and shows how the addition of extra vents can improve the air circulation.

Ventilation of Victorian terraces

Solid floor below scullery

Occasionally, a duct is laid under the solid floor to improve ventilation.

Vents are usually found under rear (french) window, under front window and, sometimes, below front door.

The shading shows the effective area of floor ventilated.

Over the years vents can become blocked with soil and debris: they can also be blocked due to changing levels of gardens and footpaths. In the image to the right, the new tarmac is just starting to encroach on the vent. If the tarmac does not 'fall' away from the vent, water can run through into the holes during heavy rain.

The paving here has been built up and blocks half of the relatively small vent. The plant growth will further restrict air flow.

Adding extra vents is a fairly straightforward task though ducting may be necessary where some of the ground floor is concrete. Levels of ventilation should be at least equal to 1500mm2 per metre run of wall (equivalent to 225mm by 150mm clay airbricks every 1.5 metres or so).

When inspecting a property, some of the problems mentioned above may be obvious. Where the house has no effective DPC, where vents are blocked, or non-existent, floor problems can be expected.

A distinctive 'mushroomy' smell inside the building will warrant further investigation because of the risk of wood-rotting fungi (see Chapter 15). A damp meter (see Chapter 16) will confirm the moisture content of the floor. If the levels are higher than 20 per cent, there is an increased risk of rot, and lifting some of the floorboards to enable a more detailed inspection would be a wise move. If the floor feels soft and spongy when jumped on, or if there is a gap between the bottom of the skirting and the floor boarding, the joists or wall plates may already be affected by rot.

Movement

Although rot may often account for movement in a floor, there are other causes which are related to the movement of the structure. If the house has joists built into the external walls, differential movement of the external and sleeper walls will result in an uneven floor. In houses built in or before the 1950s, there is an internal load-bearing wall, usually the wall between the dining room and lounge, carrying the load from purlins (via one or two pairs of struts) in the roof. These internal walls often have inadequate foundations and may settle. This can be a particular problem where roofs are recovered with heavier materials. If this is suspected, it is worth closer examination of the partitions at right angles to the suspected load-bearing wall. If the load-bearing wall has dropped, the doors will be out of square with the frames; a tapered gap above the door is a typical manifestation of this type of failure.

Joist problems

If the floor feels quite springy when jumped on, there could be a problem with undersized joists. Furniture or appliances which do not sit level or crockery rattling in cabinets whenever someone walks across the floor both suggest a floor structure which is undersized. Another possibility is that remedial or improvement works to wiring and heating installations have, over the years, resulted in excessively deep notches which reduce the effective depth of the joists. It is also possible that the mortar in which the joists were embedded when they were built into the wall has fallen away and the joists are moving within the wall 'socket'.

Floor boarding

Dampness apart, there are a number of defects which occur in floor boarding. Square-edged timber boards were common until the 1950s. These can give sterling service. Although square-edged boards do give rise to draughts because of the gaps between the boards (in centrally heated houses shrinkage can be quite high), they are easy to lift for maintenance purposes and generally of substantial section. However, if incorrectly fixed, the boards can damage surface coverings; a rippled appearance to a carpet or sheet covering is usually evidence of this.

Floor boards can be converted from timber logs in two main ways: flat sawn or quarter sawn. The former is more efficient at using all the timber but these boards readily twist and curl. Square-edged boards were traditionally fixed with brads (two nails to each joist) but, when lifted for rewiring or other purposes, they are often refixed with lost head nails. These are not as effective at restraining board movement.

Flat sawn

Floor joists

Cut floor brad

Quarter-sawn boards are less likely to twist than flat-sawn boards.

Flat sawn boards are more likely to twist or curl and can 'shell-out' if laid upside down.

Victorian floorboard nails-known as 'brads'.

Tongued and grooved boards are less likely to twist and curl, and if correctly cramped before fixing, should not shrink to a point where gaps occur. However, if tongued and grooved boards are lifted, the tongues will be broken off; subsequent renailing of the boards may lead to deflection or twisting and damage to the surface covering. Not all boards are square-edged or tongued and grooved – there are a number of other types; none of them very common.

Square edge   Cross tongue   Rebated

Tongued & grooved   Rebated , tongued & grooved   Splayed, rebated, t&g.

Square-edged boards, (left) tend to be wider than tongued and grooved boards (right). If correctly fixed and problems of rot can be avoided, the boards should last indefinitely.

Chipboard

Chipboard is common in modern construction. It is made from timber waste products, at the point of timber conversion, bound together in a resin and formed into large sheets. It is available in a range of grades. Standard grade chipboard is very sensitive to moisture. It can therefore be very vulnerable in a ground floor, not just because of the risk of rising and penetrating damp,

Loose edges should be supported by noggings.

End joints should always occur over joists.

Gap of at least 10mm required between wall and edge of chipboard for expansion.

Sleeper wall

Chipboard should run across the joists, laid with staggered joints.

Moisture resistant chipboard is more durable than standard floor grade. It can usually be identified by a red/green stripe.

Wire nails can work loose and will damage boarding; 3mm ring shank nails or size 8 screws are better (length = 2.5 times board thickness).

Boards can be damaged if they cantilever more than about 50mm beyond support.

Board thickness/joist centres:
18/20mm up to 450mm centres
22mm up to 600mm centres.

Boards without t&g edges will require noggings under both long sides.

Strand board should be fixed with 3mm annular grooved nails.

Nowadays it is recommended that all particle boards are glued to joists to prevent squeaking.

but also because of spillage and leakage. A leaking washing machine can ruin chipboard in a few months. If the floor remains wet for any length of time, the chipboard swells and breaks apart. Because of this risk, some organisations insist that where chipboard is used in houses, it should always be a moisture-resistant grade. Whatever the grade, chipboard is likely to expand as its moisture content increases. Moisture movement can be as much as 6mm on a 2400mm board. This is more than enough to crack the floor finish above if it is glued in position, e.g. vinyl tiles.

There are other problems related to chipboard, most of them due to poor-quality workmanship. Most of these problems are related to inadequate support. These are summarised in the previous diagram. Strandboard is similar to chipboard but made with shavings rather than fine particles.

Insulation

If the floor feels uneven but slightly resilient (usually with accompanying creaks), it may be because an insulation quilt has been laid over the joists. This is most likely to occur where older timber floors have been upgraded. However, quilts draped over the joists prevent the floor from lying properly and this uneven surface can soon damage surface coverings (very soon if nails work themselves loose). More importantly, there is a risk of cold bridging. This occurs where heat can escape through some areas more easily than others (see Chapter 16). It encourages condensation and is often confused with rising or penetrating damp. Where timber floors are insulated, a better method is to support the insulation on synthetic netting as shown below. Even with this method there is a risk of cold bridging at the perimeter or edge of the floor.

Potential cold bridge

Quilt supported by netting, or insulation boards on nails.

The risk of cold bridging (at the wall adjacent to the floor) depends on the nature of any wall insulation and the insulating properties of the internal leaf.

Resilient layer (quilt) laid over joists. This can lead to cold bridging; it is also difficult to get an even, creak-free, finish.

Defects in ground-bearing concrete floors

Concrete floors can be very durable. However, in certain circumstances their failure can be quite dramatic and very expensive to resolve. Fortunately, failures of this type are comparatively rare. The BRE has established that many of the problems are caused by poor site practice. Where appropriate, these practices have been summarised.

Settlement of hardcore and sub-soil

Settlement or consolidation of hardcore can occur where the hardcore has been inadequately compacted. Where the hardcore is over 600mm thick, compaction can be difficult even when carried out in thin layers. Particularly vulnerable is the edge of the floor where the hardcore may run down to the foundation and can therefore be of considerable depth. On sloping sites the settlement may be uneven, thus accounting for differing size gaps under skirtings either side of rooms. Partitions built on the floor can also crack.

Where poor-quality, soft ground is not removed before commencing the floor structure, the risk of settlement is increased. However, this would normally manifest itself fairly quickly, possibly even before the building is complete.

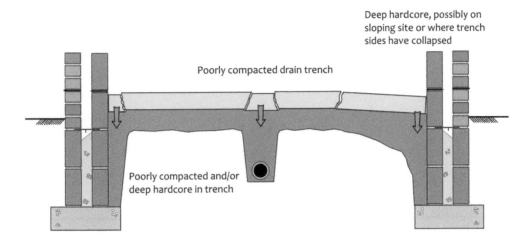

Deep hardcore, possibly on sloping site or where trench sides have collapsed

Poorly compacted drain trench

Poorly compacted and/or deep hardcore in trench

Swelling ground and hardcore

Some hardcores will swell in moist conditions. These hardcores such as steel slags, which include unhydrated lime, and colliery waste which can contain large proportions of clay, are comparatively rare. In many cases suspected problems of swelling turn out to be sulfate attack of the concrete slab. Swelling can also be confused with ground heave. This occurs in clay soils and results in expansion of the clay as it absorbs water.

Any DPM will inevitably be torn or damaged.

Moist hardcore expands - deep hardcore suffers highest rate of expansion.

Ground heave commonly occurs where large trees have been removed from clay soils. The zone of swelling is concentrated around the tree roots and, at its worst, can lift the surface of the ground by as much as 150mm. Ground heave can also be caused by changes in water courses or broken drains. Heave caused by frost may also affect floors in unheated outbuildings but should not be a problem in heated dwellings.

Ground heave in clay soils

Heave breaks slab and can also affect shallow foundations.

When trees are removed the ground will slowly start to swell

Zone of swelling – size of zone depends on nature of tree roots

The house shown here was built in the 1950s. In the 1990s a large tree, situated just to the left of the outhouse was felled. Within 12 months the outhouse floor cracked and lifted (see left-hand photograph). The main floor of the house is suspended timber and there were no signs of movement when this photo was taken.

Slab failure

In many cases, defects in the slab are caused by errors during the construction stage. These include excess water in the concrete, exposure during curing to frost, inadequate thickness or incorrect mix. These problems may not manifest themselves in any tangible form because of the relatively low loads on the slab. A much more serious defect is sulfate attack.

Sulfate attack

The phenomenon of sulfate attack has been explained in an earlier chapter. Sulfate attack occurs when sulfates in the hardcore, or possibly the ground below, migrate in solution to the underside of the slab, react with a by-product of the cement in the slab and form tricalcium sulfo-aluminate. The resulting crystal expands as it forms. The slab lifts slightly and domes, leaving a small gap between the hardcore and slab. In extreme cases the resultant expansion can break the back of the slab and even cause lateral movement of the surrounding walls.

The process of attack and expansion is quite slow and can take several years to visibly manifest itself. It is unlikely to occur where DPMs are effective as considerable amounts of moisture are required to complete the chemical reaction. A number of hardcores are known to be responsible for sulfate attack and these include colliery shale, old bricks containing gypsum plaster (calcium sulfate) and a variety of industrial wastes. Sub-soils which are liable to cause sulfate attack include London clay, Oxford clay and Keuper Marl.

It can be avoided in the first instance by careful choice of hardcore, the use of sulfate-resisting cement in the slab, isolating the slab from the hardcore with a sheet of polythene, and good sub-soil drainage. Once it has occurred, the only effective solution is to break up the slab and renew the hardcore.

Membrane failure

Membrane failure is most likely caused by poor site practice during construction or damage caused by floor or hardcore movement later in a building's life. Poor site practice includes the following (Table 7.3).

Table 7.3 Membrane failure due to poor site practice

LIQUID MEMBRANES	SHEET MEMBRANES
Inadequate thickness (or no. of coats)	Torn during concrete or screeding
Poor joint with DPC	Poor laps and joints
Dusty slab surface prevents good bond	Using DPMs of inadequate thickness
Areas omitted	

Fortunately, minor defects will not necessarily result in damp penetration. This is just as well as, in practice, few polythene DPMs can be regarded as impervious membranes. They are designed to resist capillary action, not direct water pressure. This is why they are usually positioned just above ground level. Permeable surface finishes such as carpets will allow small amounts of moisture to evaporate safely through the slab and disperse into the atmosphere of the room. Sheet vinyl, on the other hand, will resist the passage of water vapour, resulting, possibly, in moisture collecting below the sheet covering. Testing with a damp meter will give some idea of the level of water present.

Any of the structural problems related to heave and settlement are likely to cause failure of the damp-proof membrane. These are relatively easy to identify, if not diagnose. Perhaps more difficult is the damp patch which appears on a floor screed mysteriously and with no obvious associated defects. The following drawings show some typical problems.

Where vinyl tiles are starting to curl and lift, the problem may be a failed membrane but, in new buildings, it could be due to tiles being laid before the screed is dry.

Slab is often hollow around edges.

With liquid membranes it can be difficult to ensure a good joint with the DPC.

Polythene should be laid on sand or stone-dust blinding (25mm) to reduce risk of hardcore perforating membrane.

If the slab settles, the polythene can tear at junction with wall.

A polythene membrane is difficult to join effectively and damage is always likely during concreting.

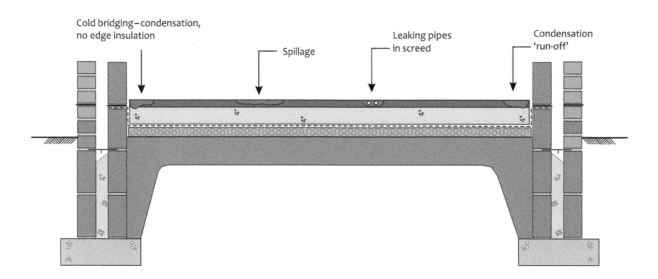

Cold bridging – condensation, no edge insulation

Spillage

Leaking pipes in screed

Condensation 'run-off'

Screed failure

Although, as explained in an earlier section, some timber finishes are found with ground-bearing slabs, most slabs are finished with a floor screed. The screed is a fairly delicate element and will normally be laid towards completion of a building. It is easily damaged and the slightest of imperfections can affect the appearance and adhesion of any thin sheet covering. A number of defects can occur at the construction stage. These include insufficient thickness, inadequate mix, excess water, poor compaction, and incorrect aggregate (sand too fine – encourages shrinkage and therefore cracking). Screeds should be allowed to dry slowly (possibly under a polythene sheet in hot weather) if they are to achieve their maximum strength and successfully resist any shrinkage.

Problems of ground heave, swelling hardcore or sulfate attack will obviously have a dramatic impact on the screed. Vertical movement in all or part of the floor screed, or wide cracks in the screed, are almost certainly caused by problems from below.

Parallel long cracks in a screed, over a block and beam floor, suggest that the floor has not been grouted properly (see the section on pre-cast floors for more detail). Similar cracking can occur if a screed is laid over insulation boards without first taping the joints or where pipes have been bedded in the screed.

A rough dusty surface suggests that the screed was poorly compacted, laid too dry or contains insufficient cement.

If a light tamp is used to level the concrete the surface may rised in the middle – the thin screed is prone to cracking.

Frequent, fine and irregular cracks suggest the screed is too thin, contained too much water when laid, or was over-trowelled.

Regular straight, and possibly, parallel cracks suggest cracking caused by movement in a badly grouted suspended concrete floor, or screed laid on incorrect grade of insulation board.

Wider cracks and an uneven surface suggest problem with slab or hardcore.

In some parts of the country, asphalt screeds are common. When laid in two coats (25mm) on a well-prepared slab, an asphalt screed provides a durable and smooth surface, quite suitable for carpets, etc. However, if they are laid too thin or on a wet or dusty base, they are likely to lift in places: they will sound hollow when tapped or walked over.

Chipboard failure: ground-bearing slabs

Chipboard flooring is a popular alternative to traditional floor screeds. As observed, chipboard is very sensitive to moisture and this has obvious implications for the type and position of a damp-proof membrane. If the chipboard is covered with an impervious material such as sheet vinyl, minor defects in the membrane can lead to wetting of the chipboard. Trapped construction water in the slab can cause similar problems which is why the chipboard manufacturers recommend an additional membrane above the slab (unfortunately this is often omitted). The position of this membrane requires careful thought. This is explained in the diagram below. With laminated boards, i.e. where insulation is bonded to the chipboard, the additional membrane has to be positioned directly on top of the slab. If insulation is separate, a better membrane position is above the insulation, this will also prevent condensation from above.

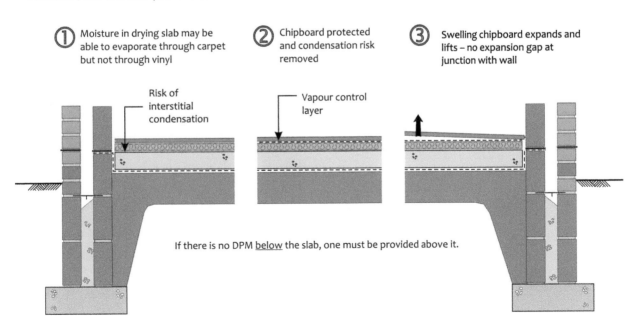

① Moisture in drying slab may be able to evaporate through carpet but not through vinyl

② Chipboard protected and condensation risk removed

③ Swelling chipboard expands and lifts – no expansion gap at junction with wall

Risk of interstitial condensation

Vapour control layer

If there is no DPM <u>below</u> the slab, one must be provided above it.

Water from leaking washing machines can quickly destroy chipboard. Even moisture-resistant grades will deteriorate if they remain saturated for long periods.

Chipboard, if it remains dry, is a durable material. Its durability depends, to some extent, on the way it is fixed. It is important that the tongued and grooved joints are tightly fitting. On site there is always the temptation to cut off a tongue which does not fit neatly into the groove. This reduces the stiffness of the floor. In addition, where boards need cutting to fit round obstructions they will need to be butt jointed and the tongued and grooved joint is lost. A stiletto heel can open up a hole where two or more boards meet and this can be difficult to repair.

Chipboard flooring will also expand slightly once laid. To allow for this expansion a gap of about 10mm should be allowed at its perimeter. If a chipboard floor has lifted and bowed in the centre of a room, this should be the first item to check.

Older floor finishes

Wood block floors

Wood block floors were common until the 1950s, not just in houses but also in schools, offices, village halls, etc. They are usually about 20–25mm thick and bedded in hot pitch (coal-based) or bitumen (oil-based) by dipping the block into a tray of adhesive and then laying it in place. The adhesive also formed the damp-proof membrane.

Wood blocks are very durable, and when the surface has worn over many years, can be sanded or planed down to provide a new, blemish-free surface. There are, however, a number of potential problems, some more serious than others:

- *Blocks which, when laid, contained too much moisture would shrink and this may account for wider than normal joints.*
- *Wetting, caused by washing machine leakage, for example, can ruin a block floor. Even if the blocks settle back into place, the bond with the adhesive will have been lost and the blocks will rock slightly. Persistent and high humidity levels can cause similar problems.*
- *Some modern polishes and sealers are too strong for block floors. They tend to build up in the joints, and because they do not have the flexibility of wax polish, they limit the ability of the floor to accommodate expansion and contraction. This sometimes means that movement is concentrated across one part of the floor (see diagram below).*
- *Most adhesives will break down after 50 years or so. This may mean re-bedding the blocks though careful use of a blow torch on the underlying pitch or bitumen may provide some respite.*

Blocks will lift if they stay wet for long periods

'Split'

Loose blocks need to be re-fixed as soon as possible to prevent adjacent blocks from lifting. Blocks should be finished with flexible coatings – oils and waxes are more durable and less prone to cracking than varnishes.

- *If the loose blocks are around the perimeter of a room (i.e. against external walls), the possibility of moisture transfer from the walls should be investigated.*

Thermoplastic tiles

These tiles were introduced just after the Second World War. They became very popular, particularly on solid floors (timber was rationed after the war and the amount of timber used in a house was limited, hence the concrete ground floor). The tiles were made from thermoplastic resin binders, mineral fillers, pigments and sometimes asbestos. In the mid-1950s vinyl tiles became available and these slowly overtook thermoplastic tiles in popularity. Their advantage lay in increased flexibility and in an improved range of colours. Both types of tile were normally laid in a solvent bitumen adhesive on a screed or trowelled slab.

At this time it was not standard practice to provide damp-proof membranes in ground floors and, over several years, evaporation through the joints of the tiles could leave behind a white crystalline material (sodium and potassium carbonates) alongside the tile joints. This can often be cleaned off, though in severe cases the evaporating moisture is absorbed into the edges of the tiles. As the moisture evaporates and the carbonates crystallise, the resultant expansion can break down the tile edges.

Quarry tile floors

These are usually laid on a screed and are sometimes found in kitchens, utility rooms, etc. There are three main causes of failure:
- *Irreversible expansion as the tiles absorb moisture (when new) from the atmosphere or from cleaning and leaks.*
- *Thermal movement – enough to break the bond with the screed.*
- *Shrinkage of new screeds.*

Failure usually manifests itself as arching along two or three rows. Before the tiles lose their bond, they may sound hollow when walked over.

Defects in pre-cast concrete floors

These floors are comparatively recent and, as such, their long-term durability is unproven. However, there are a number of problems which have manifested themselves, most of which are due to inadequate site preparation or specification rather than failure of the floor components themselves.

Beam and block floors are not usually designed to carry internal load-bearing walls. Deflection in a floor below a wall or horizontal cracks in the wall itself may suggest that over-loading has occurred. Similarly, but less significant, cracking of garage floors is commonly caused by heavy point loads, i.e. car jacks, damaging the thin concrete topping surface of the floor (even though it may be reinforced).

If a pre-cast floor is damp and the obvious problems of spillage or leakage are discounted, condensation or rising dampness may be affecting the floor. Rising damp can occur where the sub-soil is in contact with the soffit of the floor. There should be a minimum gap of at least 75mm between the underside of the floor and the ground surface, though this is obviously impossible to check without lifting part of the floor. In clays likely to heave, this gap should be increased to 150mm.

Damp patches round the edge of the floor could be attributable to condensation around the edge of the floor due to cold bridging (especially near air vents), failure of the DPC or problems with beams of incorrect length which project into the cavity. The risk of general surface condensation will depend on the type of infill blocks, the moisture permeability of the floor finish and the temperature and relative humidity above the floor. Note: these floors do not require damp-proof membranes where the minimum gaps specified between floor and ground are provided and where the floor void is ventilated.

Ventilation is provided not only to reduce the humidity of the sub-floor void but also to prevent the build-up of gases. This is particularly important where landfill gas or methane might be present.

Where beams run parallel to walls, there should be some restraint strapping to tie the wall back to the floor. In extreme cases a wall may bow slightly (see Chapter 8 for more details). Finally, narrow parallel lines in the screed suggest that the floor, i.e. the gaps between the beams and blocks, has not been grouted properly. This stage is sometimes omitted because its importance is not clear to the operatives. However, it can be very expensive to resolve and may require complete removal of all the screed.

Radon gas

Radon is a colourless gas which is radioactive. Although it is present throughout the UK, in some locations its concentrations are a health hazard. To protect occupants, a house should be sealed to prevent the gas rising up through the sub-soil and entering the building. The Health Protection Agency (which now includes the former National Radiological Protection Board) publishes guidance showing which areas are most at risk, how the gas can be detected and measured, and appropriate steps for safeguarding a house against it. A detailed examination of radon is beyond the scope of this text. However, as far as ground floors are concerned, there are two levels of protection: passive and active. Both may be required where the risks of radon gas are high. The passive system, which provides basic radon protection, can usually be achieved by increasing the 'airtightness' of the damp-proof membrane and extending it across the external leaf of brickwork. Additional, active protection, where risks are higher, usually requires some form of ground depressurisation under a ground-bearing slab, or air vents (sometimes fan-assisted) below a suspended floor slab. The diagram below shows two methods of providing basic protection below a ground-bearing concrete slab. If this membrane is damaged, either during or after construction, the barrier will not be effective and additional measures (i.e. active protection) may be required.

Damp-proof membrane providing radon barrier

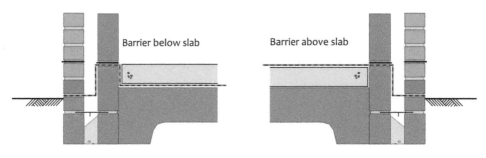

Working details will vary but the principle is to make the construction gas-tight.

Damage to radon barrier, often through service connections or later additions, etc., will breach barrier.

A heavy-duty radon-proof membrane laid over the hardcore of a new development of flats in the south-west of England.

Pipe should be insulated in roof void to prevent condensation and to maintain air flow.

Wind flowing across the top of the pipe causes negative pressure in the pipe which helps the radon-laden air rise.

Installations can include mechanical fans to ensure a fast removal of radon gas.

Warm air inside the house heats the pipe causing the gases to rise.

Sump (void) below floor slab or concrete oversite.

Cavity tray

The cavity tray and DPM separate the superstructure and substructure.

The diagram above shows a sump designed to be slid under the radon barrier. It is intended that this should collect the gas and ventilate it to the outside air. A suspended ground floor slab may be advisable to avoid damage to the slab and protective membrane through settlement.

Upper floors 8

Timber floors

Construction summary

Most houses have upper floors made from timber. The construction principles have changed very little in the last two or three hundred years though, in modern construction, pre-fabricated joists, in the form of metal web joists and timber 'i joists', have become more common – and are now probably the norm, at least on large-scale housing developments. Nevertheless, most floors still comprise a series of timber joists covered with softwood floorboards or sheet materials such as chipboard. The floor structure is usually hidden from view below by a ceiling, originally made from timber lath and plaster but, nowadays, almost always made from plasterboard. Besides concealing the floor structure, the ceiling provides a surface for decoration, improves the floor's fire resistance and reduces, to some extent, the transfer of sound.

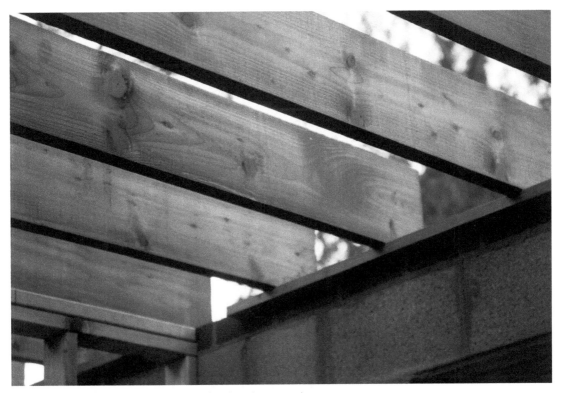

Traditional 'cut' timber joists being put in place in a new house.

Metal web joists.

Timber i joists.

In modern construction, traditional 'cut' timber joists are normally at 400mm centres (16 inches) and are usually made from imported softwood. The depth of a joist resists its tendency to bend or deform under load and the width of a joist resists its tendency to twist. Ideally the joists should span the shortest distance across a room to keep the joist depth to a minimum.

Probably the most common

400

Joists are normally at 400 centres;
300 to 500 are not unknown.

4" by 2"   7" by 2"   8" by 2"   9" by 2"   10" by 2"   12" by 3"

100 x 50   175 x 50   200 x 50   225 x 50   250 x 50   300 x 75

Logs can be cut (converted) into joists in various ways: the joist on the right will be stronger than the joist on the left. However, it is more likely to twist as a result of drying shrinkage.

Joists are commonly cut from Spruce, sometimes called whitewood or white deal; or Scots Pine (redwood or red deal). Traditionally, timber was often referred to by country of origin, eg Baltic Fir, Russian Deal.

Although joist centres of 400mm have been common for the last 150 years, they can vary from 350mm to 600mm. In earlier floors the construction is even more haphazard with joist widths and centres varying considerably.

A typical timber floor from a small Victorian house is shown below.

Late nineteenth century small terraced house

Square edged or t&g boarding

Joists trimmed around chimney breast and hearth–sometimes using wider joists.

Internal loadbearing half-brick wall

UP ———➤   Landing

Herringbone strutting; not always included

◀— Flue from room below

Joists built-in to external walls

Front

Joist size - 175 x 50mm (7" by 2")

Until the 1920s, solid walls were the norm. In small and modest houses joists were usually built into the one-brick-thick solid walls as shown below though cantilevered iron brackets and brick corbels can sometimes be found in better-quality houses. In larger buildings (in the eighteenth and nineteenth centuries) walls for the lower storeys were often more than one brick thick, diminishing in thickness towards the top of the building.

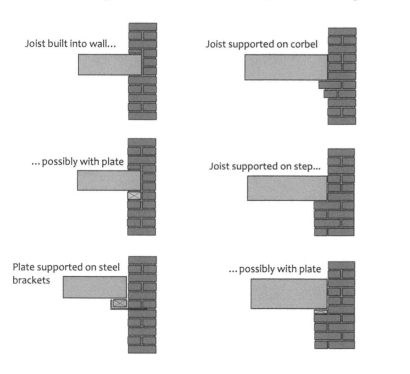

Joist built into wall...

...possibly with plate

Plate supported on steel brackets

Joist supported on corbel

Joist supported on step...

...possibly with plate

Detail showing bracket and cornice

Skirting

Pre-cast cornice

Some of these 'grander' houses had double (or even treble) floors. In a double floor, one or more beams provide intermediate support for the floor joists. These main beams are often supported on padstones to help spread the load safely into the wall.

Typical double floor – large Victorian villa

pad stones

herringbone strutting; wedged at each end

strutting →

bridging or common joists

bridging or common joists

square edged floor boards (for other options see Chapter 7)

main beams – often referred to as binders in early texts

fir lath

notch in binder

notch in joist

bridging or common joist

ceiling joists

Separate ceiling joists added to hide beams.

Timber wedges help tighten the floor.

Upper-floor joists are normally supported by the internal leaf of the cavity wall, either built into the inner leaf or supported on joist hangers. The following drawing shows the typical construction of a modern 'cut' timber joist.

Restraint and strutting

Since 1985, the Building Regulations have required joists to be tied or secured to the walls at intervals of not more than 2 metres to provide restraint for the comparatively thin external walls. Where joists run parallel to the external walls, galvanised metal straps can be fixed as shown below.

A modern timber (joist) floor

Joists are normally softwood and at 400mm or 600mm centres.

Joists can be built into internal leaf or supported on joist hangers.

In modern construction timber to timber connections are made using special joist hangers. Traditionally the timbers were jointed.

① ②

Building-in joists gives good restraint but they must be airtight. Hangers require additional straps or special hangers (at 2 metre centres).

Straps required at 2 metre intervals to provide restraint to external wall.

④

Gap of about 50mm for service runs. Gap also facilitates fixing of boarding to joist after wall finished is applied.

Stair well   ③

Trimmer joist

Trimmed joists

Strutting to prevent joists twisting and damaging ceiling. ⑤

Joist centres usually 400mm; 600mm common with 'made-up' joists.

Joists can lap at load bearing wall but joist ends should not extend more than 100mm beyond wall.

Longer projections can lift boarding due to cantilever action when joist is loaded mid span.

Timber blocking or folding wedges to 'tighten-up' floor.

Where possible joists should run the shortest distance across a room.

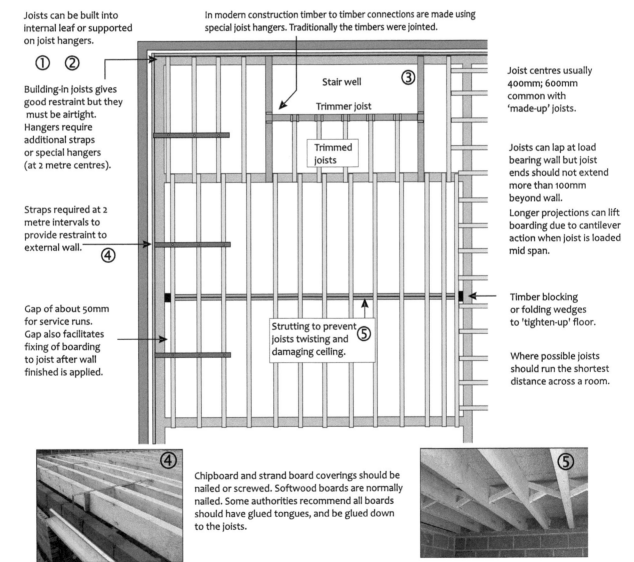

④

Chipboard and strand board coverings should be nailed or screwed. Softwood boards are normally nailed. Some authorities recommend all boards should have glued tongues, and be glued down to the joists.

⑤

Timber blocking below the strap ensures the strap also works in compression. If the joists are built into the inner leaf and have at least 90mm bearing, no additional restraint is necessary. However, where standard joist hangers are used, additional restraint is required. This can be provided by straps or by the use of special joist hangers. These restraint hangers are, again, required at 2 metre centres (usually every sixth joist). Intermediate joists can be supported by standard hangers.

In all but the shortest of spans, strutting is required to prevent the floor joists from twisting and cracking the ceiling below. This twisting can be caused by loading or by shrinkage as the new joists dry out. A single set of struts, at right angles to the joists, is normally provided mid-span. The strutting can be herringbone pattern or straight, using joist off-cuts (in practice the straight strutting is normally staggered to facilitate nailing). Herringbone strutting is probably the more traditional method though straight strutting is commonly found in Georgian buildings.

Herringbone strutting.

Traditional cut timbers built into the wall with joists stiffened by both solid strutting and herringbone strutting. Timber blocking or timber wedges, fixed at both ends of the strutting between the end joists and the wall, tighten up the floor and help minimise movement.

Openings in floors

Upper floors usually contain openings for stair wells. To form the opening, the floor needs to be trimmed as shown in the following diagram. Similar techniques were used to trim floors around chimney breasts to ensure that the joists were not directly at risk from fire.

The trimming and trimmer joists are usually of thicker section timber (or two standard joists bolted together) to cope with the increased load. Until the 1960s special carpentry joints were used to join these timbers. A tusk tenon jointed the trimmer to the trimming joist and a housed joint was used to fix the trimmed joists to the trimmer. Nowadays timber-to-timber galvanised hangers are more likely to be used.

Stair well

Trimmed joists housed into trimmer

Tusk tenon

Tusk tenon joint

A timber-to-timber hanger in the process of being fixed in place in a modern house.

Floor coverings

In the Georgian and Victorian periods most upper floors were covered with square-edged boards. These were usually made from softwood though hardwoods such as oak were sometimes found

Square-edged boarding in a Victorian house.

A chipboard floor finish being laid on the metal web upper floor of a modern house. The green covering is a temporary measure to protect the floor surface.

in the reception rooms of grander houses. At the end of the Victorian period the introduction of better woodworking machinery meant that tongued and grooved boards became more readily available. An alternative form of construction, sometimes found in better-quality housing, used grooved boards (grooved along both edges) secured with continuous metal tongues. In the twentieth century, square-edged boarding and tongued and grooved boarding were both common though, since the 1970s, tongued and grooved chipboard has become the most popular material.

Ceiling finishes

Until the 1930s ceilings were usually made from lath and plaster. Thin strips of softwood were nailed across the underside of the joists and finished with three coats of lime plaster. Nowadays ceilings are usually formed from plasterboard. The plasterboard can be self-finished, plastered with gypsum plaster or finished with flexible proprietary materials.

The photograph shows a typical mid-Victorian working-class house. The joists are quite thin, probably about 38mm. There is no ceiling so fire protection and sound insulation are minimal.

Concrete floors

Construction summary

Low-rise purpose-built flats from the 1950s onwards usually have upper floors made from pre-cast concrete. The floors in flats separate dwellings and, as such, require higher levels of fire protection and sound insulation than their timber counterparts in houses. Concrete floors are also capable of greater spans than timber floors and are ideally suited to flats where large floor areas may be required without the complication of unnecessary internal load-bearing walls.

For the past 20 years or so, plank floors and beam and fill floors have been the two most common forms of construction. In both cases the main load-bearing members are usually made from pre-cast, pre-stressed, reinforced concrete. In a pre-stressed beam the reinforcing wires are stretched to a specific tension before placing the wet concrete in the moulds. When the concrete has cured, the tension in the wires is released and this produces a very strong but lightweight beam. The beams are usually pre-cambered; in other words the tensioned wires cause the beams to arch slightly (usually no more than a 350th of the span). When the beams are under load, the camber should not be noticeable.

Once in position, the floor is usually covered with a brushed grout to tighten up the floor and fill any gaps, thus improving fire protection and sound insulation. Concrete floors usually have sufficient mass to provide adequate resistance to airborne sound (noise from radios, etc.) but require additional surface coverings to reduce impact sound (the noise from people walking across the floor). This is usually achieved by creating a structural break in the floor. One option is to use a sand cement floor screed laid on a resilient layer of paper-backed mineral wool. Another option is to use tongued and grooved chipboard laid on insulation boards. Note that in 2004 sound insulation requirements became more onerous for new buildings – some beam and block floors are no longer suitable.

Beams can span up to 8 metres or so.

All internal partitions are non-loadbearing and are supported by concrete floor.

Pre-stressed concrete planks

Pre-stressed concrete beams

Heavy partitions may crack ceiling finish if beams are not doubled up.

Floors should be grouted to fill gaps; concrete may be needed where beams sit side by side.

Framework for ceiling finish

Restraint straps will be required where beams run parallel to external walls.

Depending on the nature of the floor structure, the soffit can either be plastered directly or can be made from a timber or metal framework with plasterboard fixed to its underside.

From the 1920s to the 1950s, traditionally built (i.e. with load-bearing external walls) low-rise and medium-rise flats often had intermediate floors constructed from in-situ concrete. The floors could be constructed in a variety of ways; three are illustrated below.

In-situ concrete floors 1920s–1950s

Solid reinforced slab   Sand/cement screed

Reinforcement rods or mesh

The slab may be supported by the internal leaf or both leaves.

Hollow pot floor

Clay or concrete hollow pots

Concrete filler joist floor

Primary beam

Temporary formwork provides a platform on which to pour the in-situ concrete.

Reinforcing rods

Secondary beams

Most of these floor types also act as diaphragms, in other words, they help to distribute wind loads from walls running in one direction to walls running in the opposite direction. To do this successfully, the joints or connections between walls and floors are of vital importance. If correctly built, in-situ concrete floors provide good load-bearing capacity, good fire resistance and high resistance to airborne sound. With appropriate detailing and surface finishes they will also provide good resistance to impact sound. Many early concrete floors, in fact, had no provision for resistance to impact sound and the floor finish often comprised a sand cement screed or quarry tiling laid directly onto the floor structure.

The photograph above shows an early clay pot and steel beam upper floor.

Common site errors during the construction process

Upper floors in houses and flats should be trouble-free. However, there are a number of common construction errors, most of them caused by incorrect site practice or poor site supervision (see Table 8.1). At best, these defects have cosmetic implications; at worst, they could affect the health and safety of the occupants.

Table 8.1 Common site errors during the construction process

PROBLEM AREAS/ STAGES	TYPICAL SITE ERRORS	IMPLICATIONS
Joist support.	Inadequate bearing, usually caused by joist hanger not tight against wall or joists too short. Joists should be nailed to hangers.	Possible floor collapse.
Junction of walls and joists.	Incorrect joist hangers, straps omitted. Most concrete beam and block floors require straps along parallel walls.	Lack of restraint may permit lateral wall movement.
Joists generally.	Undersized joists, incorrect centres, strutting and/ or blocking omitted, holes and notches incorrectly cut. Incorrectly sized trimmers, etc.	Deflection causing cracking in partitions, gaps under skirtings, cracking in ceiling, 'bounce' in floors.
Joist protection & strutting.	Wet joists, lack of strutting and/or blocking at strutting ends.	Warping joists, damaged ceilings, uneven floors.
Damp protection.	Joists bridging cavity, or partial bridge and mortar droppings.	Rot and damp penetration, damage to decorations.
Floor coverings (boarding, etc).	Flooring undersized i.e. too thin for span, boards incorrectly cramped (t&g). Badly warped joists. Chipboard not t&g or inadequate noggings.	Uneven floors.
Fire (concrete floors in flats).	Pre-cast floors not grouted, service ducts not protected, inadequate ceiling or ceiling fixings.	Smoke or fire with risk to life and building.

PROBLEM AREAS/ STAGES	TYPICAL SITE ERRORS	IMPLICATIONS
Sound (concrete floors in flats).	No floating floors, pre-cast floors not grouted, sound insulation too thin, floor of insufficient mass, screeds too thin. Beam and block floors won't always meet new Regs.	Airborne and impact sound transfer, neighbour disputes.
Uneven floor finish (flats).	Screed problems, lack of reinforcement over resilient insulation. Poor grouting of beams/blocks. Sloping (crowning) chipboard finishes caused by camber in pre-cast beams.	Damage to coverings, possible effects on sound insulation.

Defects: timber upper floors

Introduction

There are a number of potential problems with upper floors. Disregarding, for the moment, defects in the floor coverings or ceiling finish, these problems include damp penetration, excess deflection and lack of restraint for external walls. Problems of pest attack are dealt with in Chapter 15. Perhaps one of the most common problems is inadequate support. This can occur where joist hangers are not tight against the wall or where joists are cut too short.

Hangers correctly sized for joist; joist a good fit in the hanger and securely nailed.

Joist too short.

Joist too short/ hanger not properly bedded in wall.

Where joists are built into the wall they should have a bearing of at least 75mm; 90mm if they are acting in restraint.

Damp penetration

In cavity walls, problems of damp penetration are usually due to faulty workmanship. This can occur in a number of ways. One common problem is caused by bridging of the cavity which can allow moisture to penetrate the joist ends. This can occur where careless bricklayers allow mortar to accumulate in the cavity. The problem is exacerbated if the joists project into the cavity and form a ledge on which mortar droppings can collect. The cavity can also be bridged where insulation has been fitted incorrectly, where cavity trays have been omitted or where mortar has collected on ties.

In solid walls the joist ends must always be considered at risk unless the external walls are quite thick. In the majority of speculative housing built during the Victorian and Edwardian periods, external walls are usually one brick thick. This effectively means that the protection to the joist end never exceeds 100mm or so (and is often as little as 50mm). Deterioration in the mortar or poor-quality brickwork can result in the joist ends becoming damp. In some cases,

Joist should never touch outside leaf.

DAMP

Even slight projections can form ledges which trap mortar droppings.

Water can come from saturation during heavy and prolonged rain (water running down inner face of brick), cracks in mortar, damp outer leaf.

A slight projection can also push board insulation out of place. If insulation boards touch the external leaf a path may be created for dampness. Water can cross the cavity by running over the top edge of the insulation board and down its inner face.

Mortar droppings can sit on insulation batts and create path for water.

joists are supported on timber wall plates built into the wall. If the wall plates rot, the stability of the wall itself may be affected.

Damp penetration will manifest itself in a number of ways. There may be damp patches or staining around the edge of the floor; a moisture meter may show that the moisture content of the timber is above 20 per cent. There may be obvious signs (or smells) which suggest rot is occurring, the floor may feel soft when jumped on and there may be wide gaps under the skirtings.

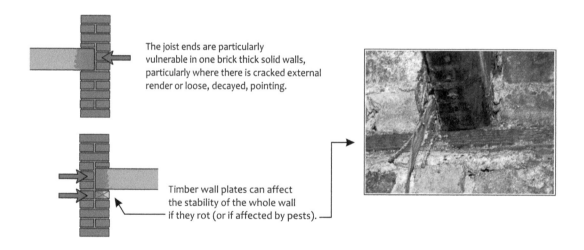

The joist ends are particularly vulnerable in one brick thick solid walls, particularly where there is cracked external render or loose, decayed, pointing.

Timber wall plates can affect the stability of the whole wall if they rot (or if affected by pests).

Sloping floors

If a floor slopes towards a load-bearing external or internal wall, movement in the main structure is likely to be the cause. Torn wallpaper or cracking in the plaster or wall itself will help to confirm this. Possible causes have been covered in the Chapters 3 and 4. If a floor sags or deflects, the problem is more likely to be related to the floor itself.

Deflection

Deflection or sagging can occur if the joist ends have started to rot. It can also occur for a number of other reasons and systematic inspection is necessary if the cause is to be correctly identified. Deflection can be caused by any or all of the following:

- *undersized joists*
- *joists at incorrect centres*
- *excessive loading from furniture or partitions bearing on the floor*
- *excessive notching for services*
- *rot or insect attack.*

New tiling can add 2 to 3 tonnes to the weight of a roof.

A roof re-covered with heavier materials can place extra, disproportionate, weight on internal load-bearing walls. Substantial settlement may affect suspended floors.

A ball may help identify slope; doors in partitions at right angles to internal wall may distort.

Differential settlement can result in deformation of the floors. If the party walls are bonded to the front walls cracking can occur in any of the walls.

In some Georgian and Victorian terraced houses the front walls are not bonded very well (if at all) to the party walls. Although this may have implications for the overall stability of the building, settlement of the front wall will not normally cause cracking in the party wall. There will, however, be stress marks or tears in the wall paper and plaster where the party wall meets the front wall.

In extreme cases deflection can affect the stability of the external walls. This occurs where the load from the sagging floor exerts lateral pressure on the wall.

Visual inspection may indicate deflection; the floor may have an obvious dip or sag. This can sometimes be confirmed by the use of a straight edge or by looking at the ceiling from the room below. In other cases the 'heel test' (i.e. stand on the floor, lift the heels and bring them sharply down) will reveal excess spring in the floor. If glass in cabinets or window glass rattles, the floor joists may be undersized (or overloaded).

Where deflection occurs, the first stage in the diagnosis should be to ascertain the depth and centres of the joists. The joist depth may be difficult to assess unless floorboards are lifted. Many joists in older houses will not conform to the current requirements of the Building Regulations though reference to them should give some idea as to their preferred size. An

Original line of wall

If joists are undersized the loads will not be vertical and will tend to push the walls outwards causing bulging.

If joists are correctly sized the loads will be transmitted safely into the walls.

alternative approach, which should only be used as a general guide, is to apply a simple 'rule of thumb' method: take the span in feet, divide by two and add two; this will give the required depth in inches. So, a span of 14ft requires a joist 9 inches deep. Alternatively take the span in millimetres, divide by 24 and add 50 to give the depth in millimetres; a span of 4000mm requires a joist depth of 216mm, say, 225mm (joist sizes are available in increments of 25mm). These rule of thumb methods assume joist centres of approximately 400mm.

In many older houses central heating may have been installed or rewiring may have taken place. Both these activities may result in careless drilling or notching of the joists. The Building Regulations contain a number of simple rules to ensure that notching and drilling do not affect the stability of the joists. These are shown in the diagram below.

Where notches lie outside these zones of safety, the effective depth of joist should be regarded as the actual depth less the depth of the notch. In extreme cases the installation or renewal of wash-hand basins or showers may result in notches as deep as 75mm to take the waste pipes. Thus a 200mm joist effectively acts as a 125mm joist.

If the depth of joist is satisfactory and notching is within the guidelines, the joist support needs to be investigated. If one or two joist ends are rotten or not adequately supported by the load-bearing walls, it will have the same effect as increasing the joist centres.

If the joist support is adequate, another defect which may cause deflection is lack of effective strutting. This could be because strutting was omitted in the original construction or, more likely, strutting has become ineffective because individual struts have been removed for service runs. Another possibility is that the folding wedges or timber blocking pieces at either end of the struts have worked loose or were omitted from the original construction (see diagram on page 232). A similar effect is caused where a joist built into a wall moves around within the wall socket – a problem is caused when the mortar 'grout' falls away, or where an in-built timber wallplate supporting the joist rots.

Some of the notches in these joists are quite deep, particularly the one on the right. The notch on the left is original – it was cut for the lead gas pipe (lighting).

This shows the empty 'sockets' where first floor joists used to sit in this party wall.

Partitions built directly onto the floor can also cause deflection. These partitions may have been part of the original construction or may have been added to change room layouts. If the floor has adequately sized joists, a partition running at right angles to the joists will not necessarily cause problems. However, where the partition runs parallel to the joists, or in some cases, in between the joists supported by the floor boarding alone, deflection may occur. In all the above cases the problems will be exacerbated if the partition is carrying heavy loads. Several rows of book shelves, for example, can increase the load dramatically. Where the partition is on the highest floor, it may be carrying loads from the roof not intended in the original construction. This may be due to structural alterations to the roof itself or the load from stored materials.

Stud partitions are quite light but the load on the floor can be high if heavy shelves are fitted or part of the roof is supported.

Partition parallel to the joists

Missing or loose wedges

Strutting will provide a some resistance to bending but only if it is complete and correctly wedged at ends.

Missing struts

A partition at right angles to the joists, and near their centre, can create even more deflection.

Joists 'doubled-up' will help to carry the extra load but if the partition is carrying load from the roof deformation may still occur.

The load from the internal partitions has caused deflection to occur in this first floor structure.

Failure of the ceiling finish

Any deflection in the floor is likely to induce cracking in the ceiling finish. Cracking can also be caused by lack of floor strutting, inadequate nailing, plasterboard of insufficient thickness for the joist centres, lack of noggings between joists and plasterboard fixed without staggered joints or with joints incorrectly taped. Wallboard, which tends to be in large sheets (often 2400 x 1200mm), is more likely to crack than plasterboard lath (not to be confused with traditional lath and plaster). Lath, usually in sheets 1200 x 400mm, does not require taped joints and, because of its smaller size, stresses can be spread evenly across the ceiling. It is rare in modern construction but was common from the 1940s to the 1980s. Failure of plasterboard, plaster skim or artex finishes are discussed further in Chapter 12.

Horizontal gaps between floor and external walls

If there is a clear gap between the floor and the wall there may be insufficient restraint to the external walls (see also Chapter 4). This is most likely to occur where the wall runs parallel to the joists. In extreme cases it is possible to stand in a room upstairs and see into the room below. The problem is most common in end-of-terrace properties, particularly those with more than two storeys, and where the floor joists on each floor run in the same direction.

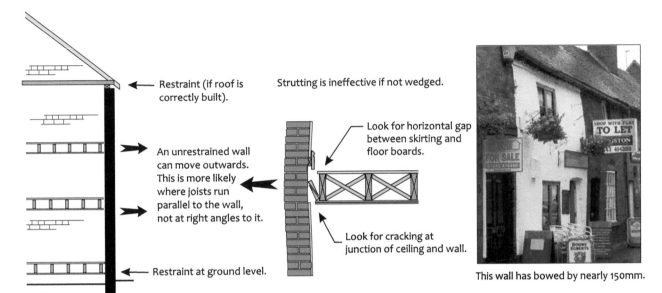

Restraint (if roof is correctly built).

An unrestrained wall can move outwards. This is more likely where joists run parallel to the wall, not at right angles to it.

Restraint at ground level.

Strutting is ineffective if not wedged.

Look for horizontal gap between skirting and floor boards.

Look for cracking at junction of ceiling and wall.

This wall has bowed by nearly 150mm.

These straps are in the right position and will offer restraint to the gable wall. However, straps should work in compression as well as tension. This requires timber blocking in between the joists under the straps.

Defects: concrete upper floors

Pre-cast floors

These floors are comparatively new, they first became common at the end of the 1950s in low-rise flats. As such, their longevity is unknown. However, some problems have come to light, and these include:

- *inadequate end bearing*
- *cracking of ceiling finishes*
- *poor sound insulation*
- *lack of fire protection where services penetrate the floor*
- *lack of restraint to external walls*
- *failure of the floor finish*
- *sloping chipboard floors.*

End bearing

Most manufacturers recommend a minimum end bearing of 100mm. Where brickwork has been badly set out or built out of plumb, there is the danger that the end bearing can be reduced. In practice, this can be avoided by careful site supervision or insisting that the beam manufacturers provide beams to suit the as-built dimensions rather than the dimensions shown on the drawings. While this will inevitably cause some delay, the associated costs outweigh the potential costs of remedying incorrectly sized beams.

Cracking of ceiling finishes

This is most likely to occur where plaster finishes have been applied directly to the soffit of the beams. It may be caused by incorrect or inadequate grouting of the floor surface, the load from heavy partitions built directly on a single beam, or the use of wet infill blocks which will shrink slightly as the floor dries out. It could also be caused by using undersized beams which suffer excessive deflection though this defect should not occur if the manufacturer's design guidance is followed.

Poor sound insulation

If the manufacturers' recommendations have been followed, modern pre-cast floors will conform to requirements of the Building Regulations with regard to sound insulation. In practice, poor site workmanship or inadequate site supervision can result in number of problems which reduce the floor's resistance to the passage of impact and airborne sound. Airborne sound resistance can be reduced if the floors are not grouted properly. Impact sound resistance will be reduced if there are defects in the floating floor. This can be caused by inadequate thickness of the separating medium (usually a paper-backed quilt under screeds or insulation boards under screeds or chipboard), or where the screed has penetrated the joints between insulation boards. Sound insulation will also be reduced if holes cut for service runs (obviously through the blocks, not the beams) are not adequately filled. This will also reduce the fire protection of the floor.

Failure of the surface finish

Screeds laid on a resilient layer will crack if they are not reinforced (the usual recommendation is galvanised chicken wire) or if they are of inadequate thickness; they should be at least 65mm thick. Screeds can also crack if they are laid too wet or are not sufficiently compacted, thus reducing their bond with the wire reinforcement.

Chipboard can fail if it becomes wet and this can occur through spillage or where wet flooring components are used and a polythene vapour check is omitted above the insulation boards (the resilient layer). As with solid ground floors the chipboard should be tongued and grooved, with glued joints, and with a minimum gap of 10mm at the perimeter to allow for moisture movement.

If a chipboard floor has been correctly laid but still appears to rise slightly towards the centre of a room, it could be due to the pre-camber of the beams. A floor spanning 8m may rise by as much as 20mm mid-span. Before laying the resilient layer the floor should have been checked for level and, if necessary, a thin levelling screed should have been applied.

Lack of restraint to external walls

Some beam and block floors require restraint straps where beams run parallel to external walls. As with timber floors, strapping is required at 2 metre centres.

In-situ floors

Three in-situ floors were described earlier in the chapter. These floor types are normally very durable and serious structural problems are rare; where cracking and deflection are evident, the advice of a structural engineer should be sought. Some structural problems such as carbonation and the use of high alumina cement are described in Chapter 17. More common are problems related to fire resistance where services penetrate the floor, and poor impact sound insulation (see above). Other defects include cold bridging and poor detailing where floors cantilever to form balconies.

Cracks in this position may be caused by lack of steel in top of the balcony slab.

Minor cracks may be hidden by asphalt.

Damp penetration

COLD

Condensation

Dampness in the screed or on the ceiling can be difficult to diagnose with any certainty. Both can be caused by condensation or damp penetration. The latter is more likely where cavity trays have been omitted above the floor slab or where cracking in the cantilevered section of the slab has torn the asphalt.

Balconies 'guarded' by railings or brick walls.

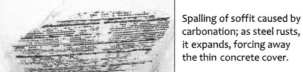

Spalling of soffit caused by carbonation; as steel rusts, it expands, forcing away the thin concrete cover.

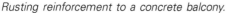

Rusting reinforcement to a concrete balcony.

Pitched roofs 9

Happily, a roof hit by a flying shark is an unlikely event in the UK. However, there are many other reasons for roof problems to occur.

Introduction

The proper performance of a roof is often considered to be more important than that of any other building element, except perhaps the foundation. This is because the whole or, more usually, partial failure of the roof inevitably has an adverse effect upon one or more elements elsewhere within the building.

A roof can fail for a number of reasons, ranging from poor design and/or poor construction to lack of maintenance. Failure may also be due to the roof having outlived its designed life expectancy or to the effect of other failed building elements upon its performance, e.g. failure of a supporting wall.

The poor state of this roof is as a result of a complete lack of maintenance over many years. The roof coverings are in an extremely poor condition allowing water penetration and leading to the failure of the roof timbers.

Construction summary

Function of the roof

The primary function of the roof is to provide protection to the building beneath from the weather. It should be constructed in a form that will meet the following five key performance criteria.

1 **Strength.** *It must be capable of supporting both its own weight and any loads that might be imposed upon it, e.g. snow, standing water, water tanks and other plant.*
2 **Stability.** *The various components should not be subject to excessive movement nor should they create instability in other elements.*
3 **Durability.** *The materials used, especially coverings, must be capable of performing for their designed life expectancy.*
4 **Weathertightness.** *The roof covering must prevent moisture penetration. This is achieved by it being either thin and totally impervious (e.g. sheet roof coverings, slate) or by the use of a thicker, less dense material that is sloping and allows evaporation plus water shedding to take place (stone tiles, clay tiles).*
5 **Thermal efficiency.** *In recent years it has become recognised that a roof must also provide good thermal insulation.*

Pitched roofs: a brief history

Protection from the elements has been essential to human survival for many thousands of years. The first roofs, once man moved out of caves, were probably formed by covering a sloping framework of branches with a covering of animal skin or vegetation – direct descendants of these forms of construction can still occasionally be found in this country, e.g. thatched roofs of reed, wheat or heather. Sloping, or pitched roofs, were found to be the most efficient form of weatherproof construction in all but the driest climatic conditions and are the traditional form in the British Isles.

A traditional pitched roof is expensive to construct. It involves the use of skilled craftsmen and a considerable quantity of materials. Earlier roofs involved cutting the individual timbers to size on site (a country-cut roof), while a modern roof is commonly constructed with factory-formed trusses. Coverings were of thatch, slates, stone or clay tiles – naturally based materials – which required steeper pitches to ensure efficiency. It was only in the twentieth century that man-made materials, such as concrete slates and tiles with interlocking details that allow much lower pitches, have been introduced.

The form of pitched roofs

The simplest form of pitched roof is the lean-to or monopitch roof that is commonly found above many rear extensions of Victorian houses. It can also be found as the roof form on a great number of housing estates built since 1960. Originally, such roofs were constructed on site from individual timbers, but the modern version is normally formed with trusses and often has an extremely low pitch (17.5 degrees or less).

Lean to or monopitch roof

A simple lean-to roof has a maximum span of about 2.4 metres.

2.4 metres

This span can be increased to about 3.8 metres if ceiling joists and intermediate support to the rafters are provided.

Purlin

3.8 metres

In practice there are lots of variations. The roof on the left has a series of struts, fulfilling the same function as the purlin. The roof on the right has a principal rafter supporting the purlin mid-span. This may be necessary on long roofs.

A lean-to roof can only span a short distance, but a simple double pitch or couple roof can span a wider length. Incorporating a collar to act as a tie at about one-third of the height enables even larger areas to be successfully roofed over, as it reduces the tendency for the roof to spread.

Couple roof

A simple couple roof has a tendency to spread outwards.

3.3 metres max.

Collar roof

A high collar is not effective.

$\frac{2}{3}$ Not effective

$\frac{1}{3}$ Effective

A good joint between the collar and rafter is vital to ensure the roof does not spread.

4.5 metres max.

A raised collar enables greater internal head height and is common in Victorian housing.

Collars are sometimes nailed (above); bolting is better (below).

A collar roof can still only span a relatively short distance, but is at its most efficient when the collar is positioned at the feet of the rafter. It is then known as a closed couple roof. Even greater spans are achieved by the inclusion of purlins, struts and braces. Such construction is typical of roofs formed during the last 150 years.

Closed couple roof

5.0 metres maximum span (approx)

Three purlin roofs

c.1800

Double or purlin roof

purlin

strut

Hanger and binder support ceiling joist

Hanger is in tension and prevents ceiling from sagging. Struts are in compression and prevent purlin from sagging.

Internal load-bearing wall

1890s

1930s

The purlins span laterally between the party or internal structural walls that support the loads involved. Where no such wall was available, one or more roof trusses, of the type known as King Post trusses, would be constructed at intervals. These were formed of large timbers which acted as intermediate supports to the purlins and hence to the rest of the roof structure.

King Post truss

Common rafter

Purlin

...with eaves

Strut

King Post

Tie beam

Wrought iron staps (optional)

Boarded finish (optional)

Principal rafter

...with parapet

Nineteenth century King Post truss

The trusses are spaced at equal distances throughout the roof (say, 4 metres) and enable increased clear spans.

Often, the roof space would be required for accommodation or storage purposes. This could be achieved by the use of an open truss known as a Queen Post truss or, for smaller spans, a single mansard roof.

Historically, the quality of roof design, workmanship and materials was variable. The higher the social status of the occupant (and the greater the income) usually meant the better the quality of the building. Speculative building was undertaken for all social levels and this often resulted in poorer-quality construction, notwithstanding the eventual occupant's social status.

Mansard Roof with King/Queen Post truss

King Post truss

Purlin

Curb plate

Ceiling joists

A modern Queen Post truss

Wall plate

Floor joists

Queen Post truss with bolted and strapped connections

NB. It is possible to construct a simple mansard roof without the provision of King or Queen Post trusses.

However, it is generally recognised that traditional construction, in particular of the roof, achieved its highest quality during the years between 1900 and 1914.

Shortages of materials and skilled labour, combined with post-war economic problems, resulted in there being little domestic construction in the 1920s. This was followed by a period of greater activity until 1939 in both the public and private sectors. Pitched roofs were still of traditional 'cut' construction, but the quality, though generally good, was usually lower than the peak reached just before the First World War.

The Second World War and the years immediately after were a period of little building activity except for some types of system building (see Chapter 17). Between 1945 and 1954, timber was subject to rationing and this led to the extending or stretching of the centres of rafters and trusses in order to save materials.

Domestic construction commenced again in earnest in the 1950s. A great number of the roofs of the period between 1950 and 1970 were of 'cut' construction, but this period also saw the introduction of modern trussed roof construction, so that by 1970 almost all roofs of low-rise domestic housing were formed with trussed rafters. The advantages of modern trusses are that they are relatively cheap to manufacture, easy to install and do not require skilled site labour. They are also lightweight and are capable of spanning wide distances. It is now uncommon to use anything other than trussed roof construction for new housing.

Modern trussed rafters

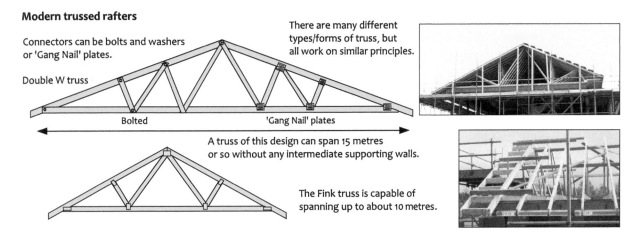

Connectors can be bolts and washers or 'Gang Nail' plates.

There are many different types/forms of truss, but all work on similar principles.

Double W truss

Bolted

'Gang Nail' plates

A truss of this design can span 15 metres or so without any intermediate supporting walls.

The Fink truss is capable of spanning up to about 10 metres.

In a simple Fink truss roof, five binders are required to tie the trusses together; a raking brace on each slope keeps the structure rigid. The lower end of the brace should be secured to the wall plate.

Where binders are not of sufficient length, they can be joined by lapping them over two trusses.

Ends of binders should abut gable walls.

Wall plate strap

Gable strap

Gable ties are required every two metres along the pitch; on high roofs they are also required at ceiling level. The need for wall plate strapping depends on pitch, load etc.

The majority of low-rise domestic pitched roofs are of timber construction. However, since the Second World War some use has been made of pre-cast reinforced concrete members and steel members, though these are often found to be installed as supports to what is essentially a timber roof (e.g. reinforced concrete trusses and purlins supporting timber rafters).

Defects in pitched roofs

Poor design

Roof failures can occur as a result of designer error. Raised standards of education and professional training, increased knowledge of the performance of structure and materials, and regulatory control of building design, all mean that (in theory, at least) roofs have become increasingly less likely to fail. However, the introduction of new and not properly understood design approaches, the use of new materials, or of traditional materials used in new ways, even the loss of knowledge of how traditional materials perform, can each result in the roof structure and/or its covering failing to perform as required.

It also needs to be recognised that, historically, the standards of roof design before (and even after) the introduction of model by-laws in the nineteenth century were variable. This is particularly noticeable in the buildings of cheaper, and probably speculative, domestic construction which were erected in the period of extensive urban expansion of the nineteenth century.

Generally, the older the building, the less likely that there will have been any scientific design criteria applied when considering structure and materials. The designer, or more likely, the master builder, would have based his construction on the accumulated knowledge of generations of previous similar builders using materials that had been 'proved by the test of time'. Furthermore, it was not uncommon for corners to be cut or design implications to be misunderstood, even in so-called quality work, resulting in the roof structure performing at the limit of, or possibly beyond, its proper structural capability (Table 9.1). While not necessarily leading to total failure, it does mean that the roof will be over-stressed and, almost certainly, showing signs of distress.

Table 9.1 Common defects in pitched roofs at construction stage (new build)

POTENTIAL PROBLEM AREAS	IMPLICATIONS	DEFECTS / MANIFESTATION
Untreated timbers	Attack by wood-boring insects	Weakening of timbers due to infestation
Rafters, purlins and other roof components inadequately sized	Roof structure too weak to cope with roof covering and/or imposed loads and/or wind	Bowed and/or split timbers Deflection of overstressed roof members Dishing of roof
Rafters, purlins and/or other roof members omitted	Structural integrity of the roof compromised	Bowed and/or split timbers Deflection of overstressed roof members Dishing of roof
Roof trusses inappropriately stored or mishandled (physical damage, distortion, water saturation)	Physical damage or distortion leading to inadequate structural performance Water saturation and risk of timber rot	Bowed, twisted and/or damaged trusses Evidence of timber decay (wet or dry rot)
Inadequate bracing of roof trusses	Trussed roof likely to move or tilt sideways as a unit	Diagonal movement (racking) of roof Gable walls leaning (leading to possible failure)
No allowance for weight of water tanks and other plant installed in roof space	Overloading of roof structure	Deflection of roof structure/components Deflection and cracking of soffits and ceilings below
Too small an eaves or verge overhang	Lack of full weather protection to top of external wall	Penetrating damp affecting top of wall and wallplate Damp staining on face of wall
Half tiles instead of tile-and-a-half tiles used at verges and valley gutters	Half tiles difficult to fix securely Likelihood of loss due to wind especially on verges	Raised/missing tiles Penetrating damp
Sarking felt improperly installed	Too limp leads to water ponding Too tight leads to tearing Poor attention to detailing leads to water penetration	Damp penetration of roof
Parapet and valley gutters with incorrect detailing (e.g. extended bay lengths, insufficient step height).	Gutter bays will not cope with effects of thermal expansion Risk of capillary action at steps	Penetrating damp into roof void and ceilings below Risk of wet or dry rot
Inadequate flashings/soakers at abutments with parapet walls, chimneys	Weather integrity of the roof compromised	Penetrating damp affecting roof structure and adjoining internal elements and components

Note: No reference is made in this table to ventilation and insulation problems. These are discussed in Chapter 16.

The following are examples of poor design:

Inadequately sized members. The width or depth of the timber, concrete or steel member is too small to prevent it bending, bowing or buckling. This is a common problem of traditional timber roofs constructed before approximately 1890. Nineteenth-century roofs often contain very narrow rafters, while earlier roofs have extremely broad timbers which are often very shallow. In both instances, such roofs have a tendency to deform and this is frequently exacerbated by replacement modern tiles that are heavier than the original roof covering.

Insufficient members. The individual members are too widely spaced apart and do not fully support any applied load.

Overloading of the structure. The roof covering or contents are sometimes too heavy for the structure to support properly. This may be caused by the original design being inappropriate for the roof covering and other loads actually applied. Alternatively, the overloading may be due to subsequent maintenance or remedial works, e.g. slate or clay tiles being replaced by much heavier modern concrete slates or tiles, or the addition of water tanks and other plant within the roof void. Post-construction over-loading became such a problem that, since the 1991

Building Regulations, it has been mandatory to submit structural calculations to the relevant local authority Building Control Department where a change of roof covering is being considered.

Inappropriate roof covering. The covering can be unsuitable for a number of reasons, e.g. pitch is too shallow or too steep for the product used; tiles are too heavy for the structure. In some cases, the roof covering (tiles or slates) is 'stretched', i.e. the gauge or distance between the battens is elongated above the recommended distance to reduce the number of rows used.

Left: These tiles have been 'stretched' – the gauge is too long and the lap too little. Right: Cheap and cheerful, but not for the long term – this Victorian butterfly roof with a central valley gutter has had both slopes recovered in the recent past with 'stretched' clay double roman tiles, and a defective lead gutter was given a felt gutter lining.

Incorrect detailing. A great number of roof problems are caused by insufficient attention being paid to the design of 'minor' details, i.e. the mass of more intimate construction that needs to be carefully designed if the roof is to perform efficiently. Inadequate design is not a problem solely of the past. It still occurs in both new roofs and those that are undergoing refurbishment. It should be less likely in modern roofs because they are subject to greater design knowledge and regulatory control, e.g. the Building Regulations. However, roofs of older buildings undergoing refurbishment have been subject to much less strict controls until quite recently, e.g. the submission of a mathematically calculated structural assessment of such a roof has only been required since 1991.

Left: This roof has had extra timber fillets inserted into the original rafters in an attempt to help remove the 'bow' in the rafters. Right: This roof is late Georgian and additional rafters are being installed to strengthen the roof. The original rafters are being left in place, not for structural reasons, but in an attempt to conserve the original material.

Poor construction

The problems of poor construction are often similar to those of poor design:
*   *incorrectly sized members*
*   *incorrect fixing*
*   *insufficient members*
*   *incorrect detailing*
*   *overloading of the structure*
*   *poor workmanship/supervision*
*   *inappropriate roof covering.*

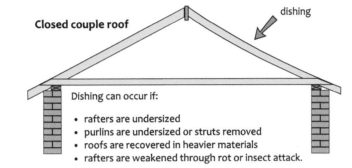

Closed couple roof

dishing

Dishing can occur if:
*   rafters are undersized
*   purlins are undersized or struts removed
*   roofs are recovered in heavier materials
*   rafters are weakened through rot or insect attack.

Roof alterations

purlin

purlin, hanger and binder removed

The owner of this roof (photo below left) removed the purlin strut, hanger and binder to provide an uncluttered roof space. Unfortunately, the ceiling started to sag - hence the hastily added hangers. Purlin deflection has also occurred and the roof has dished.

The photo on the right shows a trussed rafter roof. Two of the tension members have been removed to create space in the roof. A collar has been added to each truss in an attempt to compensate.

Roof spread

Each rafter of a pitched roof is subject to horizontal and vertical forces that result in it having a tendency to move outwards. This is overcome by the inclusion of horizontal members that tie the rafter to another element of the structure or to another rafter, e.g. a collar or ceiling joist. The closer the horizontal member is to the foot of the rafter, the more effective it is in preventing the outward movement or spread of the roof.

However, it is not always possible to install a horizontal member at the base of the rafter, e.g. where the design of the building incorporates a habitable space at roof level. In this circumstance, roof spread can develop. It can also be a problem where ceiling joists are inadequately connected to the rafters and fail to provide restraint.

Simple couple roof

Thrust along the rafter

Horizontal thrust

Vertical thrust

If the horizontal thrust is not restrained, the structure is not stable.

Ties should be well nailed or bolted to the rafters.

The lower the tie, the more effective it will be.

Collar

Ceiling joist

In 'lean-to' roofs, movement can also occur at the top.

Collar roof

Collar is too high to restrain rafter feet.

Habitable space

Roof spreads

If the horizontal thrust is not restrained, the structure is not stable.

Floor is too low to provide restraint.

Spread can occur if collar is too high or if joist is poorly fixed to rafter.

Roof spread may push wall plate outwards and may also affect the wall itself.

Two examples of poor construction. The photo shows the rafter nailed on top of the joists. The right-hand drawing shows a detail where the joist and rafter are not connected at all.

The effect of damp penetration

The roof is particularly susceptible to the effect of weather and, if there is a failure in preventing damp penetration, both the roof itself and adjoining elements of construction may be affected. Initially, it can often be difficult to precisely identify the original defect causing the dampness, as moisture has a habit of travelling out from the original point of failure over a period of time. This may occur on the surface of, and within, adjoining materials. The precise cause can frequently be confirmed only after opening up of the roof construction has taken place.

Further problems may be caused where timber or steelwork is built into the adjoining structure, especially where the walls are of solid construction. If they are of any age, it is unlikely that the timber or steel will be protected by any form of damp proofing. Where the covering structure also lacks thickness, the timber or steel truss, purlin, beam, etc. may be adversely affected by damp penetration of the wall. This situation can be exacerbated in particularly highly exposed walls.

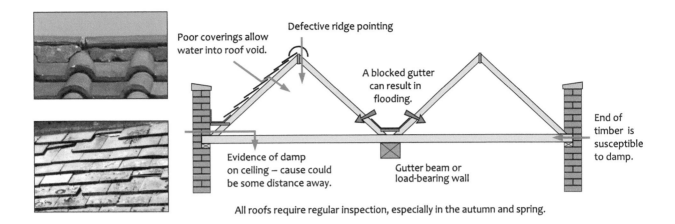

Poor coverings allow water into roof void.

Defective ridge pointing

A blocked gutter can result in flooding.

End of timber is susceptible to damp.

Evidence of damp on ceiling – cause could be some distance away.

Gutter beam or load-bearing wall

All roofs require regular inspection, especially in the autumn and spring.

Little protection from the weather is offered by this ancient stone tiled roof. Stone slates (as opposed to natural slate) are not always impervious to water. Heavy rain and cold winters can lead to frost damage.

Gutters should be appropriately sized so that they are capable of fully discharging all the water received from the adjoining roof surfaces into rainwater outlets. However, an outlet can become partially or fully blocked by debris such as autumnal leaves or a tennis ball. Occasionally, a flash storm can result in more rainwater than the original designed capacity of the gutter. The result is that the water will rise above the top of the gutter and may flow into the roof itself. Unfortunately, one of the effects of climatic change is an increase in the number, regularity and intensity of storms and this is leading to increasing problems of existing gutters being unable to cope with roof rainwater run-off.

Snow can also be a problem. Wind-driven snow can find its way through the smallest opening, e.g. beneath tiles and slates, and can be a serious problem with unfelted roofs.

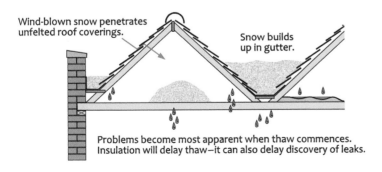

Wind-blown snow penetrates unfelted roof coverings.

Snow builds up in gutter.

Problems become most apparent when thaw commences. Insulation will delay thaw—it can also delay discovery of leaks.

Inadequate pitch

The pitch of a roof is dictated by the type of roof covering. Recommended pitches, particularly for tiles, vary greatly and the manufacturer's recommendations should always be followed.

Modern inter-locking slates and tiles can be laid at much lower pitches than their traditional predecessors. However, if either type of covering is laid at too low a pitch, rain and snow may penetrate the roof covering, especially if the roof is in an exposed situation. Even with reclaimed slates and tiles that pre-date the issue of any manufacturer's advice, sensible adherence to traditional knowledge of appropriate minimum pitches is necessary. Laying the covering on steeper pitches (up to the vertical) is not normally a problem, though all tiles need to be nailed, as sole reliance on any nibs that are provided to give adequate fixings is insufficient.

Laying roof coverings on too shallow a pitch (Table 9.2) or extending the gauge, i.e. reducing the lap (see next sub-section) of tiles, slates and other coverings, are common problems. The reason for these approaches is either because of ignorance of what should be installed or, more likely, to save money as fewer tiles and slates are needed. Unfortunately, there is also reduced performance because such a roof is much more prone to water penetration and wind damage.

Table 9.2 Examples of minimum recommended pitches for a range of roof coverings

MATERIAL	TYPE/SIZE	MINIMUM RECOMMENDED PITCH
Traditional slates	Large slate	21 degrees
	Small slate	33 degrees
Fibre-cement slates (man-made)		20 degrees
Redland 'Cambrian' slates (man-made)	Single standard	15 degrees
Traditional clay tiles*	Pantiles	25 degrees
	Single-lap tile	35 degrees
	Plain tile	40 degrees
Redland 'Regent' tiles (man-made concrete)	Through-coloured tile	12.5 degrees
	Granular finished tile	22.5 degrees
Cotswold stone tiles		55 degrees
Shingles	Western red cedar	25 degrees
	Oak	45 degrees
Thatch		45 degrees

*Note: Recent advances in manufacture have led to the introduction of interlocking clay tiles which can be installed at lower recommended minimum pitches, e.g. Marley Sandringham clay pantiles (22°) and Sandtoft New Generation 20/20 clay plain tiles (15°).

The effect of inadequate pitch

If the pitch is too low for the covering, wind-blown rain and snow will more easily penetrate the slate or tile covering.

Pitch of rafter (degrees)

Some roofs have sprocket ends fixed to the rafters for aesthetic reasons. These can further reduce effective pitch and lead to water penetration.

Note: effective pitch is always less than that of the rafters.

The rafter pitch on this roof is about 12 degrees – the minimum pitch for these artificial slates should be about 20 degrees.

Gauge and lap problems

The gauge or distance between the centreline of the battens determines the headlap (lap) or cover that a tile or slate achieves when laid above another tile or slate. The amount of headlap is important – too little and it will not prevent weather penetration. Unfortunately, many roofs suffer from this problem, either because of ignorance by a contractor or because an extended gauge/reduced lap can lead to savings in material, time and cost.

Occasionally, exactly the opposite may occur. The gauge may be reduced. This will increase the lap and probably give greater weather resistance, but will also result in an excessive load on the underlying structure.

Gauge is the distance between batten centres. Margin is the part of each slate/tile exposed to view (gauge = margin).

Headlap (lap) is the distance the tail of a tile overlaps the head of the course below (next-but-one course below for double lap slates and tiles). The steeper the pitch, the less the lap needs to be to prevent water penetration.

Left: The margin on these double romans is too large. The subsequent reduction in lap may result in water penetration and increased risk of wind lift. In addition, the lead gutter should be stepped (and laid in three or four sheets) to avoid problems of ridging and splitting due to thermal movement. Right: The lead gutter has been laid in separate sheets to accommodate thermal movement, but the tiles have been 'stretched' to reduce the number of rows required.

Failure of roof coverings

Roof coverings for pitched roofs fall into two distinct types:
1   *natural materials, e.g. clay tiles, stone tiles, natural slates, timber shingles;*
2   *man-made materials, e.g. concrete tiles, fibre slates, metal or fibre sheet.*
Below are some of the problems associated with specific roof coverings.

Natural slate

Slate-covered roofs can be found throughout the British Isles where they were extremely common in the nineteenth century. They can range in colour from red/black to green/blue, depending upon the source which, historically, was within the UK – Cornwall, Cumbria, Devon, parts of Scotland and Wales. However, the popularity of slate roofing declined in the first three quarters of the twentieth century, although there was a revival in the last quarter. This led to the re-opening of some formerly closed quarries to meet the increased demand. In recent years, foreign slates, in particular from Brazil, Canada, China and Spain, have also become commonly used because of their relative cheapness and greater availability. Some of these have been of poor quality and have caused problems, but many have performed acceptably.

Natural slates come in a number of sizes, with the larger being used on shallower pitches or at the bottom of steeper roofs. They are produced in a range of thicknesses; the larger the slate, the thicker it is. Fixing is by means of head or centre nailing, the latter on larger slates or shallower pitches.

Left: A number of slate-roofed houses. Right: This roof shows the result when the fixing fails. One defective slate has already been supported by a tingle – an s-shaped length of lead or copper.

Life expectancy generally is up to 100 years, though it can be considerably longer. It depends upon source, thickness, cutting technique, etc.

The following are some of the particular problems of natural slate.

Reliability. Natural slates are of varying degrees of durability and strength depending upon the following factors:

- *source – country, quarry, bedrock*
- *method of manufacture – roofing slates are produced by hand-splitting large blocks of slate (riven slates) and then each slate is dressed, either by hand or, increasingly, by machine*
- *installation – the grain should run parallel to the length.*

Problems can be caused when slates contain impurities that accelerate decay, e.g. calcite, pyrite, or expanding clay materials.

Delamination. Slate is a laminated material comprising thin layers of rock bonded together naturally. Poorer-quality slate may contain carbonates or clay mineral impurities. With the former, a chemical reaction with the acids in rainwater creates calcium sulfate, the crystals of which force the thin laminations apart. In the case of the latter, water gradually penetrates between the layers and subsequent frost action can force these apart. Deterioration of this type more readily affects the underneath of the slate, where moisture has been drawn by capillary action. Each

Examples of how slate deteriorates can be seen on this slate-hung elevation – the problems include nail sickness, delamination, surface erosion, broken and slipped slates.

shedded layer weakens the slate and reduces durability. It can be identified by cracking between the layers appearing around the edge of the slate, and is more likely to occur in slates of inferior quality.

Problems with torching and bedding. Torching is a traditional process that consisted of filling the gaps between slates and battens with a mortar pointing. This was commonly a mixture of lime mortar and animal hair. It was used to prevent rain or snow penetration into the roof space of inferior-quality work where there was neither boarding nor felt. It lost favour between the two World Wars when it was realised that it was better to allow ventilation to the underside of the roof slates (or tiles) and to the battens.

Torching was usually a mix of lime mortar which helped reduce the incidence of draughts and the ingress of snow and dust. However, it also reduced ventilation, resulting in premature failure of the battens.

Bedding (or shouldering) involved sealing between the tail of one slate and the slate above (often where loose) with mortar to provide both a weather-tight seal and a bond between the adjoining slates. Both torching and bedding are still found to be in occasional use as cheap 'repairs'.

The mortar in both cases tends to shrink and fall out, leading to water penetration and loose slates. Both techniques can lead to inadequate ventilation of the underside of the slate and encourage speedier deterioration of the roof covering because of moisture problems.

The batten adjoining any torching is also more likely to be adversely affected by moisture.

Split slate. Slate is a naturally brittle material and this means that it can have a tendency to split down its length (along the 'grain'). This is often along the line of a nail hole. Each slate should be checked by tapping before installation, but the defect may only appear after fixing.

Surface erosion or weathering. Slates often suffer erosion of the surfaces that are covered by the overlap of each slate. This reduces the longevity of the slate. It is caused by surface water being drawn up between the slates by capillary action. Sulfur dioxide in the atmosphere mixing with rainwater creates sulfuric acid that affects the surface of natural slate and causes disintegration. Initially, this may appear as powdering, flaking or blistering of the surface. It is more likely to occur with poorer-quality slates that contain carbonates in varying quantities.

Weathering occurs on top of the slate and underneath (caused by capillary action).

Splitting can occur along the length (grain) of a slate. This is often induced by the nail holes.

Tingle

Delamination occurs when the individual layers of slate separate (lose their natural bond).

Slipped or missing slates result in incomplete integrity of the roof covering.

Temporary repairs made by the use of lead or copper tingles.

Cracked slate

Lead or copper tingle re-secures slate to batten.

Tingle

Bedding of mortar applied between the slates to bond them together and seal the joints.

Torching of mortar applied to the underside of the slates to seal any gaps from water penetration.

Weathering may also affect the exposed upper surface of the slate for the same reason. Erosion may also be due to impurities in the slate, e.g. iron pyrites, in which case staining may be apparent.

Slipped slates. Slates may partially slip or become completely detached from their original position. The cause is normally 'nail sickness' (see later roof sub-section). Sometimes an attempt is made to re-secure such slates by the provision of lead or copper clips (often known as tingles), but this approach is unsightly.

This roof has a nail sickness problem. The slipped slate clearly shows its centre-fixing holes whilst, to its right, the (bent) slate tingle indicates the problem has occurred before. Delamination of the slate to the left can also be seen.

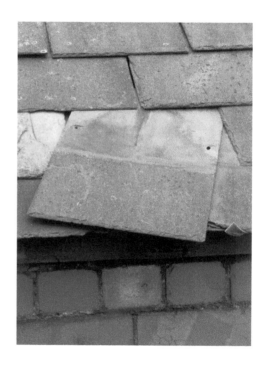

Broken slates. Broken or cracked slates may be due to a number of reasons, e.g. foot traffic, inherent defect. However, head-nailed slates are particularly prone to cracking as they are more susceptible to the effect of wind-lift.

Water penetration (or leakage). Well-laid slates lie close to each other. This encourages water seepage due to capillary action.

On the other hand, wide joints between the slates will often allow water penetration because the side lap is too small for the length. This problem is reduced as the pitch of the roof increases and water is shed more quickly. On flatter slopes, water tends to spread at a wider angle between the slates due to capillary action and may reach the nail holes of the course beneath (causing nail sickness) or even penetrating to the battens, felt, etc.

Verge failure. This is often associated with incorrect selection of slates. Slate-and-a-half slates should be used rather than half slates, as the latter are extremely difficult to secure properly against wind uplift, which is often greatest at the verge of a roof.

Further problems can be posed where no over-sail (i.e. overhang) is formed, so that water cannot drip away from the face of the underlying wall, or the under-cloak is poorly formed, allowing moisture into the structure.

The tiling or slating at the verge should be tilted-up to ensure water drains back over the roof and not down the wall.

Battens are too close to verge edge – rot is likely.

All tiles/slates along verge need to be nailed or mechanically fixed.

Interlocking 'slate'

Slate-and-a-half-slate (not half-slate unless at least 150mm wide) to avoid wind lift problems. A similar detail is needed for plain tiled roofs (right).

Verge pointing formed with a strong cement mortar is more likely to suffer from shrinkage cracking. At worst, the mortar may loosen and fall out.

Undercloak normally formed with a thin but strong material such as slate or moisture resistant fibreboard. Older roofs may have sheet asbestos undercloaks.

Verge should oversail wall by a minimum of 38–50mm to avoid shedding water down wall face.

Verge not tilted-up – water can discharge over wall.

Ridges and hips. Forming a ridge or a hip in slate is extremely difficult. Ridges are rarely so formed, but hips are occasionally constructed with mitred (cut) slate-and-a-half slates fixed over lead soakers. Securely fixing such a detail is difficult and failure is common, especially if the pitch is too low (less than 40 degrees). Ridges and hips are better formed with clay components or leadwork.

Left: Poor detailing at parapets will inevitably lead to rain penetration. Middle: This slate roof is nearing the end of its life. There are also half slates used at the verge and damage to the verge pointing. Delamination can be seen on many of the slates. Right: The clay ridge tile is bedded in cement mortar which is breaking down, leading to slipped and broken slates and necessary replacements.

Bitumastic oversealing. A slate roof that is approaching the end of its life is sometimes given an extended life by the application of a proprietary over-sealing coat. The process normally involves re-fixing/replacing any loose/missing slates, applying a coat of hot bitumen to the whole of the roof followed by a layer of reinforcing fabric and finally a further top coat of more hot bitumen. This process will give some further years of life to an old slate roof. It is, however, limited in durability/longevity and will result in the eventual replacement of the complete roof covering including any good slates, as it will be impossible to distinguish these from those of poor quality.

Left: This slate roof has been over-sealed. This type of repair is not uncommon, but it has a limited life and impedes ventilation. Right: The over-sealing is of some age and has weathered in places revealing the slates beneath. Full renewal of the roof must now be considered.

Stone slates

Roofs covered with stone slates are confined to those areas of the country where there is an easily accessible supply of either limestone (as in the Cotswolds) or sandstone (as in parts of Yorkshire).

They are sometimes referred to as tilestones. It is not unusual for such roofs, when properly installed, to last for 100 years without major overhaul. Even then, the slates are often perfectly capable of being re-used for another 100 years.

Limestone slates are normally riven and laid at a pitch of around 50–55 degrees, while those of sandstone may be riven or sawn, and are laid at pitches of between 24 and 50 degrees. Both types of bedrock are porous, and this means that the slates need to be of considerable thickness (12–25mm) in order to allow evaporation of absorbed moisture and to avoid water penetration into the roof void. This results in an extremely heavy roof covering.

Below are some particular problems relating to stone slates.

Defective mortar fillets. Traditionally, the junction between stone slates and the adjoining walls and chimneys was covered with an angled mortar fillet rather than a flashing. This would have been of lime mortar, but in recent years cement mortar is often used instead. Lime mortar is softer and will slowly weather, leading, eventually, to disintegration. Cement mortar has a tendency to shrink and crack, often within a very short space of time, and this can lead to penetrating damp problems.

Slipped slates. Stone slates were originally hung on oak or deal pegs inserted through holes near the head of each slate. Sometimes brass screws were used and more recently galvanised or stainless steel nails. Once the fixings to a slate fail, slippage will occur.

Left: The lichen on this lean-to roof will impede ventilation. Middle and right: Mortar fillets at the junction with a vertical abutment. Both have been protected – in the middle by a splayed weathering and in the right-hand photo by the addition of slate drips. These traditional techniques are very effective.

Delamination. The thin layers of stone making up the slate are often prone to delamination in much the same way as natural slates.

Left: These stone slates are clearly delaminating. Right: The condition of this old stone roof that is heavily covered in lichen and ferns can only be ascertained from the bottom slates, where delamination is taking place.

Man-made slates

These are a cheaper alternative to natural slates and have been manufactured since the Edwardian period, when asbestos cement slates were first introduced. These consist of asbestos fibre woven into a fabric which is then encased in Portland cement mortar. The slates are very light and extremely brittle. There are two types: 'tiles' which are square and laid diagonally; and 'slates' which are rectangular and laid as ordinary slates. For much of the twentieth century these were the only artificial slates manufactured, but since the late 1970s asbestos-free mineral-fibre slates have replaced them. This was, in part, a response to the health-related problems associated with asbestos-based products and the advice to discontinue their use.

Left: Man-made 'tiles' may still occasionally be found on the roofs of houses, more especially bungalows, of the early/mid-twentieth century. Notice the moss and lichen. Middle: Some of the original asbestos 'tiles' have been replaced with modern fibre 'tiles' secured with lead tingles. Right: Artificial 'slates' from a more modern house. They are showing signs of bowing and curling.

Artificial slates are often nail fixed or riveted, though many manufacturers of mineral-fibre reinforced slates recommend the use of stainless steel hooks as these slates are thick and lightweight in form and tend to be more prone to wind-lift problems. Such slates come in a range of colours and shades and many of the earlier slates do have an artificial appearance. Many of the more recent types have a more slate-like appearance. In terms of anticipated life, manufacturers' guarantees are generally for 15–30 years, but 50–60 years can normally be expected.

Below are some particular problems of man-made slates.

Dimensional stability. Man-made slates often have a tendency to bow or curl after installation. A slight defect of this nature may merely cause aesthetic concern over the appearance of the roof, but bowing or curling of greater magnitude can lead to problems of water leakage or loosening of the fixing nails/hooks.

Discoloration The slates are artificially coloured and in some cases they may become discoloured. This may affect the edges of the slate and/or the remaining exposed surface. The cause can be the effect of sunlight, which leads to fading, or it may be for more complex reasons, such as water penetration of the material over long periods. The problem may eventually lead to performance difficulties and necessitate replacement.

Cracking or splitting. The thinness of man-made slates, together with their relative lack of strength, means that they are prone to cracking. This can sometimes occur along the line of the fixing holes and can be induced by clumsy installation. This type of slate is also prone to impact damage as it ages because of the material hardening.

Bowing or curling of the slate
sometimes occurs after installation.

Discoloration can occur around the edges of the slate or over its surface.

Splits can occur along lines of weakness – often induced by fixing holes.

Most new slates are centre nailed; riveted tails prevent wind up-lift.

Verge problems. The verge detail on a roof formed of man-made slates can be constructed in a similar fashion to that described earlier for natural slates. The problems experienced are generally similar.

The 1980s saw the introduction of dry verges. These are dry-fixed components formed in man-made slate which wrap over the edge of the verge to form a water-tight detail. They avoid the problems associated with traditionally formed verges, e.g. shrinkage of the mortar bedding, inadequate over-sail, but can experience difficulties if the mechanical fixing, normally a nail or rivet, fails.

Future performance unknown. More recently, slates formed with a mixture of crushed slate, resin and reinforcing material, which more exactly replicate the appearance of a real slate, have been produced. One major manufacturer indicates a probable life expectancy of 60 years for its slate of this type. The manufacturer's advice for a new product is given in all good faith, but is normally based on laboratory and other controlled tests plus experience accumulated from the use of other similar products. The conditions of actual use are outside of its control (and liability). This means that the future performance of any new product is, to some degree, unpredictable.

The potential health hazards of asbestos slates. Asbestos-cement products present no health problems as long as they are stable and not 'shedding' fibres. The microscopic fibres can become a hazard to health if ingested as they can lead to a form of lung cancer and other lung problems. Broken or cracked asbestos slates/tiles can present a potential health hazard, but a more serious health problem is likely when the surface of the tile becomes roughened and allows fibres to be released into the air.

A roof containing asbestos slates should be monitored very carefully to assess whether it does present a health hazard. There is probably no danger to health where the components are stable and the roof may be retained, even if some of the slates are already cracked or broken. However, total removal and replacement of the complete roof area, rather than partial repair/ replacement, are recommended if there is evidence of active deterioration occurring.

Note: The removal of any asbestos product is now required to be carried out under a very strictly regulated process (under the Control of Asbestos Regulations 2012) by a licensed contractor.

Clay tiles

Clay tiles have been used extensively in England and Wales for several hundred years. Originally introduced by the Romans, they fell out of use until the thirteenth century. Since then they have never been out of production. In recent years, increasing numbers of foreign-made clay tiles are being used.

They have been manufactured in a great number of shapes and sizes ranging from the small and simple, e.g. plain tiles, to the large and more complex, e.g. double roman. Tiles such as the former merely butt up to the sides of each other, but are heavily overlapped vertically, in a similar manner to slates, to form a weather-resistant covering. Other types of tile, like the pantile and double roman, are more intricately shaped to overlap each other at each side as well as the head. The side lap increases weather resistance and this was further enhanced by the introduction of interlocking side details in the second quarter of the twentieth century. Shaped tiles for ridges and hips are produced for use on any roof, while shaped valley tiles are also manufactured for use with specific tile profiles, e.g. plain tiles.

The tiles are either hand-moulded or machine-pressed from better-quality brick clays. A good quality clay tile should be capable of lasting a hundred years or more, though a modern

Plain tiles (double lap) – showing individual tiles side-butted up to each other. A tiny gap, 2-3mm, allows for thermal expansion.

This tile is lapped by B and by A - hence the name 'double lap'.

nibs

Slight curve to limit capillary action.

These are known as single lap tiles. If the lap is less than 75mm, rain may be blown under tile.

Most modern tiles interlock rather than lap.

Double (two rolls in each tile) Roman tiles showing overlapping of the tile at the side.

Tiles can be flat or profiled.

manufacturer's guarantee may be as short as 30 years. In the second half of the nineteenth century and the first half of the twentieth, a great number of low- and medium-cost houses were roofed over with double roman or other large-profile clay tiles, whilst many better-quality houses were finished with more expensive plain tiles. In the 1940s and 1950s, clay tiles were, to a great extent, superseded by concrete tiles, and, though the former have undergone a renaissance since the early 1980s, they probably only account for about 20 per cent of all current new roof installations.

Some of the particular problems with clay tiles are set out as follows:

Delamination. Clay tiles are not watertight. They absorb moisture, but work on the principle that it evaporates back out through the upper surface before the moisture can penetrate completely through the tile. This means that the tiles are susceptible to the effect of frost, especially if they are of poorer quality. This increases the possibility of a laminated structure. Frozen moisture trapped in air pockets will expand and fracture the tile or its surfaces, particularly the uppermost. Delamination of the top surface is a common problem with older clay tiles and careful attention should be paid to these if they are being considered for re-use.

Pitted, cracked or broken tiles. Defects such as pitting, cracks or breakage may occur for a number of reasons. Pitting of the top surface may be due to delamination (as described in the previous section) or to weak spots in the clay resulting in variable weathering over a prolonged period. Cracked or broken tiles may be as a result of weakness in the tile, often due to its age, or external activity, e.g. hit by a heavy object or walked over.

Left: A double roman tiled roof. One of the tiles is showing signs of premature deterioration and the ridge pointing is disintegrating. Right: Damaged single roman tiles. Most of the defects are caused by frost action, but the broken tile could also be the result of clumsy maintenance work.

Cambered/non-cambered tiles. Machine-pressed tiles are generally 'flat', i.e. non-cambered, across their width. This means that they can lie close to each other and this increases the likelihood of capillary action. Other tiles, including certain plain tiles, have a slight cross-camber, i.e. they have a concave profile across their width, with the result that driving rain or snow is more liable to penetrate a roof laid with such tiles.

Slipped tiles. Most tiles are formed with nibs along their upper edge which enable them to be hung onto the battens. The tiles need to be nailed to the battens at all verges and where adjoining ridges and hips. All tiles should be nailed when the roof is in an exposed position or has a steep pitch. Failure to carry out the additional nailing may result in tiles being lifted and moved by the effect of the wind.

A slipped tile may also be caused by the failure of the nibs or, with some types of plain tile which have no integral nibs, by failure of the fixing nail (see later section on 'Other influential factors on the roof covering' which discusses 'nail sickness').

Typical problems of plain tiled roofs – slipped, cracked or broken tiles plus evidence of torching and bedding. Note the cross-camber of the industrial tiles (right) and the verge poorly formed with half tiles (left).

Torching. The torching of tiles, in a similar fashion to that described for natural slating (see p. 251), is sometimes discovered. The reasons for torching and its problems are similar for both tiles and slates.

A plain tiled roof in a 1930s house. Torching has blocked the ventilation of the roof space and can also lead to rot in the battens. Note the insulation at ceiling level. Interstitial condensation due to lack of ventilation is a real danger in these conditions.

Efflorescence. See next section on concrete tiles.

Verge failure. In traditional verges formed with lime or cement mortar infilling, the construction detailing is similar in principle to that of a slate roof except that tile-and-a-half tiles will be used and a tiled under-cloak is common. The causes of failure may be due to problems similar to those referred to earlier in the section on natural slates.

In recent years, dry verge systems have been designed and manufactured for use on clay tile roofs and the comments made in the earlier section dealing with man-made slates apply equally here.

The mortar of the verge to this plain tiled roof has disintegrated and is exposing the ends of the roof battens. It was found to be of too weak a mix.

Ridges and hips. Ridges and hips are commonly formed with clay components, although they may be constructed with other materials. Clay ridges are generally of a half-round or inverted 'V' shape and many older ridge tiles also had fancy detailing along the actual ridgeline. Hips are often covered with ridge tiles formed in both the previous shapes as well as a variety of purpose-made shapes, e.g. bonnet hip tiles, which are more like an inverted dish.

Traditionally, ridge tiles are installed by bedding each tile in black or lime mortar, although in recent years cement mortar has often been used. The former types of mortar are inherently weaker and more flexible than the latter, resulting in greater tolerance of differential movement. However, they lose their ability to bond to the ridge tile and the underlying structure as they age, leading to loose or missing sections of ridge. On the other hand, cement-based mortar is more prone to shrinkage cracking and this may allow moisture penetration of the ridge line. It also bonds to the ridges and adjoining tiles. As a result, they cannot have any attached mortar cleaned off.

Top: These ridge tiles have been bedded in cement mortar, which has shrunk and cracked. Bottom: These ridge tiles have been bedded in black mortar which is breaking down. It will be more difficult to re-use the ridge tiles bedded in cement mortar because this type of mortar is hard to clean off clay products.

Clay hip tiles bedded traditionally in mortar suffer from problems similar to those affecting ridge tiles. It is also normal practice to install a galvanised metal hip iron or hook, screwed or nailed to the bottom of the hip board. This provides extra security in case the mortar bedding fails. It may work loose or be missing or might never have been originally installed.

Dry ridge and hip systems are also now manufactured in clay, for use with any type of roof where it is possible to install them. Correct installation in accordance with the manufacturer's recommendations is essential. This may be difficult on some hips, where there can be difficulties in marrying up the hip components and the tile profile, due to the angle or pitch of the actual hip not meeting the manufacturer's installation criteria.

Left: The black mortar bedding of these clay hip tiles is starting to break down. In time, the hip tiles will become loose. Right: These modern concrete bonnet hip tiles have been installed with no mortar bedding and no hip iron at the base of the tiles. Their performance is based solely on a nailed fixing which does not always work due to their exposed position on a roof.

Dry systems usually rely upon the integrity of their neoprene components to be watertight. Penetration may occur where these are damaged during installation or are missing, especially the sealing washers to nails/screws. There have also been questions raised about the life expectancy of the neoprene.

Valleys. Many tile-covered roofs have valleys formed by installing a metal lining of lead or zinc, which is laid over a timber or plywood board. This involves cutting the adjoining tiles and bedding them in mortar above each side of the valley lining. Often, these details are poorly formed, with clumsy cutting and crude pointing. The metal should be laid over a small tilting fillet at its junction with the tiles, and also have a welted edge. These are often missing and, if the mortar to the tiling fails, there is an increased likelihood of water penetration of the joint. (See Chapter 10 for discussion of defects in lead and zinc roofs and flashings, etc.)

Purpose-made clay valley tiles are available for certain tile profiles (e.g. plain tiles). These are specially moulded for a range of pitches and plan shapes. Such a valley presents few problems, if correctly installed. Each tile should be nailed and the roof felt properly dressed around the valley. Problems may arise if the battens are cut short and do not fully support the tiling.

Concrete tiles and slates

These were first installed on roofs before the Second World War and have subsequently superseded clay tiles to such an extent that, for the majority of the period since 1945, they have captured approximately 80 per cent of the market for tiles. They are machine-pressed, manufactured in a wide range of traditional and modern profiles and colours, and can be sand-finished, smooth-finished, or through-coloured. The majority are formed with interlocking edges, though some, such as plain tiles, are not.

Concrete tiles are normally much heavier than comparable clay tiles, although there are specially manufactured lightweight versions of some profiles. Life expectancy for earlier concrete tiles is probably 50–75 years, but for more recent tiles it may be longer.

Below are some of the problems associated with concrete tiles:

Overloading of the roof. The extra weight of a concrete tile means that due consideration must be given to the dead load imposed on the roof structure. This is a common problem on many older houses in which the original roof covering had reached the end of its life and was then replaced by concrete tiles for reasons of cost or ignorance.

Overloading of a roof inevitably results in bowing and buckling of the roof structure, giving it a distorted appearance. In time, individual timbers can crack, split or break under the stresses created. Total roof failure is unlikely, though it may occur in extreme circumstances. The tiles on an overloaded roof will be adversely affected when the movement induced in the structure results in cocking and lifting of individual tiles and failure of the interlocking detail, all because of the unevenness caused in the roof. This can lead to leakage of the roof covering.

This roof is covered with concrete tiles and is obviously over-loaded. Possibly, it was already over-stressed before these replacement tiles were installed.

Overloading of roofs by replacing traditional slates and clay tiles with concrete products became such a problem in the last quarter of the twentieth century that Building Regulations were introduced that now require all roof recovering work to have structural calculations submitted to the relevant local authority.

Loss of surface/colour. Originally, the finish of concrete tiles, whether smooth or of sand/granules, was only surface deep. Once it is weathered away or it becomes chipped/damaged, the concrete beneath is revealed. This is unsightly, but does not impair the performance of the tile. However, replacement tiles will not match the faded appearance of the original roof covering. For the last 25 years or so, completely through-coloured concrete tiles have been manufactured as well and these avoid this problem.

Efflorescence. Earlier concrete tiles sometimes contained soluble salts, due to quality control problems during manufacture. These led to efflorescence on the underside of the tile, especially around the nibs that are gradually eroded. The presence of a white powdery deposit on the surface of any ceiling and felt beneath the tiles is an indicator of this problem. Note: this problem may also occur with under-burned clay tiles.

'Delaminated' tiles. Incorrect manufacture can sometimes result in a tile that is not resistant to the effects of moisture and frost. This results in the gradual disintegration of the tile as the concrete softens and crumbles due to the action of the weather. The problem will normally affect the manufacture of a complete batch of tiles so, once detected, the complete roof should be examined.

Cracked and broken tiles. These may be due to an inherent weakness in the tile due to poor manufacture or to a defect caused by poor installation. They may also result from external factors such as impact damage from foot traffic or balls.

Slipped tiles, verges, ridges, hips and valleys. Refer to earlier section on clay tiles.

Shingles

The use of wood shingles in the British Isles dates back some two thousand years to the Romans. Originally of oak, they fell into disuse as this particular home-grown timber became scarce. Re-introduction in the 1920s owed much to their continuing use in North America and the Scandinavian countries, though now they are more often formed from imported Western Red Cedar rather than oak.

Shingles are rectangular in shape (about 400mm long x 100–325mm wide) and vary in approximate thickness from 10mm for cedar to 20mm for oak. Both types are best laid on pitches exceeding 45 degrees, though cedar shingles can perform successfully on pitches down to 30 degrees or less. The appearance changes after a few months on exposure to the weather, both oak and cedar shingles developing a silver-grey colour.

This late twentieth-century house has a shingle roof and vertically-hung shingles on the elevation.

Durability of shingles is much greater than commonly supposed. Split (or cleft) shingles (split shingles are also known as shakes) are more durable than those that have been sawn, as splitting ensures continuous fibres throughout the length of the shingle. Durability is also increased by installing the shingles with a 5–12mm side gap rather than the normal 3mm. This encourages efficient water run-off, quicker drying and allows for expansion. It also reduces the likelihood of buckling.

Perfectly sound examples of both oak and cedar roofs of 40–50 years of age can be found, though other materials, such as larch, may only last a maximum of 20 years. However, in this country, cedar shingles need to be treated with preservative if they are to be ensured a 50-year life, otherwise they have been known to fail within 10 years of installation. This practice has generally been followed since the early 1960s.

Problems experienced on shingle roofs include the following.

Splitting. Older shingles were not pre-holed for nailing, the nails being driven directly into the wood about 150–200mm up from the butt (or bottom). This meant that the shingle could

The shingles have been installed at different times – those on the left are the most recent and in good condition, the vertical shingles are older and showing signs of weathering, while those beneath the lead flashing are the oldest and starting to develop surface deterioration.

sometimes split when being installed and, if this was a hairline crack, it might not become apparent until later. Modern shingles are often pre-drilled to avoid this problem.

Loose or missing shingle. These problems may be caused by the effect of wind lift. Shingles are extremely light in weight and require very careful nailing – it is normal for them to be twice nailed. A more secure and effective fixing is provided by ring-shanked nails, which have larger heads. (See also the next sub-section on nail problems and the later section on nail sickness.)

Nail problems. Corrosive acids in both oak and cedar will attack any zinc galvanised steel nail, resulting in 'nail sickness' and loss of the shingle. This may be avoided by the use of non-ferrous nails, though there may be some staining from those containing copper or brass. (Zinc gutters, flashings and ridge cappings should also not be used for similar reasons.)

Excessive dampness. Earlier reference has been made to the need to treat shingles with preservative in order to prolong durability. Great Britain, particularly the western part, is renowned for its wet climate and any timber roofing material is particularly susceptible. The shingles are secured to battens that traditionally were fixed above an unfelted roof. The modern approach is to install a felt underlay that can restrict air circulation under the shingles and prolong drying out. This may lead to more rapid deterioration of the shingles. In some cases, air circulation and drying out are improved by the installation of counter-battens.

Surface deterioration. The exposed surfaces of the shingle will weather and gradually erode over a period of time.

Left: This shingle roof is showing signs of its age. Split, twisted and up-lifted shingles are evidence that replacement should be considered. Right: The upper surface of the shingles is weathering and delaminating.

Lichen. Lichen may be found on the exposed upper surfaces of the shingles. It grows on the moist timber and, though it does not directly attack the timber, it is unsightly and can adversely affect the efficiency of water run-off, leading to accelerated deterioration of the shingles.

Effect upon metals. The water run-off from shingles is acidic in content and this may have a corrosive effect upon any metal gutters and flashings affected. Other materials should be selected for these components, where they come into contact with such run-off.

Thatch

In the British Isles, there is evidence of the use of thatched roofing for some 2500 years. For a period, during the Roman occupation, clay tiles were also used, but thatch remained the roofing material until after the Norman Conquest. Large-scale fires in London and other cities in the twelfth and thirteenth centuries, which resulted in thatched roofs being banned in many cities and towns, were followed by a gradual loss of popularity of thatch generally in urban areas over the next 300 years. Many thatched roofs in rural areas were replaced with other materials in the nineteenth and early twentieth centuries, but remained popular in some parts of the country, e.g. Devon and East Anglia. Thatched roofing has recently undergone a revival in rural areas and there are still almost 100,000 thatched roofs in the UK.

The main thatching materials are:

- **Water reed** – *generally used in East Anglia. This is the most durable thatching material, giving a useful life of 50–80 years for locally grown water reed. Imported water reed has about the same longevity as wheat straw.*
- **Wheat straw**, *which is prepared as either:*
    - *long straw with a 15–25 years life. Its use declined in the 1960s and 1970s, but it is increasing in popularity.*
    - *combed wheat reed with a 30–40 years life, commonly used in the West Country.*

Useful life varies because of regional/localised climatic and construction differences as well as any variable quality of both the materials and the installation. Natural deterioration is inevitable, however well constructed the thatch roof, due to the presence of moisture retained within the thatch. This can be delayed by increasing the pitch (it should never be less than 45 degrees) and encouraging more efficient water run-off. Roofs in drier or more exposed areas of the country will also tend to last longer. Life expectancy figures drop noticeably the further west the thatched property is situated and/or the lower the pitch.

The thickness and quality of workmanship in laying the thatch also play an important part in the performance of such a roof. A properly constructed roof will eventually need replacement or repair because of natural deterioration, but it will frequently be found that the straw base coat of the thatch is still in perfect condition and can be retained. In southern England, some thatched buildings have base coats that are 500 or more years old.

Left: The steep pitch can be clearly seen in this. Right: This roof unusually incorporates a lead gutter.

Some of the problems experienced by thatched roofs are listed below.

Atmospheric conditions. The life expectancy of a thatched roof is determined by the rate at which the material breaks down. This is due to micro-biological activity which tends to be more rapid where the climate is wetter and more humid, e.g. the West Country. The roof of a building in a sheltered position, or under trees, or a roof with a shallow pitch will also suffer from accelerated deterioration. This is because the lack of air movement (wind), shade and/or inability to shed moisture before it penetrates the thatch will increase the rate of decay.

Fungal attack. Degradation of a thatched roof commences as soon as it has been installed due to the effect of fungi that are produced within the materials used. Increased fungal growth can be expected as the roof ages and also the further west the building is situated, as warmer, wetter and more humid climatic conditions encourage activity.

Gullies. These are valleys running down the slope of the roof and are caused by water run-off wearing the surface of the thatch unevenly. The extent of the wear can often be determined by examining the depth of the valley at the eaves. If dealt with early enough, they can be patch repaired to enable the roof to reach its anticipated life-span.

Left: Gullies in the thatched roof formed by water run-off. Right: In this roof, vegetable growth can be seen at the junction with the chimney. Note the lack of any rainwater gutter on both properties. This is quite normal. The thatch should oversail the external elevation by a considerable distance to avoid damp penetration.

Moss, lichen and algae. Moss, lichen and algae can grow on the surface of the roof, especially if it is situated in a damp and shaded position, overhung by trees. Re-growth of a previously treated outbreak can often occur in such a situation. It is debatable whether the actual growth has an adverse effect upon the thatch. However, moss and algae indicate the presence of relatively high levels of moisture (which will accelerate decay of the thatch) and all three will prevent the free run-off of water as well as affect the appearance of the roof.

Weathering of the ridge. Whatever the thatching material used, the ridge tends to deteriorate at a fairly rapid rate. A new ridge of straw or sedge needs to be added every 10–15 years. At the same time, the whole roof can be cleaned, redressed (i.e. repaired) and any accumulated moss removed.

Rodent and bird damage. Rats, mice and squirrels can all have an adverse effect upon thatched roofs. Birds, particularly in spring when they are seeking nesting materials, can cause considerable damage. Generally, the result does not immediately encourage water penetration, but can lead to a weakening of the roof covering and increase the rate of deterioration. The presence of protective wire netting can initially help deter pest invasion, but, if they get beneath it, the result may be even greater damage as they attempt to escape.

The cause of the damage can normally be identified by examining the straw or reed on the ground below the affected area. Short, chewed lengths are left by rodents, while birds tend to pull out whole lengths.

This thatched roof has been netted to prevent pest invasion.

Fire This is an obvious risk with a thatched roof, but there is no evidence of it occurring any more frequently than with other roof coverings. Great care needs to be paid to the condition of internal electrical wiring in the roof void and any accumulation of stored materials or roof debris in the void should be avoided.

Traditionally, lime-wash or lime and clay slurry was applied as a fire retardant (note: a thatched roof can be fire-retarded, but not fire-proofed). Modern thatch is chemically treated, although this may reduce the longevity of the thatch, i.e. the process can affect durability, and may be washed out in a relatively short time (10–15 years). All thatched roofs should have smoke detectors in the roof space.

Chimney stacks and central heating flues need careful construction and/or installation to ensure that heat within the flue does not affect the thatch. Regular maintenance of the flue is also essential, e.g. sweeping. Dirty flues causing sparks have led to thatched roof fires (as have garden bonfires). The increased use of wood-burning stoves in the last few years has increased the risk because there are greater deposits of tar left on the inside surfaces of the flue. Spark arrestors are sometimes fitted to the top of a flue or chimney, but they need regular cleaning to avoid clogging, which affects performance and also increases fire potential.

Other influential factors on the roof covering

The overall performance of any roof covering will also depend upon factors other than the material itself. These include the following.

Nail sickness. Man-made and natural slates, clay and concrete tiles and timber shingles are normally secured by nails (though man-made slates can be secured with stainless steel hooks and tiles can be hung on their nibs alone). Ferrous nails inevitably rust in the presence of the moisture always present between the roof covering and the sarking felt. Slipped or missing slates/tiles are evidence of such a problem. Always use non-ferrous fixings for securing roof coverings and battens. (See photographs in previous sections on slates and tiles.)

Batten failure. Modern roofing battens should be of preservative-treated softwood, though untreated softwood may still be erroneously used. Untreated softwood and, occasionally, hardwood battens may be found in older roofs.

Moisture leakage through the roof covering may adversely affect the battens and their fixings in even the best-constructed and maintained roofs. Over a prolonged period of time, this may result in the battens being infected with wet rot. Any ferrous nails securing the battens to the rafters may suffer nail sickness.

The size of the batten is also important. Battens that are not wide enough will split when they are nailed. On the other hand, too thin a batten will tend to 'spring' as the nail is driven in. Since 2003, BS 5534 has increased the minimum batten size for single-lap tiling on 600 rafter centres from 25 x 38 to 25 x 50. Roofs with rafters at 450 centres are required to have 25 x 38 battens. Many earlier roofs may be found to have under-sized battens.

Battens may be of insufficient length to fully span the full width of a roof. Butt jointing all of the battens above one particular rafter must be avoided as it will create a structural weakness because the battens will not provide effective lateral bracing.

Sarking felt (underfelt). It became the practice in the twentieth century to provide a pitched roof with secondary weather protection by means of an underfelt. This prevents both rain and snow penetration (especially when wind-blown) of the roof void. The felt is secured to the top of the rafters by the roofing battens. The provision of an underfelt was a gradual process and did not become a standard practice until the 1940s/1950s.

Care needs to be taken during the installation of the underfelt to ensure that it is laid with a slight sag between the rafters. This allows any rainwater/snow penetration of the tiles or slates to drain away and not collect against the battens. On the other hand, the felt needs to be properly supported at the eaves to avoid sagging, as this would allow any run-off moisture to pond. The trapped moisture can lead to wet rot in the battens, reduce the life of the felt and, if the felt fails, penetrate the underlying structure.

The felt should be laid with a slight sag between the rafters. This allows any leakage of the slates or tiles to run away to the eaves.

This felt is tightly drawn; any moisture trapped above the batten encourages wet rot in batten, and sometimes in the rafter.

Most felts before the 1990s were bitumen-based; they can become very brittle with age and are easily torn.

Absence of a tilting fillet at the eaves leads to a sag in the felt. This may result in ponding of any moisture and water leaking into the building as the fabric deteriorates. The eaves edge piece (right) acts as a tilting fillet; it also protects the outer edge of the underlay from ultra-violet light and birds.

Prior to the twentieth century only better-quality buildings had any secondary protection and this was by means of close boarding laid over the rafters. The boarding also increased thermal insulation, but was prone to moisture attack and did not fully prevent moisture penetration. To assist moisture run-off/evaporation, counter-battens were often fixed between the boarding and the roofing battens. Alternatively, on some roofs, slates (but not tiles) were fixed directly to the boards. In cheaper housing and in the less important and ancillary roof areas of better housing, it was normal practice to make no provision of any form of secondary protection.

A number of different materials have been in use at various times. They include the following.

Hairfelt. This was manufactured out of animal hair and was in use in the early part of the twentieth century. It provides no real weather protection and becomes extremely brittle with age. Very rarely found now.

Unreinforced bitumastic felt. No longer in common use, this tears easily and becomes brittle with age. Occasionally found now.

Reinforced bitumastic felt. In common use in the second half of the twentieth century, this is considerably stronger than the unreinforced variety, although it will still tear if mishandled or hit. The bitumen content becomes brittle and increasingly susceptible to damage with age. Still occasionally found in current use, this type of roofing felt will still be found in many pitched roofs formed after 1950.

A common problem of the bitumastic type of felt is its tendency to become brittle and disintegrate at the eaves and other vulnerable positions, such as valley gutters, after a relatively short period. A short piece of plastic DPC or vapour-permeable plastic felt installed at the bottom of each slope will avoid this problem.

Left: The photograph shows three types of roofing felt in one roof. Hanging down is an older unreinforced felt; on the right is reinforced felt; top right is vapour-permeable synthetic felt. Right: The tiles of this roof have been removed to reveal an unreinforced felt which has broken down immediately above the gutter which is full of dirt and moss. The observant will note that the gauge of the tiles has been overly extended.

Vapour-permeable plastic and synthetic membranes. These appeared on the market in the 1980s and allow low levels of trapped vapour (which would otherwise condense within a roof void) to escape through micro-pores contained within the material. The pores are too small to allow water droplets (e.g. from rain/snow) through from the outside. The plastic materials are tear-proof and rot-proof so, if laid in accordance with the manufacturer's instructions, should perform properly. These membranes are now the norm in domestic pitched roof construction. They received BBA Certification for warm roof construction in 1982 and Certification for cold unventilated pitched roof construction in 1999.

Some designers and contractors erroneously believe that this type of 'breathable' felt is a full substitute for proper roof ventilation as required by the Building Regulations (e.g. eaves and ridge ventilation). However, this can lead to condensation problems. There can also be a problem of premature aging of these membranes which can be caused by over-exposure to UV light during installation (i.e. due to not being covered quickly enough – within three months) or by temperature peaks throughout a roof's lifetime. These factors may lead to brittleness or material breakdown. Slope orientation needs to be considered carefully when such problems are experienced. In normal conditions, life expectancy should be comparable with that of a traditional felt underlay.

Valley and parapet gutters

The more complex the roof design, the greater the likelihood of valley gutters to drain abutting roof slopes. Often, the design will call for the roof to be 'hidden' behind a parapet wall, creating the need for a parapet gutter to drain any adjoining roof slope to a suitable discharge point. Both types of gutter are commonly found in Victorian and earlier roofs.

Any gutter should be designed to drain the maximum anticipated water run-off from ALL surfaces within the roof catchment area. Account must be taken in the design for both normal rainfall patterns and short duration exceptional rainstorms, which in the past have happened rarely, but are now occurring much more frequently. If the design is inadequate, or the construction not properly carried out, the gutter will not be capable of efficiently receiving and carrying away the water run-off. Unfortunately, recent climatic change means that many existing gutters that were appropriately designed at the time of installation for the worst anticipated circumstances likely will no longer fully cope with the increased intensity of current and future rainfall.

A variety of cover materials are used in the construction of valley and parapet gutters. The problems associated with each type of material are commented upon as follows:
- *shaped clay/concrete tiles: see the appropriate material in this chapter.*
- *sheet materials, e.g. lead, zinc, mineral felt: see the appropriate material in Chapter 10.*

Flashings and soakers

These are discussed in Chapter 10.

Wind-related damage

The annual cost of wind-related damage to buildings in the UK runs into many millions of pounds. A high proportion of this damage is to roofs. Wind may adversely affect both older and newer roofs; the former because both design and construction almost certainly made insufficient allowance for its effect. Increasing knowledge of the effects of the wind has led to better-designed roofs (in particular, coverings and fixings) and to more carefully controlled construction. However, future climatic change may result in existing roofs suffering greater failure as wind speeds are tending to increase and weather patterns change.

The effect of the wind is influenced by a number of general factors:
- **Exposure.** *The exposure of the building to the prevailing wind, which in the UK tends to be from the south-west.*
- **Height.** *The height of the building above sea level. Buildings on higher ground tend to experience higher wind speeds.*

- **Wind speed.** *In general, this varies, depending upon the region, but tends to be greater in coastal areas than in inland districts, e.g. the south coast experiences winds that are up to 25 per cent greater than those at a similar altitude in the South Midlands. It also increases the further north the building is situated within the UK; the strongest winds are experienced in north-west Scotland.*
- **Gusting.** *Air movement is subject to turbulence that causes rapid changes in its speed and direction. Such increases/decreases in wind speed are infinitely variable. However, the effect of wind gusts increases as the size of the element or component it crosses decreases. Thus, a small roof is more affected than a larger roof and, similarly, a small component, such as a tile or slate, suffers more from gusting than a roof covered with sheeting material. The type and quality of fixings used (or not) need to be assessed when investigating a roof adversely affected by the wind, as they play an important role in how the covering performs.*
- **Local topography and adjoining buildings.** *Valleys, hills, ridges, escarpments and ridges all have an influence on air movement and may directly influence wind behaviour. Similarly, the size, shape or height of an adjoining building or group of buildings may affect wind performance and patterns.*

Wind speed increases as it passes over a roof. Wind also creates both positive and negative pressures (suction) as it moves over a building. Roof pitches of 30 degrees or lower are always subject to suction forces. However, though leeward slopes (i.e. those on the opposite side to the wind) are always subject to suction, which increases as the roof pitch decreases, windward slopes are subjected to positive pressure once the pitch exceeds 35 degrees.

Normally, suction is a greater problem to a roof than positive pressure. The latter may result in great force being applied to the roof surface, but, as long as the roof structure is performing adequately, wind usually only becomes a danger when it is able to get underneath the actual roof covering and lift it. More common is the damage caused by the effect of suction, which may lift, or completely remove, the tiles or slates on both the windward and leeward slopes of a pitched roof, particularly those situated along ridges, eaves and verges, where higher suction forces are experienced than on the more central areas of a roof slope. It is for this reason that it is normal practice to mechanically fix roof tiles on these 'exposed' areas of a roof. Similarly, all tiles on the entire roof in an extremely exposed location should be mechanically fixed. It is the failure to follow these practices (which are part of the manufacturer's recommendations with new tiles) that often leads to tile displacement. All slates should, in any case, be mechanically fixed, but even the best installed slate or tiled roof may suffer wind damage in extreme conditions.

A storm in October 2002 with winds of up to 80 mph caused this damage to houses in suburban Bristol. Note: apart from one broken tile in the centre of the left-hand roof, the damage is around the perimeter of both roofs, i.e. classic storm damage.

Damage from aircraft wake vortices

This phenomenon affects the roofs of properties situated beneath, or close to, the flight path of a busy airport by dislodging the tiles or slates. All aircraft trail powerful air currents, or 'vortices', behind them and, in general, these are at their strongest when the aircraft is heavy and flying slowly – exactly the conditions when a modern passenger plane is taking off or landing. The likelihood of damage is greater when an aircraft is landing, as it is closer to the ground for a longer period than when taking off. Damage occurs immediately beneath the aircraft, and also to each side, as the vortices tend to drift downwards and then sideways, just above ground level.

The highest risk property is that situated within a 2km-long area immediately before the runway. There is a medium risk area between 2 and 4km from the runway and a low risk area 4–6km away, where the likelihood of damage is slight, but still a possibility. The footprint of the affected area is cone-shaped, widening at an angle of 10 degrees either side of the centreline the further away from the runway that an aircraft is flying.

Damage tends to be towards the centre of any affected roof slope rather than at, or near to, ridges, hips or eaves (i.e. wind-related damage). It has been found to be in one of two distinct patterns:

1   *one, or more, small, localised area of loose or dislodged tiles;*

2   *longer, often diagonal, band(s) of dislodged tiles across the slope of the roof, commencing just below the ridge.*

It is not uncommon for at least some of the dislodged tiles to remain on the roof surface. This residual effect is different from that of wind damage where there is a tendency for any loosened tiles to be blown off the roof surface.

Conclusion

Good standards of design and construction allied to regular inspection and maintenance are essential if a roof is to perform properly. Unfortunately, many building owners and occupiers (and some property managers) fail to meet even the most perfunctory standards of post-occupation inspection and maintenance. This is particularly the case with a roof, which is often ignored until its failure has impacted adversely on another element of the building.

All roofs should be given a full external and internal annual inspection as a matter of routine and, where problems are observed, immediate remedy undertaken. Such a policy will protect both the roof and the building beneath.

Neglect due to poor or non-existent maintenance plays a major factor in roof failure. Another major reason for roof failure is poor workmanship, in some cases due to ignorance, in others to rogue builders. Some examples of roof problems that were due to poor maintenance or workmanship follow.

Left: Lack of maintenance together with general long-term neglect has allowed moss to spread over the whole roof. This has impeded drainage and resulted in extensive frost attack of the tiles. Middle: This plain tiled bay roof of a Victorian house has received no recent maintenance – tiles are missing or broken and something bad has happened to the plastic gutter! Remedial action is urgently required. Right: The mortar fillet is failing. Attempts at repair have been made – the insertion of a length of lead fillet and the use of a bitumastic seal – both repairs must be considered as temporary solutions only.

Left: The cement mortar repairs to the original lime mortar verge are a short-term solution at the best and will, inevitably, lead to further failure. A stronger mortar should never be applied over a weaker one. Middle: A future failure in the making – the gauge of these clay double roman tiles has been extended much too far, leaves too little head lap and will result in rain penetration. Right: These clay pantiles have been recently used for re-covering a roof. Unfortunately, they are showing signs of delamination and should not have been re-used. Whoever laid them also has no idea how to form a water-tight verge.

Finally

Finally, it would be impossible to predict how any building would perform in the conditions experienced below. In fact, the main structure of the house has survived surprisingly intact. Unfortunately, there is very little left of the roof.

*19 December 1946, a
Dakota plane crashes
onto the roof of a house
in Angus Drive, near
Northholt Aerodrome in
Middlesex.*

Flat roofs 10

Introduction

A flat roof is a roof with a pitch of less than 10 degrees. It has been a common alternative to the pitched roof for the past 60 years or so. The decision to use a flat roof is often based on its cheaper initial capital cost. This is as a result of its easier construction and use of less material than a pitched roof of comparable plan size. However, a flat roof may not be the cheaper option when long-term costs are taken into account. Some of the covering materials used on flat roofs are not as durable as those used on pitched roofs and the extremely low pitch means that a flat roof is less efficient at draining water from its surface. Flat roofs also tend to be much more prone to leakage and water-related problems than pitched roofs, though recent advances in material technology have addressed some of these concerns.

A number of materials are used to form the structural element of a flat roof but, in the case of dwellings, though a reinforced concrete deck is not uncommon, the majority are probably constructed of timber. The waterproof covering can be any one of a number of materials, e.g. asphalt, single-ply sheet, multi-layer mineral felt, metals such as lead, zinc, copper or aluminium, or even glass reinforced plastic (GRP).

The failure of a flat roof, as with a pitched roof, may have a serious effect on other building elements and therefore any assessment of a flat roof should consider both its current condition and its future performance. Any inspection of a roof should include the adjoining elements, as it may be in these that the first signs of failure within the roof become apparent. If those first signs (in roof or adjoining element) are ignored or misread, the longer-term effect may be an increased likelihood of serious failure. Inaction may also lead to 'knock-on' effects, e.g. fungal attack leading to structural failure.

Construction summary

Function of the roof

This is discussed in greater detail in Chapter 9. The primary function is to provide weather protection in a form that is:
- *strong*
- *stable*
- *durable*
- *impervious*
- *thermally efficient.*

Flat roofs: a brief history

Both economic and skill factors have played an important part in the increased use of flat roof construction in the past 80 years or so, although, in some instances, such roofs were installed on purely aesthetic grounds. In the UK, dwelling houses constructed in the several hundred years up to the 1930s generally had pitched roofs. There were exceptions. A considerable number of grand houses of the Georgian and Regency periods were constructed with large areas of lead flat roofing and many artisan houses built in the eighteenth and nineteenth centuries had ancillary parts formed with flat roofs over such elements as bay windows and small extensions.

Left: A small extension to a Victorian houses has a lead flat roof. Right: A typical lead-covered flat roof to a bay window in a Victorian terrace.

Left: The original small pitched roofs over bay windows are quite often replaced with flat roofs as shown in this example of a Victorian terraced house. Right: Many Victorian terraced houses had a ground-floor bathroom extension added in the 1970s, or 1980s, most with a bitumen felt flat roof. Note the Second World War air raid shelter in the foreground.

In the 1920s and 1930s, the influence of the Modern Movement resulted in an increase of the popularity of flat roofs. This influence continued to some extent into, at least, the 1950s and 1960s.

Top left: The 'purist' ideals of modernism influenced the design of these houses from the 1930s that were built with flat roofs. Top right: An in-situ concrete house from the early 1950s. Right: a 'pre-fab' from the period just after the Second World War.

Defects in flat roofs

Poor design

A flat roof is generally considered to be an unsophisticated and simple approach to providing a building with a roof. However, flat roofs are more prone to failure than pitched roofs. This is frequently a result of poor design, particularly as simplicity of form does not mean that efficient performance is simple to achieve in technical terms; for example, while a pitched roof sheds water quickly, a flat roof does not. Another aspect of poor design, and indeed specification, is the failure to properly understand how materials (especially new covering materials) perform in the extremes of climate experienced in the UK. A further important factor is a lack of thought given to the proper design of the mass of small detail that needs to be carefully worked out if the roof is to perform properly.

Left: The large flat roof is performing properly – draining efficiently with no water held back. Right: This roof has a major problem. Action is needed urgently to remedy the problem.

Examples of poor design of flat roofs include:

- *inadequately sized structural members (timber, steel, reinforced concrete);*
- *insufficient structural members (too widely spaced);*
- *overloading of the structure (often when service plant is placed on top of a roof, or by replacing the roof covering with a heavier material, or by allowing general access, such as creating a roof garden, when the roof was designed for maintenance access only);*
- *selection of an inappropriate substrate (e.g. chipboard decking which disintegrates when wet);*
- *incorrect selection of the covering (e.g. zinc in contact with copper – on account of the significant risk of electrolytic corrosion of the zinc);*
- *poor specification (often because a product and its performance are not properly understood).*

The sub-section on the poor design of pitched roofs included in Chapter 9 sets out some examples of pitched roof problems that may also be experienced by flat roofs.

The form of flat roofs

There are a number of different forms of flat roof and they are defined essentially by the position of the thermal insulation. They are referred to as either a cold roof or a warm roof – with warm roofs being divided into two categories: sandwich and inverted (or upside-down).

Both cold and warm roofs should be designed to avoid interstitial condensation. Interstitial condensation occurs when warm, moist internal air from the building passes through the roof structure (because the internal air pressure is higher than the cold external air) and on meeting any colder air, or colder structure within the roof, the warm air condenses. (See Chapter 16, which discusses the principles of condensation in more detail.)

Cold roof

This was the normal form of flat roof design until, perhaps, 40 years ago. The insulation is placed immediately above the ceiling and between the joists. The void above the insulation needs to be fully ventilated to prevent moisture condensing within the space. It is also essential to install a vapour check on the warm side of the insulation.

Solar reflective chippings

Roof covering (felt in this example)

Timber firring to provide slight fall

Insulation – with vapour check beneath

Because vapour checks are often breached by service fittings, a cold roof relies upon adequate ventilation to remove moisture vapour which may enter the roof space. In practice, effective ventilation is often difficult to achieve particularly where the roof is sheltered and/or the provision of cross-ventilation is not possible.

Unlike a warm roof, the deck of a cold roof is exposed to the heat of the sun (warm roofs have thermal insulation above the deck). This may increase the likelihood of thermal movement.

Warm roof: sandwich construction

In a warm roof with sandwich construction, the insulation is placed above the decking and beneath the weatherproof roof covering. This results in the roof construction remaining at (more or less) room temperature and, as it is warm rather than cold, this reduces the likelihood of condensation occurring within the structure. Once again, a vapour check must be provided on the warm side of the insulation, but the roof is not ventilated as this will draw in cold air which will reduce the temperature of the roof construction and increase the likelihood of condensation occurring.

It is important to ensure that the thermal insulation layer does not cause any damage to the waterproof layer through thermal movement. To counter this possibility, an isolating membrane (known as a slip layer) can be placed between the insulation and the waterproof layer. An alternative approach is to rely on the selection of a relatively stable insulation material.

Another potential problem with this roof type is that the waterproof layer is exposed to a greater range of temperatures than other roof designs, as it cannot lose heat to the structure because the insulation is immediately below it – and it receives no heat from the building for the same reason.

Warm roof: inverted (upside-down) construction

With this type of flat roof, the insulation is installed above the waterproof roof covering which is placed on the decking. No vapour check is required as the roof covering performs that function as well. The insulation is held in place by ballast or pavings positioned above it. The photo shows ballast in the form of large pebbles.

It is important that the insulation selected for this roof type is able to resist frost, but its positioning also means that the structure and the waterproof covering are subjected to less thermal movement than with the other flat roof designs.

Condensation

It can sometimes be difficult to clearly differentiate between condensation and damp as the cause of moisture-related problems within a flat roof. In both cases, there may be high levels of moisture and decayed material, e.g. rotten timber, saturated plasterboard or concrete. The investigation should include an assessment of the potential causes of interstitial condensation as shown in the diagrams below.

Potential causes of interstitial condensation

COLD ROOF

Air flow reduced/blocked by lack of adequate gap or obstruction.

WARM ROOF

No insulation, or insufficient insulation above the vapour check.

Lack of vapour check.

Splits or gaps in the vapour check, especially around pipes, flues and light fittings.

Any ventilation/flow of air through the roof void will lower the temperature and can lead to interstitial condensation.

Consideration must also be given to evidence that the cause of the dampness may be condensation, such as:

- *Black mould, which is indicative of condensation.*
- *The type of human activity in the area beneath the roof, e.g. bathing, cooking or clothes washing.*
- *The difference in temperature between the inside and outside of the roof. The higher the internal temperature, the greater the potential for condensation within the roof element, especially if it is accompanied by high internal moisture levels and incorrect roof construction.*
- *Internal dampness without accompanying evidence of water penetration of the roof covering or adjoining structure (i.e. the cause is NOT penetrating damp).*

The subject of condensation is covered in detail in Chapter 16.

Poor construction

Many of the problems experienced by flat roofs would have been avoided if correct construction standards had been achieved. Even the best-designed roof may fail if the installation is poorly carried out. (Reference is also made to poor roof construction in Chapter 9.)

Reasons for construction-related failure include:

- *poor workmanship and/or supervision*
- *lack of co-operation/co-ordination between the parties involved (designer/contractor, contractor/ specialist sub-contractor)*
- *inappropriate selection/combination of materials, including fixings (e.g. copper nails with zinc sheeting)*
- *unfamiliarity with products and processes (often when they are new on the market)*
- *following trades/works causing damage to the newly installed flat roof (e.g. decorators, scaffolders)*

- *working in unsuitable weather conditions (e.g. wet weather leading to trapped moisture/blistering of the covering or a lack of adhesion to the substrate; frosty weather causing cracking of the material).*

Entrapped moisture

This is a problem suffered by flat roofs that are wholly, or partially, formed with in-situ cast concrete or topped with a cement and sand screed. These construction processes require considerable amounts of water and the slab/screed should be allowed to fully dry before being covered over. Alternatively, the concrete or screed may be drained through the deck, via drain tubes, or ventilated through the roof covering by means of breather vents, so that the evaporation process can be properly completed.

Cast in situ roof

Insulating screed — Mushroom vent

Vapour check — No vapour check

Migration of trapped moisture may take many years. Initial appearance is often around light fittings, vertical pipes or low points in the roof.

General problems

There are a number of defects that are not strictly related to a particular form of flat roof construction, nor the actual covering. These general problems include the following.

Poor falls

Rainwater and melted snow should drain efficiently from all areas of a roof. A 'flat' roof should always be self-draining, i.e. constructed with a slight fall (or slope) to ensure the proper flow of any water that collects on its surface. The fall is created by installing an upper substrate with a slope – this may be achieved by sloping the entire substrate (e.g. decking or insulation laid on sloping timber fillets) or by sloping the top surface of the substrate (e.g. shaped insulation or screed). The minimum fall for any flat roof should be 1:40 (i.e. a fall of 25mm in every 1.0 metre). The fall can be greater than this, and indeed it is normal practice to consider a roof to be 'flat' at any slope of less than 1:10 (i.e. a fall of 100mm in every 1.0 metre). Some guidance advises a fall of 1:60 or even 1:80. The danger of such a shallow pitched roof is that it will tend to drain more slowly. Also, the real fall may be even less than intended if construction is not carried out extremely carefully. Such a shallow fall is also more likely to experience ponding problems, as the slightest unevenness will have an adverse effect upon the flow of water across its surface (see diagram and photographs on the next pages).

The fall of the roof should be towards the drainage outfall. This may be a valley/parapet gutter or an eaves gutter draining the section of roof concerned. Valley, parapet and other gutters should also be laid to an appropriate fall for the material being used. Alternatively, the roof may be drained towards a single opening (or scupper) in a parapet wall, or a sump or gully positioned above an internal rainwater pipe. The slope should be shaped to direct water to the outlet.

Ponding

Ponding is an extremely common occurrence in flat roofs. Any unevenness in the covering, or the substrate beneath, may prevent free drainage occurring properly and can lead to ponding, i.e. trapped water collects in a puddle on the surface of the roof covering. The depth of a pond can vary immensely, from a millimetre or two to many millimetres (75–100mm are not unknown), depending upon the amount of unevenness in the structure or covering. Gradually, there will be a build-up of dirt deposits on the surface of the covering at the bottom of the pond. This will be apparent even after a dry spell which has caused the water to evaporate.

A flat roof should have an even fall. This should be not less than 1:40 to ensure that it is self-draining and to reduce the incidence of ponding.

Ponding.

Ponding is normally due to unevenness of the covering or sagging of the substrate/decking. A roof badly affected by ponding is much more likely to suffer from long-term failure. Other causes include blocked or raised outlets, or the fall being too shallow.

The effect of the ponding is to set up irregular stresses in the covering at the edge of the water. For example, one effect will be on the surface temperature of the covering, because the water in the pond will absorb heat. Over a period of time, these stresses can lead to cracking, especially of built-up felt. Ponding may also attract insects and larvae which in turn may attract birds whose pecking has been known to damage fabric-based coverings. A further potential hazard is that a leak in the 'pond' area, for whatever the reason, will result in a greater amount of water penetrating into the building than from a properly drained roof area.

Out-fall problems

The correct size and positioning of the rainwater outlet are critical; if it is too small for the surcharge of water, or incorrectly placed to properly drain the whole of the required roof area, this will cause problems. The effect of a blocked or inadequate outlet may be water building up on the roof gutter and eventually leaking into the roof void, or flooding over the ineffective outlet and down over walls.

Flat roof draining into internal valley/parapet gutter.

Outlet to a parapet wall. Also known as a scupper.

Outlet blocked with ball, etc. or too small – flow discharges over top of outlet and down wall beneath.

Ball causes blockage.

Ponding due to debris.

Outlet should have gravel/leaf guard installed.

Horizontal and vertical outlets can become blocked with dirt, gravel, leaves, balls or other debris. All outlets should be regularly inspected (at least twice a year, especially in the autumn if the roof is overhung with trees).

Outlet component set too high to enable efficient flow of water – result is ponding & build up of dirt, cracking of roof/gutter material.

Left and middle: Damaged or displaced rainwater hoppers result in water being shed down the face of the building. Right: Ponding and leaves around an outfall caused by the outlet being higher than the roof.

Damage by foot traffic and/or human activity

Most flat roofs are not designed to be walked upon, except for maintenance purposes. Where access is needed for maintenance and repair, it may be necessary to use crawling boards. Examples of damage include gouges, rips, tears and indentations caused by human movement itself, or the placing or dropping of heavy objects on the surface. A roof covering that is already blistered or rucked, or brittle through age, will be particularly prone to damage of this nature.

Organic growth

Lichens, mosses and algae may appear on the surface of the covering, especially where it is overhung by trees or is in permanent shade. A heavy build-up of organic growth may indicate inefficient run-off of rainwater. It may also be indicative of a lack of regular inspection and maintenance of the roof.

A build-up of moss is a common problem on older, mineral felt roofs, especially those that have a poor fall and are finished with chippings. Not only is it unsightly, but its presence indicates a level of retained moisture on the felt roofing which may cause problems.

Lead roofing and guttering can be corroded by the acidic run-off from an adjoining roof that is covered in lichen. The resulting acidic attack may be in the form of narrow, clean-cut grooves on the surface of the lead, or there may be round spots beneath the adjoining roof's drip-off points. Similar corrosion problems can be experienced by zinc roofs and gutters from the run-off from both mosses and lichens, as well as from timber shingles. In each case, the strongly acidic content of the run-off can rapidly corrode the metal.

Thermal and moisture movement

Failure to take account of movement is a common cause of flat roofing problems. Problems occur where a cycle of expansion and contraction is set up through either changes in moisture content or changes in temperature.

Many of the common materials used for flat roof structures and coverings will undergo thermal expansion and contraction. The amount of thermal movement will depend upon the coefficient of expansion of the materials affected, and the external temperatures (sometimes the internal temperatures) involved. Thermal movement (expansion and contraction) may occur throughout the day as, for instance, the sun shines, is then obscured by clouds, then reappears, etc. Sudden outbursts of rain on a sunny day will also cause a drop in the temperature of the roof. This cycle of heating and cooling, leading to expansion and contraction, can damage the roof covering.

The tendency to be affected by moisture movement is related essentially to the porosity of the material. In comparison with thermal movement, moisture movement tends to occur in less frequent cycles, except perhaps where the source of the moisture is the result of a significant amount of condensation generated from intense activities within the building.

The movement cycle in the covering material or the underlying structure may subject the waterproof covering to damaging stresses. Often, the initial effect is to ruck up or ripple the covering. Over a period of time, the continual expansion and contraction will weaken the material and eventually lead to it splitting. When the substrate suffers from thermal movement, it invariably results in the covering being adversely affected as well.

Thermal movement

Expansion

Covering

Supporting substrate

Structural element

Contraction

All components of a flat roof are subject to thermal movement. In a flat roof the effect of such movement tends to be greater than in a pitched roof.

The covering will normally suffer more readily than the other components as it is fully exposed to seasonal or diurnal changes in temperature.

moisture

The covering may ruck or ripple. It will eventually split. Alternatively, it will merely split as it is slowly weakened by movement.

moisture

The covering splits as the substrate moves.

Secured end of covering is loosened over a period of time.

Sometimes, expansion detailing/jointing is inadequate to deal with the stresses involved (or the components are improperly fitted together).

Left: Movement in the felt, or perhaps the substrate below, has led to rucking (rippling) which may eventually lead to splits in the felt. Right: Here, the rucking has created a gap at the side lap through which water can penetrate (there is also evidence of ponding). In both examples, the undulations in the felt will impede the flow of water from the roof.

Differential movement

A flat roof may be affected by differential rates of movement between the roof itself and an adjoining element of construction. The element may be at the same level as the roof (e.g. a parapet wall or another area of roof, perhaps of a separate construction) or below it (e.g. an internal wall). The effect upon the flat roof will be horizontal and/or vertical movement stress which can manifest itself in a number of defects such as:

* *bucking or rippling*
* *stepping, unevenness or sagging*
* *splitting or cracking.*

As with thermal and moisture movement, the problem may affect only the roof covering, but it is more likely to have an effect upon the substrate as well.

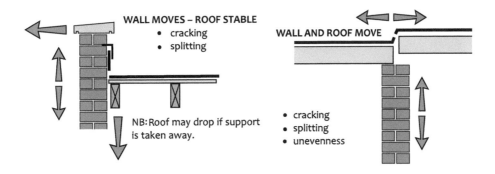

Defects at upstands and skirtings

The junction of a roof with adjoining walls and with penetrations such as pipes, flues, access hatches and rooflights is a common point at which failure occurs. This is because the roof covering is required to be turned vertically through 90 degrees in order to form an upstand (also known as a skirting) that is watertight.

Failure can occur for a number of reasons including:

* **Splitting along the 90° turn in the covering.** *This occurs because it is either weakened by the initial formation of the bend in the material or by subsequent differential movement between the roof and the adjoining element. A tilting fillet should be used to reduce the angle of change (see p. 286).*

- **Insufficient height of the upstand.** *The waterproof covering should be dressed up the wall or component for a minimum distance of 150mm. If the distance is less than 150mm, the upstand may be surmounted by standing water left on the roof/gutter as the result of a blocked outlet or an exceptional rain-burst. This height is also considered sufficient to prevent the effects of splashing from rainwater as it hits the roof surface.*

This felt upstand is splitting along the upper turn. The roofing felt at the lower level is cracking badly and the bitumastic paint repair provides a temporary fix only.

A cover flashing, ideally in lead, should be used where a felt or metal roof covering is dressed up a wall or other vertical component. The felt should not be attached to the wall in order to avoid the consequences of differential movement between the roof structure and the wall. Asphalt, however, if laid on a concrete base can usually be bonded to the wall and dressed into a suitable chase, in which case a cover flashing is not normally necessary. However, if asphalt is laid on a timber deck, or there is a risk of differential movement between deck and wall, the asphalt should be finished with a freestanding upstand which should not be bonded to the wall; a separate cover flashing is then necessary.

90 degree (acute angle) too severe in both places

Vent or soil pipe, etc.

Where the roof meets a wall or pipe flue, hatch or rooflight, the junction to the roof needs to be properly formed with an upstand. Often, turning the roof covering through 90 degrees creates unacceptable stress in the material and it cracks. (Cause may be weakness induced by installation or thermal movement.)

Moisture penetration

min.150mm

Benefit of angle/tilting fillet

Cracks/ splitting   No cracks/ splitting

90 degree (acute angle) too severe in many cases

Asphalt laid over tilting fillet at junction.

Asphalt is often dressed into a chase in the background. It has a tendency to slump where chase is not provided or is inadequate.

Top of upstand (min. 150mm) should be protected with cover flashing – with asphalt and mineral felt but this is often omitted – leading to potential water ingress problems.

Mineral felt should not be chased in because it is vulnerable to tearing through differential movement. It is better to cover it with a separate lead flashing.

Upstands around outlets (often called scuppers) in parapet walls are frequent sources of water penetration. Often this is because they are poorly formed as it can be difficult to properly shape or dress the roof/gutter material through the various vertical and horizontal planes.

Top left: The lead flashing has been omitted from a section of the upstand detail. It is possible that the builder assumed the coping would provide sufficient protection. As a consequence, the roofing felt has had to be attached to the upstand; this may cause problems due to differential movement. Top right: Cracks in the asphalt at the junction of the decking and the upstand could be caused by differential movement or by the lack of a tilting fillet. Bottom left: The asphalt has slumped because it has not been properly attached to the upstand.

Decking problems

In domestic flat roof construction, the structure is normally formed with a reinforced concrete slab or by timber joists. The former will usually be finished with a cement and sand screed topping which is laid to a fall or with insulation boarding incorporating an integral sloping top. The latter will normally be covered with board or sheet decking and the roof slope is formed by either installing the decking on firring pieces fixed to the joists or by shaped insulation boarding.

There are a number of materials that may be used for the decking, but the following are the most common.

Chipboard. This is manufactured in a number of grades, including moisture-resistant which is intended for use in roofs. However, some leading authorities have expressed concern over the use of any grade of chipboard in a roof, as they are all capable of absorbing moisture under certain conditions, particularly those where high levels of humidity are present. If this occurs, the chipboard loses its structural integrity causing it to sag under its own moisture-sodden weight and it will, eventually, disintegrate.

Plywood. External-quality plywood or marine plywood sheets should be used for roof decking, as the glues in these are water-resistant. Plywoods of internal quality are adversely affected by moisture and may expand, disintegrate or develop fungal infection (wet or dry rot) if the roof leaks.

The felt roof covering was torn off in a storm about two years ago. The resultant deterioration of the plywood decking and the timber fascia can be seen.

Softwood timber boards. These are the traditional decking above a timber structure. They should be at least 19mm thick, tongued and grooved, and fixed with two nails to each joist. Straight-edged boarding is often used on the grounds of financial economy, but it is more prone to warping, either due to initial drying out or the effect of subsequent water penetration of the covering. Both types may shrink as a result of initial drying out.

Excessive warping or shrinkage may result in the roof membrane being affected. Warping will produce ridges in the covering, while shrinkage results in dips or mini-valleys between the individual boards. In a badly affected roof, either may be a substantial problem because of the considerable number of board joints that may be involved.

The mini-valleys formed by the shrinkage of the boarding beneath this flat roof can be identified by the water in the resultant channels.

Wind-related damage

There are a number of potential problems that may be experienced by flat roofs because of the effect of the wind. They are discussed below. (The general background to this subject is discussed in Chapter 9.)

The effect of suction. As with pitched roofs, negative pressure (suction) is a greater problem than positive pressure and affects the edge of the roof (i.e. the eaves) more than the central area of a flat roof. This is because maximum suction occurs at the edge, though flat roof coverings that are particularly light in weight may be prone to lifting by suction across their whole surface area. Any weakness in the (edge) fixing is exploited by the wind and may lead to the covering being lifted. Multi-layered roofs can experience successive removal of the individual layers of the roof, i.e. the covering, insulation, decking, sub-structure. At its worst, it is not unknown for a complete flat roof to be physically removed as an integral unit from the top of a building and deposited some distance away.

Scouring action on protective dressings. The wind can have a scouring action on any gravel covering used as a protective dressing on a flat roof. Loose ballast of less than, say, 20mm diameter on inverted roofs may be particularly affected, leading to damage of the underlying substrate. The movement of the dressing may also result in the exposed roofing material being degraded by ultra-violet

The effect of the scouring action of wind on a protective dressing can be seen on this flat roof in a Cornish fishing port.

light, drainage out-falls becoming blocked, injury to pedestrians or damage to adjoining glazed areas, especially where there is a lack of any raised parapet walling.

Pressure on laps, joints and steps. A wind speed of 26 metres per second is not unusual in the UK and is often exceeded (this wind speed is described as 'storm' on the Beaufort Scale, which indicates wind force). Such a wind speed, which is becoming more frequent due to climatic change, will support a 50mm column of water. Therefore, any welted or folded joint or seam, any upstand or step, that is less than this height must be viewed with suspicion as there is every possibility that it will leak under high wind conditions.

Failure of roof coverings

A wide range of materials can be used to cover flat roofs. The following discussion concentrates on those that are most commonly found on domestic flat roofs.

Built-up mineral felt

This roof covering usually consists of two or, preferably, three layers of felt, bonded with hot bitumen. The exact number of layers depends on a number of factors including whether a partially bonded first layer is needed to accommodate substrate movement; the type of reinforcement in the individual layers – polyester felt membranes normally consist of two layers, glass and other fibre membranes of three. The top layer is normally finished with stone chippings or, sometimes, reflective paint to provide protection against solar radiation (which can rapidly degrade the material).

Mineral felt has been in use as a flat roof covering (and, occasionally, as a gutter lining) since before the Second World War. It became very popular in the 1960s and 1970s. This was because of its relative cheapness and ease of installation. It has since declined in popularity, because of its relative lack of durability. In the 1960s and 1970s, the life expectancy of mineral felt roofing was thought to be 15–25 years, but it rarely achieved this, often suffering failure in less than 10 years. This was due to a mixture of poor installation and inadequate materials (e.g. a core of rag fibres, asbestos fibres or glass fibres coated with oxidised bitumen). However, recent advances in material content (e.g. the introduction of polyester fleece cores – especially when coated with polymer-modified bitumen to form a 'high performance felt') and better-quality installation techniques should ensure a longer life.

Apart from general faults covered in the previous sub-section, built-up mineral felt can have the following problems.

Blistering. Blisters are quite common in felt and asphalt roofs. The blisters are usually formed through the expansion of trapped air and moisture underneath the covering (and/or between sheets in the case of felt). The blisters usually form as a result of the trapped, moisture-laden air expanding as it is heated by the sun. The blisters will tend to expand and contract as the trapped air is heated and then cooled (a process sometimes referred to as 'thermal cycling'). The expansion and contraction cycle can result in the felt being permanently stretched and not returning to its original state. The stresses can cause a split – though often the damage is caused through a mechanical impact on the blister, e.g. a footfall.

Blisters in flat roofs: In this example, the roof is above a kitchen which is likely to be the source of the moisture-laden air. Solar reflective chippings or reflective paint help in reducing the risk of blisters forming because they absorb some of the heat from the sun.

Eaves and verges. It is normal practice to finish the mineral felt by dressing it down over the face of the eaves to form a drip. The same detail may be formed at a verge, though it is better to create a weathered edge (i.e. a raised kerb), to direct water away from the verge (and avoid it running down the face of the wall below). It is also an alternative common practice to incorporate an extruded aluminium edge trim along a verge (but not the eaves) to provide a mechanical fixing that is resistant to wind uplift.

A) Eaves detail

Felt dressed through 90 degrees to form drip to gutter.

Min. 50mm turn down

B) Verge detail

Raised kerb to direct water away from verge.

50–75mm high

38–50mm oversail

C) Verge with trim

Trim screwed to substrate and felt taken into recess in trim.

Verge formed with aluminium trim – the raised section of the trim makes it unsuitable for use at the eaves.

Problems that may occur at the eaves include:

Batten throws water away from fascia and into curve of gutter. (Otherwise, it will trickle over face of the fascia or even miss the gutter.)

Note: Older felts may crack as they are not reinforced with more resilient, flexible materials.

Ponding at ridge

Wind suction across top of insecurely fixed covering – edges are particularly weak if not secure.

WIND SUCTION

Water may not flow across covering at eaves due to detailing. It may result in small amounts of water being held back. This does not normally affect performance.

Wind may get under the covering once it has torn.

Problems that may occur at the verge include:

A)

Lack of weathered detail/kerb allows water to run straight over verge.

B)

Aluminium edge trim – fixed in overlong lengths (exceeding 1300mm) and/or without 3mm expansion gap between lengths.

Aluminium or plastic edge trim insecurely fixed – releases covering and allows water penetration. Weaknesses may lead to wind damage.

C)

Expansion can cause crack or split in covering.

Trim may be dislodged.

No allowance for thermal movement in edge trim.

Loss of surface protection. An even spread of solar chippings, or the use of solar reflective paint, over the complete roof area is essential. This is to prevent the bitumen content oxidising and prematurely aging the covering. The cause is the damaging effect of ultra-violet light from exposure to the sun. The result of the loss of any surface protection will be the general deterioration of the felt, usually accompanied by embrittlement. Once this occurs, water can migrate through the material and cause damage to the roof structure and boarding beneath. If left unnoticed, it may affect internal areas of the building. The penetrating water may also freeze in cold spells and cause further damage to the roof covering.

Protective ballast (small chippings or grit) should be dressed over the entire surface of the felt to a depth of 10–15mm, but they can become displaced for a number of reasons including the fall being too steep and allowing the chippings to be swept by rainwater to the bottom of the 'slope' or sagging of the substrate.

Chippings can be moved around by wind or rainwater or simply brushed to one side during repair works and then not swept back in place.

The use of solar reflective paint as an alternative to chippings will provide protection against the damage caused by ultra-violet light. Chippings, however, can also work to some extent as a 'heat sink' absorbing the heat of the sun and, in so doing, potentially reducing thermal movement; paint will not provide this benefit.

Solar reflective paint is applied to a roof surface that is free of any debris or encumbrance, such as chippings or grease. It requires frequent renewal at approximately 2–5-year intervals, depending upon weathering and exposure, to ensure that its integrity is maintained and the felt is fully protected from the damaging effect of ultra-violet light.

Problems with jointing of the top layer. The joints in the lower layers of a mineral felt roof are formed by butting together the individual sheets/pieces, ensuring that the joint in each layer is staggered to avoid a through joint with the one above. For the top (or weathering) layer, the adjoining sheets are overlapped and then bonded together with hot bitumen.

Patched seam (A) and tear (B)

Side joint along the fall – defects may be serious, but not so readily subjected to penetration.

FALL (AND FLOW)

chippings may fall down slope.

moisture drains into defective joint.

TOP LAYER   overlapped and bonded joint
BOTTOM LAYERS   joints butted and staggered

Blister has lifted side lap.

Bottom joint across the fall – likelihood of increased problems if the joint is defective as there is a greater/more regular flow of water to exploit any defect.

A joint may fail because it was, initially, poorly bonded, or it may pull apart subsequently, perhaps as a result of blistering. The failure of any joint in the top layer is a serious matter, but it is particularly so when it is across the fall of the roof or a gutter, when any weakness will tend to be rapidly exploited by water flowing across the surface.

Felt should always be lapped OVER the lower sheet.

Flow

Gutter

Single-ply membrane

Recent years have seen the introduction of single-ply roofing membranes from Europe and the USA. The introduction of these membranes into the UK took place in the late 1960s, but they had not been fully developed and did not perform successfully. In part, this was due to the thinness of some of the membranes – less than 1mm thick. Subsequent research and development have resulted in materials of much greater integrity and durability.

Generally, they have been installed over commercial premises, but there have been some examples of their use on domestic roofs. They are manufactured from various plastics or from synthetic rubber-based materials. Sheets of the former can be softened and welded together by heat or solvent, whereas the latter must be joined by adhesive. Site welding is thought to produce a more satisfactory joint than gluing, although with both types of material installation is a skilled operation and requires carefully controlled site conditions. Manufacturers claim that the materials are easily repaired if damaged.

The membrane should be mechanically fixed to a warm deck roof by screw fasteners and disc washers. Each fastening should be set into a seam, or be beneath a spot patch on a larger roof. On an inverted warm roof, the membrane is loose laid and covered with a layer of ballast.

Evidence to avoid problems includes the following:

- *Some manufacturers claim that their product may be laid on an absolutely flat roof without any drainage fall. However, this inevitably leads to standing water on the covering and this will result in an increased likelihood of water penetration if the membrane suffers from a defect.*
- *Certain single-ply membranes may be loose laid with only an edge fixing. These are more susceptible to the effect of wind lift if the manufacturer's installation requirements are not precisely followed.*
- *For those materials requiring a mechanical fixing, the size of the roof and its degree of exposure may dictate closer spacings than normal for the fixing points, otherwise the roof may suffer from wind-lift.*
- *Careful selection of adjoining materials is important. Some plastics/rubbers are not suitable for use beneath the extruded expanded polystyrene sheeting commonly used on inverted warm roofs for insulation.*
- *The membranes should never be laid over a previous roof covering. At the very least, a layer of plywood or similar material should be laid first.*

Mastic asphalt

Mastic asphalt is an extremely durable material with a life span of 50–60 years that has been used as a waterproof covering for flat roofs since about the beginning of the twentieth century. Mastic asphalt consists of fine and course inert aggregates which are mixed with bitumen. It is laid over a separating layer, normally of black sheathing felt, that isolates it from any movement in the substrate.

Mastic asphalt comes in the form of solidified blocks that are reheated on site until it becomes soft enough to be applied by trowel. This is usually carried out in two layers, to a combined thickness of 20mm. The work is carried out in bays of around 2.5–3 metres and, while the proper jointing of each bay needs to be properly carried out, the fact that an essentially seamless surface can be produced is partly why mastic asphalt roofs can be so durable.

The action of applying the mastic asphalt by trowelling commonly results in a bitumen-rich skin to the top coat. This skin is susceptible to surface crazing in cold weather. To avoid the formation of this skin, a light dressing of sand is rubbed in to the final coat after it has been applied. The surface should then be protected by solar reflective paint or a 10–15mm thick dressing of small chippings.

Asphalt has a tendency to slump when incorrectly laid on a slope or vertical upstand. It also tends to crack when formed into a sharp angle such as occurs at an upstand, and so should be laid over an angled fillet (see earlier sub-section on Defects at upstands and skirtings).

Left: The asphalt has slumped. Right: It has cracked along the top of an upstand.

An older asphalt roof (40 years +) will tend to deteriorate by developing cracks as elasticity is lost. This is a result of the effect of continual changes in temperature. Newer roofs may be less prone to these stresses, due to the introduction of polymer-modified asphalts.

Apart from those defects that were discussed in the earlier sub-section dealing with general flat roof problems, the following problems may also occur.

Adverse effect of the substrate. Mastic asphalt is probably more dependent upon the condition of the substrate than any other covering material. Any movement or unevenness in the substrate tends to be transferred to the asphalt and will frequently induce major cracking/splitting. A multi-jointed decking, such as timber boarding, can be particularly uneven if the boards are insufficiently fixed and/or suffer curling. Sheet decking or insulation board may have uneven or widely spaced joints. All forms of decking may be poorly supported by the structure, resulting in dipping and sagging. Any movement in the substrate will be readily transferred to an asphalt roof unless adequate precautions (such as thermal or differential joints/details) have been incorporated.

Surface crazing. Surface crazing (see left-hand photo above) together with wrinkling, pimpling and crocodiling are normally the result of solar radiation, and are often seen on older roofs. They may also be caused by ponding. These cracks are not considered serious as they are usually of little depth. However, those cracks associated with ponding can, over a period of time, develop into a deeper and more serious problem.

Blistering. Blistering (see right-hand photo following) in asphalt has the same cause as that in felt (see earlier explanation). Because of the inherent durability of asphalt, blisters tend to cause fewer problems than they do with mineral felt roofs.

Left: This blister has not yet affected weather-performance. Right: This blister, however, needs repair as it has 'burst'.

Lead roofing

Lead sheet is commonly used as both a pitched and a flat roof covering. Lead is a well-proven and long-lasting material, which is extremely malleable and easily dressed into shape. As it ages, the surface dulls as a coating of lead carbonate is formed due to contact with the atmosphere. This coating protects the lead from further deterioration. The traditional method of manufacture, cast lead sheet, which is formed by pouring molten lead over a bed of sand, has been found to last over 400 years. A small quantity of this type of sheet lead is still produced and is used for conservation and other high-quality work.

The majority of the lead used in roofing is known as milled lead sheet. It is thinner than cast lead and has a shorter life (at 100+ years). Milled lead sheet is made by running a slab of lead through a rolling mill several times, until a sheet of the correct thickness is achieved. There is a range of six thicknesses, each being given a British Standard Code ranging from Code 3 (the thinnest and lightest) to Code 8 (the thickest and heaviest).

The type of use will dictate the thickness selected. Large exposed areas of lead (such as are found on flat roofs and in valley/parapet gutters) are subject to considerable thermal expansion, caused by exposure to the sun. The thicker/heavier sheets (Codes 6–8) are normally selected for such positions as they are not so easily, or so adversely, affected by extremes of temperature. They can also be installed in larger panels or bays. The thinner grades (Codes 3–5) are used where there is a lower level of exposure involved, e.g. for flashings and soakers, and for smaller gutters such as those found on pitched roofs. Selection of a particular Code will also be influenced by the life-span sought and the shape and/or profile of the area to be covered.

Many of the problems associated with lead roofs are caused by inadequate design and/or construction. Such problems include the following.

Corrosion. The design and construction of the roof should ensure that there is no problem with interstitial condensation. It has long been known that condensation may affect the underside of leadwork (where it is not protected by the lead carbonate patina that forms on its topside.) This can lead to corrosion because the condensate is distilled water, which slowly dissolves lead.

A more recently recognised problem is that sometimes called 'summer condensation'. This was first identified in the late 1980s and affects the underside of a lead covering or gutter above a warm roof construction. Its cause is condensation created by the following, previously unknown, phenomenon in warm roofs.

In summer, rainwater falling on a hot roof covering and rapidly cooling it results in sub-atmospheric pressure being created in the zone between the vapour barrier and the covering (i.e. within the insulation). Rainwater is then drawn into this area by the lower pressure through the expansion and other joints in the roof covering. The trapped water subsequently vaporises under further hot weather conditions, migrates through the substrate and condenses on the underside of the lead as further reductions in temperature take place, leading to corrosion.

This phenomenon also occurs with other roof coverings such as zinc, copper, aluminium, asphalt and mineral felt, though the effect may be different.

Excessive bay size. Lead is extremely sensitive to changes in temperature. In summer, leadwork that is continuously exposed to the sun may be subjected to daily changes in temperature of up to 40°C. It is important that the individual pieces of lead sheet or guttering do not exceed maximum recommended sizes, as there is an increased likelihood of failure when they do. This is due to the expansion/contraction cycles involved and the effect will be rippling and cracking of the lead. Overlarge bay size is the most common cause of failure of lead roofing. Guidance on the maximum recommended dimensions for lead covered roofs and gutters is given in the Rolled Lead Sheet: The Complete Manual, published by the Lead Sheet Association.

The maximum size of each bay is determined by the thickness (code) of the lead. For flat roofs, Code 6 lead (2.65mm thick) is suitable for bays up to 2250 x 675mm. Code 4 lead (1.80mm thick) is suitable for bays up to 2000 x 500mm. When using Code 4 or 5 lead, bay length may need to be a bit shorter depending on exposure to sun.

To avoid premature failure, the lead must be free to move – nowadays, a geo-textile membrane is specified below the lead.

Incorrect fixing. Correct fixing of leadwork is essential in order to avoid premature failure. The fixing nails should have a life equal to that of the lead. They should be sufficient in number, and fixed in the correct pattern, to properly support and secure the lead, while permitting thermal movement. Fixing nails should also be of the correct type – either copper or stainless steel ring-shanked clout nails. Other materials, including aluminium or galvanised steel, are now considered to be inappropriate. Common fixing defects are shown below.

Side nailing should be to top third only of one side. Further fixing leads to problems of thermal movement.

If fixing holes are too close to the edge, there is a danger of the sheet splitting as its weight can cause drag on the restraining section.

Bottom edge should never be nailed.

FALL

The bay should always be fully fixed along top edge. Unless the fall exceeds 1:60, this may be the only fixing required.

Fixing both sides of a sheet can ruck/split the lead – the sheet cannot accommodate thermal movement across the fully restrained axis.

Incorrect flashings. Lead flashings are to be found on all types of flat and pitched roofs, not just those that are covered with lead. This is because of the ability of lead to be easily beaten into any shape, together with its durability and its compatibility with a wide range of other materials.

It is normal to use Code 4 lead for flashings and a properly installed flashing will perform for 80–100 years, often longer, without any attention. Occasionally, Code 3 lead flashings may be found. These are installed for cheapness, but do not perform as well as the higher code. Most problems arise from poor installation, some of which are shown below.

Inadequate turn-in: unless turned into a mortar joint or chase with a minimum depth of 25mm and fixed with lead wedges at 450mm centres, the lead will work loose.

Bottom edge of render cracks or breaks off.

Overlong pieces: maximum length of Code 4 piece to be 1500mm to avoid thermal movement and cracking/splitting/loosening of the flashing.

Inadequate overlap: minimum overlap of 100mm to prevent moisture penetration.

Inadequate fixing: flashings nailed to face of wall tend to fail early due to thermal movement. (If unavoidable, use screws and plugs.)

Inadequate depth: should be at least 150mm.

Lead wedge

Inadequate upstand height: minimum upstand of soaker, gutter or roof covering of 100mm behind the flashing to prevent moisture penetration.

The upstand of the lead soaker is far too short. It is insufficient to prevent rainwater getting above it due to capillary action between the soaker and the stepped cover flashings.

Incorrect underlay. Both building paper and felt underlay have been used as the separating layer between lead and substrate for some considerable time. The bitumen or resin content of the felt may leach out under solar heat, leading to adhesion between lead and substrate. This will restrict thermal movement. A further problem is that even slight condensation may cause the organic fibres of the felt to rot. To avoid these problems, a needled, non-woven polyester textile underlay should be installed, though building paper may still be used where there is no risk of condensation forming under the lead.

Junction problems: drips. A drip is used to form the longitudinal junction between adjoining bays of a flat roof, or of a gutter where the fall is less than 15 degrees. It is a stepped detail that should be both watertight and able to allow thermal movement to properly occur. Its installation requires care, as any deficiency in design or construction may lead to moisture problems, usually as a result of capillary action. Failure may occur for a number of reasons including those illustrated below.

Drip problems

Too small a splashlap may lead to moisture penetration. A minimum lap of 40mm is required to prevent capillary problems. The horizontal tail of the splashlap stiffens the free edge and helps keep it in position.

- Undercloak should be minimum 25mm long. A ridge/step will be formed if the deck is not rebated (ponding may result).
- Undercloak should be nailed at 50mm centres (not less than 25mm from end).

Minimum height of drip should be 50mm to prevent water being drawn up between the sheets by capillary action and/or wind pressure. This can be reduced to 40mm if an anti-capillary groove is incorporated. Even then, a high flow may penetrate the detail.

Rebate formed in edge of boarding.

Junction problems: rolls. The junction along the length of each bay is normally formed with a roll, and sometimes with a welted upstand (see next sub-section). The roll may be hollow (see photograph on the next page) or formed over a solid softwood core. A hollow roll is only suitable for use on a pitched roof, as on a flat roof foot traffic can easily crush it.

The lead from each bay should be freely dressed over the timber roll to form a junction that allows thermal movement. Only the lower sheet should be nailed, if absolutely necessary (and at one side only). Inappropriate fixing will result in buckling and cracking of the sheet, often along the roll.

Solid roll

top piece free to move

Bottom piece may be fixed (never the top piece).

overlap

min. 45mm upstand

Hollow roll

This is liable to crushing from foot traffic.

Rolls may be solid or hollow. Lead is dressed over solid core or metal spring (subsequently withdrawn) to form a detail that will allow thermal movement, prevent capillary action and avoid penetration of any normal level of standing water.

Boards are laid diagonally to minimise effects of warping (i.e. causing ridges in the lead).

Left: Hollow rolls have been formed on this lead roof. Right: A lead door canopy with a welted joint.

Other junction problems: overlaps and welts. A welted joint may be used on the side junctions of small flat roof areas, e.g. a door canopy (see photograph above and diagram below). This is only acceptable where the flow of water is unlikely to exceed the height of the welt. A welt should never be used for lead sheet jointing on a large flat roof area, nor in any low-pitch gutter. A roll should be used in both situations. (Note: Pre-clad lead panelling on flat roofs and other areas involves a standing seam joint at the side junctions. Similar in principle to a welt, this type of joint usually requires mechanical site assembly.)

A welt

This should not be used where a large flow of water is anticipated. It is easily breached by a high water level.

FALL

Lapped joints

A simple overlapped joint is not acceptable on a flat roof, or low pitch roof, however great the lap.

Capillary action and wind blown rain will result in water ingress.

Lap: 150mm at < 30 degrees increasing to 300mm at 15 degrees.

Lap

Fixing point

Lapped joints on steeper pitches and gutters are acceptable as long as a minimum recommended lap is provided.

Simple overlapping of two adjoining sheets on flat roofs and gutters is incorrect, however wide the amount of overlap. Moisture will penetrate such a detail and get into the roof substrate and structure beneath. However, the simple overlapping of bottom joints and the use of welted side-joints are frequently found on steeply pitched roofs and gutters. Both are only acceptable practices on surfaces where rainwater is drained away quickly.

Staining of adjoining surfaces. Newly installed lead sheet produces an uneven white carbonate on its surface as a result of wet or damp conditions. Unless treated with a patination oil immediately after installation, any run-off from the affected surface may create stains on adjoining surfaces. The staining on the lead surface will eventually disappear as the surface of the lead acquires an overall patina of lead carbonate. As mentioned earlier, this patina of lead carbonate is a key factor in ensuring lead's durability.

Zinc roofing

This became popular as a flat roof covering, gutter lining and for flashings in Victorian times, because of its cheapness when compared with lead. It is also considerably lighter, but is not as durable. In recent years, this has led to a decline in popularity, even though it is now normally manufactured as an alloy – zinc/titanium for sheeting and zinc/lead for flashings – to give better performance. Its use is more popular in the rest of Europe, especially France.

Zinc can, possibly, achieve a life of up to 60 years in reasonably clean urban areas and possibly longer in rural areas, but 25 years is more likely, especially on flat roofs. In heavily polluted industrial areas, or even some inner city areas, it can fail in an even shorter period. Its durability depends, to some extent, on the gauge (thickness) of the zinc used. The five thicknesses most likely to be used for zinc roofing are from 12 to 16 gauge; the normal thickness used is 14 gauge (0.8mm thick).

Left: Zinc covering to the main slope of a mansard roof and to a dormer window. Note the screw-fixed hollow 'rolls'. Right: Zinc flashings over clay double roman tiles. No attempt has been made to shape or cut the metal to the tile profile because it is a very hard, rigid material; this may lead to rainwater penetration at the wall junction.

Zinc is a more difficult material to work than lead, being less malleable and more prone to splitting where it is turned or shaped (which is carried out prior to installation). It does have the advantage over lead of being less susceptible to thermal movement, so can be used in larger bays – up to 600mm wide and 5 metres long. For this reason, extremely long zinc gutters, having fewer individual bays and a minimum fall of 1 in 20, can be installed. This also results in fewer expansion joints, which are formed with drips and rolls, although standing seams, which are more popular elsewhere in Europe, may sometimes be found. Care still needs to be taken in not exceeding the maximum recommended bay lengths and girths, or else buckling and splitting may occur.

The zinc sheeting should be given full support by the substrate and be laid over a suitable separating layer. This is to prevent adhesion and possible thermal movement problems as well as the effects of acidic content of plywood and new oak boarding. As with lead, careful selection is needed of the material for the separating layer.

The problems associated with zinc include the following.

Generally. A number of the problems experienced with zinc are similar to those suffered by lead, e.g. failure of flashings, excessively large bays/overlong gutters affected by thermal movement.

Acids and alkalis. Zinc is attacked by both acids and alkalis and should not be positioned next to oak timberwork or where it will be affected by run-off from oak components. Acidic glues in plywood can also be a problem, so softwood boarding should be used. Cat urine is also corrosive. Zinc should be given a protective coat of bitumen where it is likely to come into contact with any acid or alkali. Alternatively, it can be embedded in mortar or have chalk emulsion applied.

Bi-metallic corrosion. Zinc is subject to electrolytic (i.e. galvanic) action when placed in close proximity to metals that are more 'noble' (e.g. copper). This occurs where there is inappropriate use of adjoining materials, especially fixings, and may lead to corrosion of the zinc. Care should be taken to avoid run-off from these materials. Its use next to lead roofing and components is acceptable. Fixings should be of heavily galvanised steel screws or nails, and clips formed in zinc.

Brittleness. As zinc ages, its surface is increasingly affected by pitting and it also takes on a crusty appearance. This is zinc carbonate, which develops as a result of the zinc's exposure to air and water. The carbonate does not (as with lead carbonate) develop into a skin of sufficient thickness or density to provide a protective coat to the zinc. The metal gradually deteriorates as it ages, becoming extremely brittle, and prone to cracking and splitting. This process is accelerated in a polluted atmosphere.

Corrosion. Zinc is less corrosion-resistant than lead or copper and, like lead, is adversely affected by both interstitial condensation and 'summer condensation' (see section on corrosion in the section on lead). In these conditions, water vapour in the roof condenses on the underside of the zinc, and reacts with the metal to create zinc hydroxide on its surface. This is known as 'white rust', because of its appearance. Zincwork should normally be ventilated on the underside unless installing sheeting with a specialist-treated corrosion-resistant underside for use over unventilated roofs.

Fire. Most metals normally melt when heated, but zinc catches fire at red heat (which creates an extra fire hazard for a designer to consider).

Joints and welts. It is possible to form changes of direction in the metal – the zinc can be folded with hand tools (though the factory pre-forming of shapes is often undertaken to ensure straightness of the folds). The folding of the metal tends to weaken it and this may be accentuated when the work is clumsily undertaken, resulting in early failure due to cracking.

Foot traffic may damage the joints, especially standing seams. Welts may also suffer from human clumsiness.

Restrictions of height often mean that parapet and valley gutters will not tolerate a stepped lining. The result is that zinc is used for the gutter lining, as it can be laid in relatively long lengths without the need for too many drips. Sometimes, sections of zinc are soldered together to form an extremely long gutter without any drips (steps), in order to deal with a severely restricted height. The lack of an expansion detail (drip) may place considerable stress on both the zinc and, in particular, on the soldered joint(s). As a result, these may fail, allowing water into the roof construction beneath.

Soldered joint failure. It is a difficult metal to bend and shape, so it is normal practice to cut the material and solder the pieces into the shape required. This may result in many such joints in a flat roof or gutter. Thermal movement of the metal tends to place soldered joints under great stress. They are common points of failure in zincwork. Great care is needed in the formation to ensure that the cutting of the metal and the soldering are properly carried out.

Left: Carefully pre-shaped zinc flashings. Right: A zinc valley gutter showing its age. It will need replacing shortly. Note the way in which the flashings on the wall above have been carefully cut to fit the profile of the tiles.

Copper roofing

Copper is a durable roofing material, light in weight with at least a 50-year life expectancy, although there are examples of copper roofing exceeding 100+ years. It is sometimes used for flat roofs and also on pitched roofs (albeit, often on quite a shallow pitch as copper suffers excessive stress when laid at a pitch exceeding 10°).

On exposure to the atmosphere, copper produces a number of chemical reactions which changes its colour from almost 'pink' through shades of brown until it eventually becomes the light green colour that is most commonly associated with it. This natural process of oxidisation adds to the durability of copper and normally takes about 8–15 years, depending on the level of sulfur dioxide in the atmosphere. A salt-laden coastal environment accelerates the process. The patina that is created as a result of this process acts as a barrier against corrosive elements in the atmosphere and aids the longevity of the roof.

Copper is a soft and malleable metal, less affected by thermal movement than other metals, fairly easy to work and solders well. Care needs to be taken with installing it as it hardens rapidly when worked and this can lead to early failure.

The panels can be formed with 0.6mm or 0.7mm thick copper sheet over a suitable substrate, commonly plywood (it is important that this is smooth). Each panel can be up to 10 metres in length depending on sheet thickness, although they are often much shorter. The width is commonly 600mm, sometimes less. Vertical joints are normally formed with standing seams, although sometimes rolls or welts are installed. The base of each sheet should overlap the one beneath with at least a double welt and preferably a step incorporating a welt to avoid water penetration by capillary action. Flashings should also be of copper to avoid the possibility of electrolytic (i.e. galvanic) action.

The light green colouring of these copper roofs is produced by the patination of copper as it reacts with the atmosphere. The undulations in the copper roof (right) are caused by condensation which has adversely affected the decking.

Relatively few problems are experienced with copper roofing, but they include the following.

Copper run-off. Traces of copper salts in the run-off may stain adjoining materials such as brick, limestone, granite, marble and other metals. The staining may be green or brown and is difficult to remove.

Corrosion of copper. Contact with a range of other materials can cause corrosion of the copper. This includes contact with acidic run-off from mosses, lichens and algae as well as from wood shingles. Masonry cleaners, ammonia and various alkalis are also a problem. The patina offers only limited protection against these materials or run-off. Salts, such as sea salt, can accelerate corrosion.

Erosion. Copper can be eroded by: constant dripping water, e.g. from a rainwater discharge; by foot traffic; by general abrasion, e.g. slight movement due to wind of a copper covering above an uneven substrate.

Galvanic action. Any run-off from copper can also hasten the deterioration of any adjoining metal which is vulnerable to any contact with copper in association with water, i.e. galvanic action. Such metals include steel, iron, tin or lead and aluminium and zinc (these last two are the most susceptible metals).

Repairs. It is extremely difficult to repair or patch existing copper sheeting effectively. Any defect that arises really means complete replacement of the affected sheet/sheets, flashing or detail.

Staining. The run-off from iron and steel products may result in dark brown or black staining of copper.

Conclusion

Because of their tendency to be hidden from view, regular maintenance inspections of flat roofs should be carried out. These should be at least annually and more regularly for many flat roofs to take account of external factors such as the effect of wind or leaves and shading from trees. Once the roof finish shows evidence of approaching the end of its life, the types of roofing materials used dictate that an earlier rather than a later decision is needed in terms of replacement.

The following photographs give further examples of reasons for having a regular cyclical inspection programme and for avoiding delay in any decision.

Left: This flat roof drains (not very well) to the top of the photograph where it is (incorrectly) channelled to the right-hand corner. However, in the right-hand photograph, it can be seen that the rainwater discharges down the face of the wall beneath because there is no overhang and no gutter.

Left: Lack of protective gravel has resulted in the felt covering cracking and splitting. Right: The felt upstand has failed and been 'repaired' with Flashband or a similar self-adhesive material. A more permanent solution involving a proper flashing, probably of lead, should be considered.

Left: Regular maintenance of this roof is essential in order to ensure that this outlet does not get blocked with leaves. Middle: Regular inspection of this asphalt gutter would have identified its problems much earlier. Now, it is too late to undertake repair and total replacement is the only option. Right: A cyclical inspection revealed that thieves had removed most of the lead flashings to this flat roof.

External rendering   11

Introduction

This chapter summarises 'good practice' in external rendering and explains a number of defects which commonly occur.

In the early nineteenth century it became fashionable to render houses to give them an appearance of stone. The renders, usually made from lime, sand or other bulk fillers, were often incised with fine lines to imitate the courses of stonework. In the Victorian period modern cement was invented. It was called Ordinary Portland Cement because, when it hardened, it looked not dissimilar to natural Portland stone (or so its makers claimed). During the latter part of the Victorian period, hundreds of thousands of speculatively built houses were erected. Many of these were rendered, partly to hide the poor-quality brickwork used in their construction, and partly to improve their weather protection (cavity walls were very rare until the early 1900s). From the 1900s to the present day, rendering has maintained its popularity as a walling finish in both new build and rehabilitation work.

The two photographs above show wattle and daub as used in traditional timber frame buildings from the medieval period onwards. Medieval Vertical staves of oak with an interwoven hazel wattle usually formed the base for the daub. The daub itself could comprise a variety of materials including lime, straw, mud, sand and dung. In later periods the wattle was replaced by bricks to which the daub would then be applied.

A traditional lime render on a masonry background.

This picture shows the main shopping street in Monmouth. By the late nineteenth century it was quite common to find whole streets of properties rendered, particularly in Wales and the West, and in seaside towns.

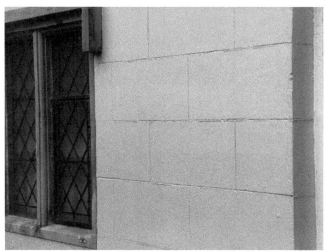

The two photographs above show a render which has been incised with fine lines to look like stonework. In many cases the brick is of poor quality.

This image shows the re-rendering of a brick wall to a Regency house. The top coat of the render will be scribed out to mimic ashlar in the manner shown in the previous two photographs.

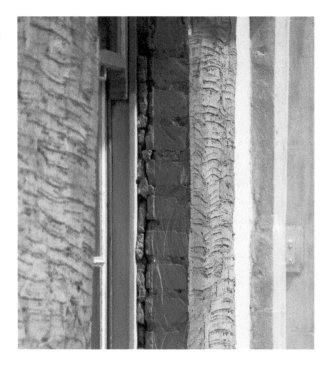

Rendering is applied to buildings for a number of reasons. These include:

* *weather protection*
* *hiding materials such as blockwork or common brickwork*
* *providing a decorative surface*
* *providing a smooth surface to take a painted finish.*

In recent years, renders have also been used to protect and hide external insulation systems.

Render finishes

Cement-based renders are normally applied in two or three coats (three coats are recommended in areas of high exposure). The manner in which the final coat is applied produces a range of aesthetic finishes which also may have an effect on the durability and performance of the render. Some of the more common finishes are described briefly below.

Smooth

Smooth finishes are popular. The finish is achieved by smoothing with a trowel. If the render is over-trowelled, this can bring a thin film of water/cement paste to the surface which forms tiny hairline cracks as it dries. This is perhaps the most difficult finish to achieve well.

Wet dash (roughcast)

This is a mixture of cement, lime, sand and small stones which is thrown and trowelled onto the undercoat to give a rough textured finish. This topcoat is quite strong which makes this finish unsuitable for weaker backgrounds. It provides a very durable surface but, in common with most coarse finishes, it will gradually accumulate dirt in an urban environment.

Dry dash (pebble-dash)

In roughcast, the coarse aggregate is mixed in with the final coat but in dry dash the aggregate is thrown (flicked) onto the topcoat after it has been applied but is still wet. The aggregate usually consists of small pebbles and/or coloured chippings.

Tyrolean

This is perhaps the most common of a number of proprietary finishes. It is usually applied by a hand-operated machine which 'flicks' the material onto the topcoat (usually with a hand-held machine) after it has been finished level. The material is usually pre-mixed and contains aggregate, cement and coloured pigments.

Some renders are quite unusual – pebble-dash is common enough but not using pebbles of this size.

This is a rather unusual render – marked out in a way that suggests an attempt to replicate ashlar stone but with exposed pebbles.

Unfortunately, render defects are all too common. Their causes are attributable to a number of factors, some of which are related to structural movement, an area which has been explained in earlier chapters. This chapter is more concerned with defects in render specification, inadequate background preparation and poor site practice. Before considering specific defects it is important to understand some simple principles relating to render specification.

Render specification

The strength and density of a render are related to its cement content. Strong or cement-rich renders are mixes of about 1 part cement to 3 parts sand. This mix produces a dense but brittle render which has excellent weatherproofing qualities as long as cracking can be avoided. The render is dense because the cement fills all the voids in the sand aggregate. A given volume of sand contains approximately 25–33 per cent air voids. By replacing the sand with the cement binder a void-free material is produced. However, renders containing such a high proportion of cement are likely to shrink on drying. As the render dries, there is consequential shrinkage which can cause the render to crack. Some strong backgrounds with a good mechanical key can resist this shrinkage because of the good bond formed with the render. However, most backgrounds do not form a strong key and cannot resist the ensuing shrinkage forces created by the drying render. The cracks which may appear form a path for rain penetration. Water entering the cracks cannot escape (because the render is dense) and may, therefore, possibly increase the incidence of damp penetration as water can only evaporate through the internal face of the wall.

Weaker mixes, which contain less cement, tend to be more durable, but a mix of, say, 1 part cement to 5 or 6 parts sand can be very difficult to 'work'. This problem can be resolved by adding lime to the mix or by using chemical additives. Chemical additives, or plasticisers as they are more commonly known, act as air-entraining agents; they reduce the internal friction of the render making it more 'plastic', with consequential improvements in its 'workability'. Lime fulfils a similar function though in a slightly different way. Lime improves the workability and the cohesive nature of the mix. It also reduces the amount of drying shrinkage and the risk of cracking. The cement and lime are usually batched to ensure that the two materials together form approximately one-third of the volume of the sand. This ensures that the voids in the sand are filled. If the voids are not filled rain penetration is more likely. Table 11.1 shows the typical mixes for render.

Table 11.1 Typical mixes for render

	CEMENT/SAND	CEMENT/SAND/PLASTICISER	CEMENT/LIME/SAND
Weak	1:7 (difficult to work)	1:2:9	1:7
Medium	1:5 (difficult to work)	1:1:6	1:5
Strong	1:3	1:0.5:4.5	1:3

Although weaker mixes are less dense, in practice, they are more effective at resisting rain penetration because they are less likely to suffer drying shrinkage and cracking. Although the render will absorb some rainwater, it will soon evaporate when the weather changes. In addition, weaker renders are more able to accommodate the stresses of thermal and moisture expansion which may occur in the render itself or in the backing material. Generally, it is accepted that weaker render mixes result in fewer overall problems than strong, cement-rich ones.

The sand used in the mix should be clean, sharp sand (coarse sand of a sharp angular nature) and should be well graded. In a well-graded sand there is a mix of particle sizes; this helps reduce the voids to a minimum. Very fine sands require a lot of water to achieve a workable mix and this will subsequently evaporate, resulting in high shrinkage and increased risk of cracking.

Some renders, by their very nature, are only suitable for backgrounds with a good strong key. Roughcast, for example, is essentially a mix of concrete trowelled onto a base coat. It is therefore a relatively strong mix and will tend to shrink as it dries. To maintain durability, the backing coat also needs to be strong to help resist this shrinkage movement. The base or backing coat, in turn, needs to be applied to a wall with a good mechanical key if shrinkage is to be avoided. Roughcast is therefore rarely successful when applied to backgrounds such as old friable brickwork or aerated concrete blocks.

Thickness and number of coats

On most surfaces two coats are acceptable though in areas of severe exposure three coats will be necessary. The thickness of the first coat should not exceed 15mm. Coats thicker than this will tend to sag under their own weight. In addition, when applying a thick render coat to a wall it is more difficult to squeeze out any air pockets; trapped pockets of air will prevent a good bond being formed with the background.

When the undercoat has begun to stiffen it should be combed or scratched to form a good key for subsequent coats. The surface scratching also provides stress relief points which help to prevent large cracks appearing as the render sets. The base coat should be left for a few days before applying further coats. Subsequent coats should be thinner and preferably slightly weaker than base coats to minimise the risk of shrinkage. A strong render coat should never be applied to a weaker base coat.

Table 11.2 presents some common defects during the construction process.

Table 11.2 Common defects during the construction process

POTENTIAL PROBLEM AREAS	EXAMPLES	IMPLICATIONS/MANIFESTATION
Background/bond/surface preparation.	Insufficient suction on smooth brickwork. Porous bricks provide too much suction. Flush mortar joints prevent key.	Lack of bond, loose and hollow render.
Render mix and number of coats.	Single coats rarely satisfactory. They are not weatherproof and cause 'grinning'. Render mix too strong.	Damp penetration caused by insufficient protection. Uneven drying after rain,
	Sand too fine, requires more cement and more water to get workable mix.	Shrinkage cracking – lack of weather protection.
Application.	Thick coats will sag. Dubbing out therefore required on uneven surfaces.	Uneven surface, usually with some loss of bond.
	Base coats must be property scratched to provide key for top coats.	Top coat partly bonded. May detach from base coat.
	Over trowelling.	Surface crazing.
Insufficient drying time.	Top coats should not be applied until base coats have completed shrinkage. Fast dry-out.	Shrinkage cracking of base coat affects top coats.
Chemical agents.	Lime bloom. In situ slaking of quicklime nodules causing cone shaped depressions in surface.	Surface finish affected.
	Incorrect angle beads, bedding beads.	Rusting of beads, sulfate attack in gypsum plaster.

POTENTIAL PROBLEM AREAS	EXAMPLES	IMPLICATIONS/MANIFESTATION
Quality of finish.	Day joints show at scaffold lifts etc., most common in smooth renders.	Aesthetic mostly.
	Bare or uneven dash finishes caused by top coat drying too quickly.	
	Tyrolean finish not confined to render (gutters downpipes etc!). Tyrolean too wet or too dry (see text).	

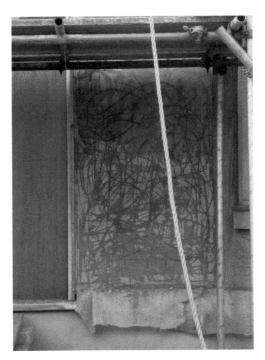

A common error on site is incorrect scratching of the base coat. Some plasterers assume this is just to provide a bond for subsequent coats. In fact, it is also to provide stress relief points for the drying (and shrinking) render. A properly scratched base coat will enable the drying render to spread these stresses across its face thus preventing wide cracks from appearing. Good practice is shown on the left, poor practice on the right.

Defects: manifestation and causes

Hollow renders

If a render sounds hollow when tapped, the cause may be lack of adhesion to the background. Hollow spots will often be accompanied by cracking (see later section) but this is not always the case. Hollow renders can occur through lack of adhesion of the base coat to the wall itself or lack of adhesion in the subsequent coats. It can be caused by incorrect mix (usually too strong), incorrect thickness of coats or inadequate surface preparation. Occasionally a hollow sounding render may not actually be defective, e.g. renders supported on metal lathing which are not designed to bond directly with the wall (see photograph following).

Hollow base coats

In order for the base coat to bond successfully to the wall, there should be a good mechanical key and the right degree of suction (absorbency). In addition, the render strength should be designed to match the nature of the background. Dense concrete blocks (or bricks) with well-raked-out joints provide a good key and the right level of suction and may even be suitable for quite strong mixes, say, 1:4, cement:sand. Aerated blocks have higher suction. Their relatively porous nature can 'suck' water out of the render mix before the process of hydration (setting) has taken place. To reduce this risk, the manufacturers recommend that dry blockwork is dampened before rendering takes place. Excess wetting should be avoided as aerated blocks will expand slightly as their moisture content increases.

Also aerated blocks have inadequate key to resist the shrinkage of dense renders and mixes should be no stronger than 1:1:6, cement:lime:sand, or its equivalent.

On smooth surfaces such as in-situ concrete or ashlar stonework, some form of surface treatment is necessary before applying the render. Surface treatments usually take the form of a spatterdash coat (a slurry of cement and coarse sand thrown onto the wall to provide a key – see photograph) or various water-based adhesives. To bond successfully, the render must be applied while the adhesive is still 'tacky'. If the render remains wet for long periods or the background is damp, these adhesives can break down after a number of years resulting in hollow spots in the render.

A spatterdash coat has been applied to this stone wall in order to provide a key.

Soft, porous stone or friable brickwork is rarely suitable as a base for rendering and metal lathing is usually required to provide separate support for the rendering.

Surface preparation of any background should include complete removal of moss, paint, loose material and soot.

Hollow rendering can also be caused by applying the render in extremes of hot or cold weather. In hot weather, the water may evaporate before hydration occurs. In cold weather, the render may freeze.

The detachment of render base coats is relatively common. In the cases shown here the fact that they are all gable ends, and therefore exposed, probably added to the rapidity of the deterioration. Note that in the middle left image (which is a close-up of the top right image), frost attack to the brickwork can be clearly seen.

The failure of the rendering on this stonework was probably aggravated by the use of too strong a mix.

The fact that the detached render is self-supporting seems to indicate that the mix is (too) strong.

Hollow subsequent coats

If the hollow spots are in the second or third coat, the likely cause is one or more of the following:

- *Inadequate scratching of base coat.*
- *Incorrect thickness; second and third coats should normally be between 6 and 10mm.*
- *Inappropriate mix – too strong a mix will shrink, too weak a mix may be difficult to apply, may not bond successfully and, in extreme cases, may not harden adequately.*
- *Render applied in weather conditions which are either too hot or too cold.*

The three images above show a detached top coat on a rendered wall (the second and third image are close-ups of the same defect).

The two images above again show a detached top coat (with the right-hand image showing the defect in detail). The problem is probably caused by inadequate background preparation and an incorrect (too strong) a mix.

The three photographs above show render which has been damaged by the rainwater soaking into it after the hopperhead to the rainwater system had been removed. It is likely that frost attack has occurred and that this has forced off the render.

Cracking: movement of the background

In many cases, renders will crack as a result of structural movement in the background. Even minor background movement can cause cracking in renders, particularly in stronger mixes. The pattern of cracking will give some clues to the cause; these have already been discussed elsewhere (see Chapter 9) but a few common examples are included below as an aid to diagnosis.

If chimneys lean like this, always suspect sulfate attack.

Cracking above a window suggests lintel failure.

Cracking which follows the brick joints is likely to be caused by structural movement.

Cracking along every course, especially in exposed positions, suggests sulfate attack.

Cracking at 450mm or so intervals suggests wall tie failure.

Note: Rendering can increase the risk of sulfate attack; there is more cement to react with the sulfates.

NB: Weak renders are more tolerant of minor movement than strong ones, but render cannot cope with the stresses caused by sudden structural movement.

Later extension

If the crack is wider at the top the cause is likely to be differential structural movement.

Cracks at high level may be caused by lateral thrust from 'cut' roofs or the lifting of modern trussed rafters which are not strapped down.

The render on this gate post is suffering from sulfate attack.

The cracking here is caused by a failed segmental arch over the window.

The cracking in the render (and the wall behind it) is caused by foundation movement.

Most of the examples above often result in fairly pronounced cracks. Finer cracks are often caused by differential movement of the walling material. This can occur in a number of situations. Concrete framed buildings with block infill panels will suffer from differential movement. The infill blocks will contract and expand at different rates from the load-bearing concrete frame; this is mostly caused by thermal movement.

Where fine cracks run horizontally and vertically, and coincide with the joints between frame and infill block, differential movement is the likely cause. At the design stage this could have been prevented by fixing a galvanised or stainless steel mesh across the columns and beams secured to the blockwork either side. A layer of building paper should be sandwiched between the mesh and the in-situ concrete to prevent the render bonding to the frame.

Cracking due to differential movement of frame and blockwork.

Special precautions are necessary to prevent renderings from cracking – see text.

Textured renders may hide cracking for a short period.

In modern construction where extensions are added, but are built from different materials, similar movement can occur. Cracking of a render finish can be prevented by fixing a flexible movement bead at the junction of the two different walling materials.

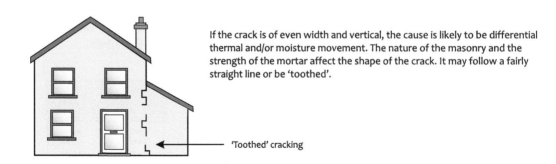

If the crack is of even width and vertical, the cause is likely to be differential thermal and/or moisture movement. The nature of the masonry and the strength of the mortar affect the shape of the crack. It may follow a fairly straight line or be 'toothed'.

'Toothed' cracking

Where terraces are constructed without provision for movement joints, cracking can occur in the walls and in the render. A building constructed with a weak mortar and covered with a weak render may be able to accommodate thermal movement; strong mixes are more likely to crack.

Rendered brick or stone terrace

If a building is constructed with fairly weak mortar cracking is less likely to occur in the wall itself. However, if it is rendered with a strong mortar, cracks may still appear between windows if there is thermal movement.

Terraces built in a strong cement mortar and with inadequate provision for horizontal movement are likely to crack due to long-term thermal movement.

The cracks will appear at the weakedt points, usually between windows. With weaker mortars any cracks in the masonry itself are likely to follow the joints; with strong renders the cracks can run through the stone or brick. In either case the cracks in the render tend to be fairly straight.

Cracking: problems related to render specification and workmanship

If the foundations and walls appear stable, problems of cracking are probably due to inadequate specification or workmanship in the render itself.

Shrinkage cracking

One of the most common problems is shrinkage cracking, usually as a result of using a render which is too strong for the background. In many cases drying shrinkage will also result in hollow spots. Drying shrinkage is normally fairly easy to identify. It is characterised by a pattern of cracking known as 'map cracking'. If dense renders are used on weak backgrounds, shrinkage cracking is almost inevitable. Rain penetrates the render but cannot escape through evaporation due to the render density.

The six images here all show shrinkage cracking.

Evidence of shrinkage cracking can be seen in this render. In this case, the render has also detached from the background.

Cracking along edges of angle beads

Vertical cracks at quoins or at window jambs are often caused by problems with angle beads. In traditional work, angle beads were rare but, nowadays, their use is widespread. It is, unfortunately, not uncommon to find angle beads bedded in gypsum plaster. This material sets quickly and is often seen as an ideal material in which to bed the beads. In wet conditions a chemical reaction between the cement and the gypsum (sulfate attack) causes expansion of the render and this will result in cracking.

Galvanised beads are easily damaged and will soon rust – particularly if they are in very exposed locations or near the sea (as is the case in the window bottom left).

In addition, rust can be caused by the use of galvanised angle beads which are easily damaged by the operative's razor-edged float. The sharp float edge can scrape off the zinc protection, leaving the unprotected steel free to rust. This can be avoided by using plastic-coated, stainless steel or non-ferrous angle beads. In 'traditional' rendering, angles were formed by working to a batten fixed proud of the wall at the quoin. This method is still occasionally used today and prevents the problems above from occurring.

Rendering over horizontal DPCs

Render should terminate just above the DPC, usually about 150mm above external ground level. Timber battens or proprietary stop beads can form a neat render 'stop'. On many houses, particularly those built in the 1920s and 1930s, the render runs down to ground level. Thermal and moisture movement will cause the wall above and below the DPC to move at different rates; the DPC providing a slip layer. This can result in cracking at DPC level. Apart from its unsightly appearance, the crack encourages rain penetration; this is often confused with DPC failure. Even if cracking is not apparent, damp problems can occur internally as ground water can bridge the DPC through capillary action.

Left: The render runs down to ground level and bridges any DPC. Right: This example has cracked, probably at DPC level.

Surface defects

There are a number of defects related to workmanship or specification which may not necessarily affect the render's long-term performance but which may affect its appearance. Surface crazing is a common form of cracking. Surface crazing occurs when the render surface (smooth renders) is over-trowelled. Over-trowelling results in a mix of water and cement rising to the face of the render. As this dries, it forms a pattern of very fine cracks. It does not necessarily affect the performance of the render itself though it can be very difficult to disguise.

Pebble-dash can be quite a durable finish. However, it is quite common to find areas of pebble-dash which are almost 'bald'. This is usually caused by inadequate workmanship and there are two aspects of good practice which need to be considered. First, it is usually necessary to add an approved waterproofer to the base coat, not to improve the weather protection of the wall, but to reduce its suction so that the top coat (butter coat) will dry slowly. While the top coat is still soft, the selected aggregate can be thrown against the wall with a scoop. Second, once applied, the aggregate should be lightly tamped into the butter coat to ensure a good bond.

Pebble-dash with some patches showing loss of aggregate.

When adding waterproofers to renders, caution needs to be applied. If the renders shrink and crack, they can allow rainwater penetration; the waterproofer restricts subsequent evaporation.

Tyrolean failures

The quality of a Tyrolean finish will depend on the suction of the coat to which it is applied. If the suction is too high, the finish will not bond properly and, if the suction is too low, the Tyrolean finish may sag or run down the wall, ruining its appearance. To be successful, it must be applied in one complete 'run'.

Scaffold lift marks

Smooth renders, particularly if they are not painted, require considerable skill if they are to provide a consistent, smooth finish. The normal practice is to start the top coat at the highest scaffold lift. The render is trowelled onto the wall working left to right (for a right-handed plasterer) and top to bottom. Before dropping to the next scaffold lift, the render is rubbed with a plastic float and possibly finished with a sponge. This exercise is then repeated on the lift below. Achieving a good joint at the junction of the lifts requires skill and fast work. It also requires the right weather conditions. If the weather is too hot, dry-out will be too fast and it will be impossible to achieve a 'seamless' joint.

Scaffold 'lifts'   Render sequence   Surface blemishes caused by poor workmanship on smooth renders

Surface popping

This problem is most noticeable on smooth renders which contain lime. If the lime contains small particles of quicklime (see Chapter 12), there is the risk of hydration (slaking) occurring after the render has been applied to the wall. The resultant expansion causes the render to 'blow', leaving small conical holes in the render surface. The first evidence of this is small circular cracks in the render about 50–75mm in diameter. After a few weeks or months a small cone of render is forced out by the expanding lime. It is often possible to see a small spot or blob of white slaked lime at the base of the cone.

The use of good-quality hydrated lime should avoid this problem. If the lime is in powdered form, the best results are achieved if it is soaked in water overnight before being added to the mixer.

Surface popping of render.

Grinning

If a faint pattern of brick or block joints can sometimes be seen on the surface of the render, 'grinning' may be occurring. Grinning usually occurs after rainfall and is often caused by using mortars of different strength, and therefore varying suction, from the bricks or blocks around them. The phenomenon is exacerbated by thin renders. The differences in suction, and therefore rainwater absorption, will result in the wall drying out at different rates.

In some cases grinning is caused by 'cutting corners'. One example is the practice of applying Tyrolean direct to fair-faced blockwork without any render base coats. However thick the Tyrolean, it is difficult to hide the minor imperfections in the blockwork and, thus, the joints 'grin' through. It is particularly noticeable on sunny days because the imperfections create small shadows which emphasise the defect.

Two examples of 'grinning'.

Lime bloom

Lime bloom manifests itself as patches of a pale, misty white powder on the render surface. It is caused by calcium hydroxide (formed during hydration) coming to the surface of a render during drying and slowly absorbing carbon dioxide to form calcium carbonate.

An example of lime bloom.

Problems with painted finishes and other surface treatments

Paint

Smooth renders and roughcast can be painted. If the correct paints are used and if due attention is paid to surface preparation, painted finishes can be durable. In addition, they will help to reduce the tendency of a wall to absorb rainwater. In practice, unfortunately, there are a number of problems which can, and do arise.

Paints should be vapour-permeable. External-grade emulsions and proprietary 'stone' paints are acceptable though conventional oil-based paints should be avoided. Apart from their impermeability (see bitumen paint below), they are not compatible with the alkaline nature of the cement in the render. They are likely to blister and flake after relatively short service.

Even if the correct paints are used, life expectancy can be quite short if the render surface is not properly prepared. Cracks should be cut out and filled, preferably with a flexible filler, and loose surface material and organic growth should be brushed, or washed off, the wall. Special 'stabilising' solutions are usually required on older renders to overcome the adverse effect of loose flecks of old paint, microscopic organic growth or other undesirable surface materials.

In some cases the painted finish is less than satisfactory in terms of appearance. This often occurs where the wall is 'marked out' into small manageable 'bays' as the work progresses. This effectively means that some parts of the wall have more coats than others. The pattern of the marking out is often visible and the only solution is to provide one, or more, extra coats.

'Marking out' can cause an uneven painted finish. Additional coat(s) will be required to hide the blemishes.

Do not confuse with 'grinning' or shrinkage cracking.

Bitumen paint

Completely impermeable coatings such as bitumen paint can cause problems more serious than the ones they are meant to eradicate. They prevent moisture escaping from a wall and can, in freezing conditions, lead to frost attack in the render. In many cases this type of coating is often specified in an attempt to resolve perceived problems of penetrating damp. However, incorrect diagnosis of the causes of dampness is common; surface condensation is often mistaken for penetrating damp. Where this is the case, external surface coatings will have no effect on the condensation. In some cases, for example, where interstitial condensation is occurring (see Chapter 16), impermeable surface coatings will prevent moisture evaporation from the wall and will be counter-productive.

Silicone

In an attempt to reduce the incidence of damp penetration, it is not uncommon to find renders coated with various clear silicone solutions. These can be effective in the short to medium term though they will be counter-productive if the render is cracked. Although these solutions are, to a degree, vapour-permeable, they do limit evaporation from a wall. Rainwater will penetrate the cracks and its impeded evaporation can lead to damp patches on the wall's inner surface. Where DPCs are non-existent or unreliable, silicone treatment of the render can lead to problems of damp penetration and efflorescence on the internal plaster just above DPC level.

Render treated with silicone solution or painted with vapour impermeable paint impedes evaporation from wall. Rising damp is more likely to evaporate into the building rather than through the render.

Imperfections in DPC (or no DPC) can lead to increased problems of damp and efflorescence.

Plastering and plasterboard 12

Introduction

This chapter describes the various types of internal wall finish found in housing and explains a number of common defects; defects, in the main, caused by inadequate specification, incorrect use of materials, poor workmanship, building movement and chemical attack.

Lime plaster

Lime plaster was common until the 1950s, though its use nowadays is largely confined to conservation work. Lime is produced by heating ground limestone or chalk (calcium carbonate) to a high temperature to remove carbon dioxide. The material thus formed is known as quicklime (calcium oxide). This material was traditionally slaked with water in large pits to produce slaked lime (calcium hydroxide). The lime pits could contain lime putty (slaked lime for later mixing with fine aggregate) or 'coarse stuff' (slaked lime and fine aggregate already mixed). The longer the material was left, the better the plaster it produced. In Roman times, it was not uncommon for lime to be left three years before use. Nowadays, there are still some specialist suppliers who slake lime in the traditional way, though most slaking takes place in factories where exactly the right amount of water is added to enable the material to break down into a fine white powder. This bagged material, known as hydrated lime, can be tipped straight into a mixer, though better results are obtained if it is mixed with water and left overnight before use.

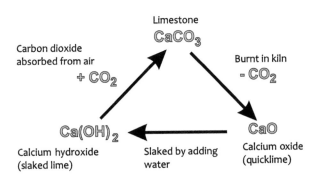

Limestone
$CaCO_3$

Carbon dioxide
absorbed from air
$+ CO_2$

Burnt in kiln
$- CO_2$

$Ca(OH)_2$
Calcium hydroxide
(slaked lime)

Slaked by adding
water

CaO
Calcium oxide
(quicklime)

Limestone burnt in open kiln to form quicklime.

Slaked lime in the form of lime putty can be stored indefinitely in the right conditions.

Slaked lime can be used in three main forms:

- lime putty
- 'coarse stuff' – mixed with sand
- hydrated lime (powdered form, sold in bags)

Lime plaster was usually applied in three coats: render, float and set – 8mm, 8mm and 3mm thick respectively. The render coat roughly levelled out the wall, the floating coat provided a true and even surface with uniform suction and the setting, or finish, coat provided a smooth finish for decoration. The base coats contained sand, or other fine aggregate, and were normally mixed in the proportions of 1 part lime to 3 parts sand. They were usually reinforced with animal hair in an attempt to resist the shrinkage action of lime as it set. Each coat would, or should, have been left for several weeks before applying subsequent coats. The setting or finish coat usually contained equal parts of lime and fine sand.

A thick base coat of lime plaster.

Lime mortars are relatively flexible and surprisingly able to adjust to seasonal movement and to subsidence, but their major drawback is their slow set. Lime plaster sets by slowly absorbing carbon dioxide from the atmosphere and returning to its initial chemical form, calcium carbonate.

The process of absorption is very slow and the plaster often forms a thin skin on its outer surface, as this is the part which carbonates first. It was quite common to leave these walls several months before applying paint or wallpaper. This delay would be unacceptable today. Carbonation can only take place in the presence of moisture and plasters which have dried out too quickly will have little strength.

Typical late Victorian specification

Lime plaster was traditionally applied in three coats: render, float and set.

The first two coats were normally reinforced with animal hair to improve tensile strength and reduce the risk of cracking.

Typical 1930s specification

Two-coat work sometimes replaced three-coat work. Cement was sometimes added to the mix to provide strength and increase the speed of set.

If the binder was cement or cement/lime, the hair was usually omitted.

In both periods, the sand was sometimes 'bulked out' with ash – a cheaper aggregate.

Lime plastering on timber lathing

Until the1940s, ceilings were usually formed from timber lath finished with lime plaster. The best practice used riven laths (split by hand) though, from the Victorian period onwards, the majority of housing used sawn softwood. The thin strips of timber were nailed to the joists. Small gaps in between the laths provided the key for the plaster.

In conservation work, an understanding of lime plaster is essential. Here a sixteenth-century ceiling from a Scottish castle is being re-created for Historic Scotland.

Three coats of lime plaster, sometimes gauged with a little cement, were used to form the finish itself. The first coat of plaster which was squeezed in between the laths was known as the 'pricking up' coat. A similar technique could be used on walls where the laths were fixed to a series of vertical battens fixed to the walls or to a free standing stud framework.

Most ceilings are attached to the floor joists – in some larger properties, particularly those with ornate ceilings, there may be separate ceiling joists.

Fir lath (nailed) and three coats of plaster.

Lime gauged plasters

Lime plasters, in their later years, were often gauged with a small proportion of cement or gypsum plaster to impart early strength.

The three-coat work to the wall dates from the late 1940s. The two base coats both contain cement, lime, sand and ash. The ceiling plaster 'pricking-up' coat is a similar material, although from the early 1930s. Note that the 'pricking up' coat has been forced so well between the laths of this ceiling that they can hardly be seen from above. These plasters only contain small amounts of cement and should not be regarded as brittle.

Cement plaster

Although the Romans discovered materials with properties similar to modern cement, the material used today has only become widespread since the early part of the twentieth century. Cement differs from lime in a number of ways. Unlike pure lime, it sets by hydration rather than carbonation. In other words, the chemical reaction occurs when in contact with water and not with carbon dioxide. Again, unlike lime, it cannot be 'knocked back' (i.e. returned to its original state) once the set has started. It is strong and brittle, virtually impermeable if mixed properly, and it develops initial strength rapidly.

Dense or cement-rich plasters (say, a mix of 1 part cement to 3 parts sand) have high drying shrinkage and their brittle nature means that they are unable to accommodate minor thermal and moisture movement; where these occur, cracking is common. They must therefore, be used with caution, particularly where strong mixes are specified (strong or dense mixes are sometimes specified to resist damp penetration). Increasing the amount of sand in the mix (say, 1 part cement to 5 or 6 parts sand) will produce a weaker, more tolerant material mix, but it is difficult to work and not favoured by plasterers.

By adding lime to the mix, greater plasticity can be obtained; the material is easier to work and, because it hardens more slowly than a cement and sand mix, there is less risk of shrinkage cracking. The cement and lime together (i.e. the binder) should be no more than one-third of the sand volume to ensure that the air voids in the sand are filled. A moderate strength mix would be 1:1:6 (cement:lime:sand), a weaker mix 1:2:9. Cement/sand plasters are usually applied in two coats (render and float) with a finish or setting coat of gypsum plaster. Cement/sand plasters cannot be trowelled to a fine, smooth finish and are rarely used as a setting coat.

Another method of producing a workable, plastic mix is to add a plasticiser. This is added as a liquid to cement/sand mixes and usually acts as an air entraining agent. It surrounds the tiny particles of cement and sand with air, thus reducing the internal friction of the mix and making it easier to work.

Cement/sand plasters are most likely to be found on the inside face of solid walls or where DPCs have been added or renewed. They are also common in basements, where the addition of chemical waterproofers can help reduce problems of damp penetration.

Most cement/sand plasters are batched on site, though some manufacturers produce pre-bagged cement/lime/lightweight-aggregate plasters suitable for use on most backgrounds (except plasterboard). Probably the most common of these is a proprietary mix called 'Limelite', of which a number of grades are manufactured to suit a variety of backing materials. They can also be used to replaster walls after the installation of new or replacement DPCs (see Chapter 16 for information on DPCs). Unlike gypsum plasters, they are not suitable for application to plasterboard.

Gypsum plaster

The mineral gypsum (hydrated calcium sulfate: $CaSO_4 2H_2O$) was formed by the deposition of salts in inland lakes during the course of many geological periods. As these lakes dried, and as the surface formation of the region altered, the mineral so formed was buried to varying depths. Pure gypsum is white in colour, but small proportions of various impurities cause the colour to vary to shades of grey, brown and pink.

The use of gypsum plaster spread to this country from France in the thirteenth century. Plaster of Paris (hemi-hydrated plaster), formed by heating ground gypsum and driving off part of the water, was renowned for its white nature and its ability to be moulded into complex shapes.

It was, however, very expensive and was reserved for the 'top end' of the market. Plaster of Paris sets very quickly and is not therefore suitable as an in-situ wall finish. It was, and still is, used to form pre-cast plaster mouldings found in decorative ceiling work.

By adding various substances to Plaster of Paris, its setting time can be delayed, and by adding bulk filler, such as sand, this material is ideal for plastering internal walls. Keratin is the chemical most commonly used to retard the set (thus giving retarded hemi-hydrated plaster); keratin is a natural protein found in the hair, horns and hooves of cattle. Much of it is imported from South America.

A cavity wall will normally be finished with two-coat work; float and set– 11mm floating coat and 2mm setting coat.

Gypsum plasters are the most common, although cement/lime-based pre-mixed plasters are also available.

Cement/lime/sand or cement/sand/plaster base coats are fairly rare, but quite acceptable.

Unlike lime, gypsum plasters set quite quickly and they do not shrink on drying; in fact, they are accompanied by a very slight expansion. The plaster sets by combining with water to form a mass of needle-like crystals which interlock and provide a set material of considerable strength. Because of their fast set, gypsum plasters can be successfully laid on a wall in thicker coats than lime plaster. Two-coat work (float and set) is normal with an overall thickness of about 13mm. Its fast set and reduced number of coats made it an obvious choice in the post-war reconstruction of Britain. A typical float coat would contain 3 parts sand to 1 part gypsum (batched on site) and the set or finish coat would be neat gypsum plaster, possibly with the addition of some lime to improve its working characteristics. Another advantage of gypsum plaster is that it can be applied to plasterboard, a material which first became popular during the Second World War where it was extensively used to help patch up bomb damage.

Gypsum plaster should never be mixed with cement. In wet conditions, sulfate attack can occur, caused by a reaction between the calcium sulfate and a by-product in the cement.

In traditional lime plastering, external angles were often formed with long strips of timber...

...or, in some cases, rounding the plaster arris and/or using a harder grade of plaster, possibly a hydraulic lime.

Nowadays, galvanised steel, stainless steel or plastic angle beads are used.

Lightweight gypsum plasters

In the last 30 years, or so, pre-mixed lightweight gypsum plasters have replaced the sanded mixes of earlier years. Nowadays, there are a number of grades available to suit almost any background. The pre-bagged plasters usually contain a lightweight bulk filler (aggregate) and only require the addition of water on site. This provides a material with reliable quality control that is both light and easy to use, and a material which, at the same time, provides improvements in thermal insulation when compared with cement-based plasters. Lightweight plasters can also provide substantial fire protection to vulnerable materials, such as steel beams and columns.

The lightweight aggregates are normally formed from exfoliated vermiculite (by expanding a mica-like mineral at high temperatures) or perlite. Perlite is a volcanic rock which is chemically combined with water. When heated to high temperatures, the combined water turns to steam, thus expanding the perlite to many times its original volume.

Table 12.1 Common site errors during the construction process

PROCESS/STAGE	TYPICAL SITE ERRORS	FUTURE IMPLICATIONS
Background, etc.	Smooth backgrounds, such as dense concrete, not primed. Mould oil left on face of concrete	De-bonding
	High suction backgrounds may suck water out of plaster	Plaster will dry before it has set properly – powdery finish with poor bond
		Efflorescence, hygroscopic salts, damp penetration
	Solid, possibly damp walls, are not generally suitable for most gypsum plasters	De-bonding
	Plaster board is not suitable for most undercoat plasters	Rust staining
	Metal lath is not suitable for some plasters	
Plaster choice (other than items above)	Incorrect grade of plaster. For example, where high impact resistance is needed, 'hardwall' plasters should be used	Surface damage to plaster
Plaster storage, etc.	Bags stored in damp conditions Use of old plaster Use of plaster straight from works	Plaster will not achieve its full strength
Mixing water	Water from butt may be contaminated with chemicals such as plasticisers and cement	Plaster will not achieve its full strength, sulfate attack
Undercoat plasters	Excessive thickness	De-bonding – it is difficult to exclude air when laying on thick undercoats
	No key for finish coats – light scratching is preferred	De-bonding
Finish plasters on plasterboard	Excess thickness – over 2mm or so, an appropriate undercoat plaster should be applied first	

Notes: For failure in cement sand plasters at the construction stage, see Chapter 11. For failure in plasterboard, see end of this chapter.

Durability and defects

Plaster

Generally

The durability of any plaster depends on a number of factors. These include:
- *background key*
- *background suction*
- *background movement*
- *strength of the mix*
- *number and thickness of coats*
- *dampness*
- *chemical attack*
- *defective workmanship.*

Left: The application of a modern gypsum-based finishing coat over the original lime plaster on the walls of the Victorian house together with further plaster repairs produced this result. Right: The effect of penetrating damp on lime mortar plaster in the corner of the room (caused by a poorly maintained leaking gutter) means that the affected area will need replacing.

Defects which occur because of the above problems manifest themselves in a number of ways. There may be cracking, either in the plaster itself or in the walling material. Surface deterioration accompanied by loss of bond may occur and, in extreme cases, partial or complete detachment of one or all the plaster coats. There may be staining, mould growth or efflorescence, each of which, while not necessarily affecting the durability of the plaster, may damage decorations. In many cases, the cause of the defect can be diagnosed without detailed investigation; shrinkage of cement/sand undercoats, for example, forms a characteristic and unique pattern. In other cases, thorough chemical or physical analysis may be necessary to pinpoint the exact cause of the problem.

Left: This problem is caused by minor structural movements. Right: This problem is caused by the shrinkage of the cement/sand undercoat.

Specification and site practice

Even with a good mechanical key and good suction of the background, bond failure, surface deterioration and cracking can still occur because of poor specification of the plaster and inadequate site practice. There are a number of common causes leading to these problems and these are explained below.

Mix. Gypsum plasters, nowadays, are nearly always factory 'bagged'. This ensures that the mix is of the correct proportions. As long as no materials other than water are added on site, the mix should not cause any problems.

However, with cement-based plasters, site batching is normally undertaken and this is often a fairly crude process. Excess water in the mix, or a badly graded sand, will lead to problems of shrinkage cracking. In addition, it is common to find mixes containing too much cement. These will shrink as they dry and, though some backgrounds such as dense blockwork can resist this, most will not. As the plaster dries and shrinks, it forms the characteristic map pattern seen in the previous photograph.

Water. Water that contains contaminants can affect the ability of the plaster to set properly. This can affect both cement-based and gypsum plasters. Water should be drawn from the mains, not from butts or drums which have been used to clean tools, etc.

Water required in the setting process can, as already mentioned, be removed by high suction backgrounds. It can also be removed by evaporation. This is most likely to occur where walls have been 'forced' to dry by using hot air blowers. A similar problem can occur where very thin plasters have been applied in very hot weather. The resultant finish, in both cases, is soft and powdery – this is known as 'dry out'. The only remedy is to strip and replaster the wall.

Additives. To improve the workability of site-batched cement-based plasters, a plasticiser can be used. This should be gauged strictly in accordance with the manufacturer's recommendations. Most plasticisers act as air entraining agents. They surround the tiny particles in air, thus reducing the friction in the mortar. Too much plasticiser, or plaster too long in the mixer, may contain too much air which will ultimately reduce the strength and bonding properties of the plaster. Washing up liquid has also been used as a plasticiser. This material produces a very workable mix, but chemical additives in the liquid will ultimately affect the strength of the plaster.

It is not unknown for milk or urine to be added to gypsum plasters in an attempt to slow their set and keep them workable for longer periods. These prevent a good finish from being achieved and may ultimately affect the bond of the plaster.

Sand. For best results, sand should be well graded and clean. Sea sand, if not washed properly, can affect the set of plasters. It can also cause efflorescence (because of the presence of chlorides) and recurrent problems of dampness (it becomes hygroscopic, i.e. acts like a sponge). Chlorides can also affect paint finishes and will corrode metals such as angle beads bedded in the plaster.

Pit sand often contains a small proportion of clay. Although this provides a very workable, loamy mix and is, therefore, favoured by many plasterers, the plaster will not achieve its intended strength.

Thickness of coats. On uneven walls, 'dubbing out' (i.e. filling in of hollow spots) will be necessary before applying the render or floating coat. Trying to compensate for the uneven nature of the wall by applying an extra thick render coat is rarely successful. Thick plaster coats need considerable pressure if they are to bond to the wall and ensure that any air pockets are squeezed out. At the same time, thick coats are likely to sag under their own weight. Thick coats are a particular problem with heavy cement/sand plasters.

If cement/sand render coats are too thin, they will dry out too quickly. This will result in shrinkage and loss of bond.

'Dead rats'. This is a problem that is sometimes found in old lime plaster. The 'dead rats' are bundles of reinforcing hair which have not been distributed correctly in the mixer. They can be a problem where hair has been added to the setting coat because it affects the smoothness of the surface finish.

Cracking

The nature and pattern of cracks will help to diagnose their cause. Perhaps the first stage in any diagnosis is to identify whether the cracking is caused by movement in the background, i.e. the wall, or is confined to the plaster itself. Some cracks are usually attributable to a fault in the building itself. These can take many forms (see Chapters 3 and 4). Diagonal cracks, whether stepped or in fairly straight lines, are usually caused by building movement. In new buildings, these may be fairly minor and may be caused by initial building settlement. Cracks which occur in later years, particularly cracks over 2 or 3mm wide, are likely to be caused by differential structural movement. Careful monitoring is required before taking any remedial action.

In this example, removal of a small area of cracked plaster reveals that the problem is caused by structural movement.

In older properties, horizontal cracking at junctions of ceilings and walls, or vertical cracks in the corners of rooms, may be the first signs of progressive vertical or lateral movement. In some cases, further monitoring and exploration will be required, though in others it may be due to minor differential movement caused by a combination of shallow foundations (possibly at differing depths) and a long hot summer. As the weather changes and the ground once again absorbs water, the cracks will often close.

In older houses with solid walls, horizontal and vertical cracking of the brick joints can be caused by sulfate attack. Cavity wall tie failure can also cause horizontal cracking of the joints (unlike sulfate attack these cracks will normally be at 450mm centres – 2 courses of blockwork or 6 of brick). This problem only occurs in cavity walls and will not usually manifest itself until a property is at least 20 years or more old. In many cases, wall tie failure will only be noticeable in the external leaf. (This serious problem is explained in more detail in Chapter 6.)

Hairline cracks at the junction of walls with adjoining walls and ceilings are usually due, in new buildings, to shrinkage caused by post-construction drying out. A wider crack at the junction of the upper-floor ceiling and wall may be caused by lack of roof strapping. If the crack is horizontal and about 50–70mm below the upper-floor ceilings, it can be caused by applying plaster directly to timber wall plates. As the plate shrinks slightly, a hairline crack is formed.

Cracks like this, usually at high level and in solid walls only, suggest sulfate attack.

Cracks every 4 or 6 courses of bricks or 2 courses of blockwork may suggest wall tie failure.

Wide diagonal cracks, and stepped or in straight lines, suggest serious structural movement.

Structural cracks will usually be visible on both sides of the wall although some wall tie failures will only be evident in the external leaf.

Cracks like these (left and right) suggest major structural failure.

Cracks which follow mortar joints (internal leaf of cavity walls only) and which occur in the early stages of a building's life are probably due to shrinkage of wet blockwork or blocks laid in too strong a mortar.

As shown below, map cracking is most likely to occur where gypsum finishing plasters have been applied to a cement/sand floating coat. The cement/sand coat will shrink as it dries, particularly if it contains a high proportion of cement. If the gypsum setting coat is applied before shrinkage of the floating coat is complete, the continuing shrinkage will crack the finish coat. This type of failure is most likely to occur where houses with solid walls have been replastered. Cement/sand render and floating coats are rarely used in new housing.

In two-coat gypsum work, fine vertical cracks above switches are usually caused by inadequate thickness of the finish coat where it hides plastic trunking containing electricity cables.

A horizontal crack at the top of the upper floor may be caused by inadequate strapping of the roof or rot in wall plate.

A less significant problem can be caused by failure to provide a mesh covering to the joint between wall plate and wall.

In recent years another problem has come to light; cracking caused by rotation. As new bricks expand and wet blocks shrink, the cavity closer rotates slightly. Flexible closers are better.

Inadequate thickness of plaster over trunking.

Cracks above lintels suggest differential movement or lintel failure.

Typical 'map' cracking where gypsum finish coat is applied over sand/cement undercoat which is still undergoing drying shrinkage.

Shrinkage of thin 'finish' coats in hot weather.

Hairline cracks following courses of blockwork may be caused by moisture movement in the blocks.

Cracking also commonly appears where two different backgrounds meet. Even if blockwork and brickwork are 'toothed' together, cracking may occur because of the differing rates of thermal and moisture movement. In housing, a common example of two different materials causing cracking is where extensions have been added. In framed blocks of flats, the problem can occur where concrete blockwork abuts concrete columns.

Bond failure

If a plaster surface appears to be in good condition, but sounds hollow when tapped, this usually indicates a failure to bond properly with the background. This may be due to top coats not adhering to base coats or base coats not adhering to the wall. Localised hollow spots do not necessarily mean that the plaster is loose and they may not warrant further action. However, where hollow spots are extensive, or where there is associated cracking, hacking off and replastering may be the only option.

Failure of a base coat to bond successfully to a wall can be caused by a number of factors. Two of the most significant are the mechanical key and the degree of suction offered by the walling material. Bond failure caused by chemical attack is included in a later section.

Mechanical key problems. Hollow or loose plaster may indicate that there is insufficient key with the background. This may be due to inadequate surface preparation or because of the nature of the walling material. Problems with the former include failure to remove loose material, paint, dust, soot and organic growth. The nature of the walling material can vary considerably. Dense concrete blocks, for example, with well-raked-out joints provide a good key for all plaster work. However, some materials such as glazed bricks, in-situ concrete, old friable brickwork, stone with wide lime mortar joints and composite surfaces, such as timber and brick, will rarely provide a suitable background.

In existing properties where plaster has been removed, much of the key will be lost because it is virtually impossible to remove all the plaster from the walls. The tiny pockets and blemishes that occur during manufacture on the blocks and bricks form an important part of the key. These will usually be filled with the previous plaster and this is difficult to remove. There are a number of techniques used to help deal with surfaces such as these including the use of spatterdash coats (see Chapter 11), metal lath and proprietary bonding agents.

Left: Metal lath has been fixed to brickwork to provide a key for plastering. Middle: The brick corner has been covered with an angle bead in order to avoid the type of damage seen in the right-hand photograph.

Metal lath is often used on weak backgrounds, but there are a number of ways in which it can fail. These include:
- *inadequate fixing (screwing is better than nailing)*
- *flat rather than ribbed meshes fixed flush against the wall, thus preventing the 'pricking up' coat from being forced round the back of the mesh to form a good key*
- *no overlaps between adjoining sheets of lath.*

Three-coat work is normally required on metal lath, as the first coat, known as the 'pricking-up coat', should squeeze through the mesh and form a strong mechanical key. Cement/sand plasters, because of their high alkalinity, will help prevent the lath from rusting. However, if gypsum plasters are used, a grade should be used which includes a rust inhibitor.

Adhesives and spatterdash (a wet, rich cement/sand mix) are more likely to be applied to smooth brick or concrete surfaces where there is little key, though there are a few proprietary plasters which will adhere well to in-situ concrete, e.g. Thistle Bonding. However, they need to be applied with care if problems are to be avoided. Water-based adhesives such as PVA (polyvinyl acetate) have a tendency to emulsify in wet conditions. They are also ineffective if the plaster is applied after the PVA has dried. EVA (ethylene-vinyl acetate), silicon and bitumen-based adhesives are often preferred.

Clean common brickwork with raked-out joints provides a good key for plaster.

This cement-based render coat has been firmly scratched to form a key for the floating coat.

The floating coat is lightly scratched to form a key for the gypsum-setting coat.

Suction problems. The suction, or absorbency of the background, can also lead to problems. Backgrounds with moderate suction are suitable for gypsum plasters and cement/sand plasters with a moderate or low cement content. Low suction surfaces are also suitable as long as there is a good mechanical key. Where suction is high, even if there is a good mechanical key, problems can occur through incorrect choice of plaster. Some aerated blocks, for example, have very high suction due to their open, porous nature. Aerated blocks, because of their high thermal insulation, are the normal specification for the internal leaf of modern cavity walls. British Gypsum produces a plaster that is suitable for aerated blocks; it contains a water-retaining agent which prevents water (required for the chemical reaction which binds the plaster together) being drawn into the blocks. If other gypsum or strong cement/sand base coats are used, the high suction of the aerated blocks will draw water out of the plaster and reduce its strength, the bond, and the quality of its surface finish.

On site, plasterers sometimes try and overcome the problem of high suction by dampening or wetting the background. If done with care and in moderation, this can be successful. If the blocks are saturated, they will expand; the resulting shrinkage will form cracking which tends to follow the block joints. This can be confused with problems in the initial construction. Lightweight blocks, for example, laid in a dense mortar, can result in cracking of the bed and perpend joints.

Loss of bond in gypsum finish (or setting) coats. In many cases, plaster failure will be a result of failure between coats rather than failure of the coat in contact with the background material. Most vulnerable is the thin finish coat which, in most cases, will be formed from gypsum plaster.

The floating coat should be lightly scratched to form a key for the finish coat, which is about 3mm thick. Inadequate key will result in the finish coat 'shelling off'. For plasterboard, the correct choice of finish coat plaster is even more vital. Plasterboard's relatively smooth surface does not provide a strong key for the finish coat and most manufacturers produce a finish coat plaster designed specifically for plasterboard. It has reduced expansion on setting to reduce the stresses which may occur at the interface of board and plaster.

An inadequate key can also occur if the backing coat (cement/sand) is over-trowelled. This will bring the cement to the surface and can result in the finish de-bonding.

If a cement/sand backing coat is still wet when the finish is applied, the excess moisture can prevent the crystallisation growth which bonds the finish plaster to the backing coat. In addition, cement/sand base coats which are wet will not have undergone their full shrinkage.

In warm weather, or when using finish plaster that has started to set, there is a temptation to add water. This can result in a soft friable finish. A similar problem occurs when using plaster that is beyond its shelf life. Plaster should normally be used within two months of manufacture. Old plaster has a 'flash set' which results in a rough powdery finish.

Chemical attack and staining

There are a number of problems caused by chemical action which may result in surface deterioration or, in some cases, disintegration or de-bonding of the plaster.

Efflorescence. Soluble salts, usually present in the wall itself (or in some plasters), can be deposited on the surface of the plaster during drying. These salts are usually carbonate and sulfate ions. They are normally harmless and can be brushed off once the plaster is dry. However, if the efflorescence continues, it may suggest a problem of damp penetration that warrants further investigation (see Chapter 16). More serious are problems caused by nitrates and chlorides. The former are carried up the wall in solution and indicate a problem of rising damp. The latter may be present in the ground or in some building materials (unwashed sea sand, for example). These salts tend to be deposited at the peak of the rising damp. They are hygroscopic (they absorb water) and this prevents them from crystallising. (Chapter 16 explores this phenomenon in more detail.)

Left: An example of efflorescence caused by problems in the brick, i.e. not the plaster. Right: The efflorescence has been caused by long-term penetrating dampness through a solid wall.

In some cases, efflorescence can be caused by chemicals in the plaster itself. Lime which contains a small proportion of clay (hydraulic lime) may contain the soluble alkalis of potash and soda. These can effloresce in the form of sulfates or carbonates. Some gypsum plasters are also affected. The photo above (left) shows efflorescence which is derived from chemicals in the bricks and not the plaster. This is often a temporary phenomenon in new, drying brickwork and should not be confused with the staining cause by long-term rising or penetrating dampness (right).

Magnesium sulfate attack. This problem only occurs in cement or lime-based plasters (i.e. alkaline plasters); gypsum plasters are not affected. It causes a loss of bond due to the formation of needle-like crystals of magnesium sulfate below the surface of the plaster. It occurs because magnesium sulfate, which is present in some bricks, is dissolved by water in the wet plaster and reacts with slaked lime (calcium hydroxide), forming a deposit of magnesium hydroxide. In gypsum plasters, crystallisation of magnesium sulfate occurs on the surface of the plaster and appears as efflorescence.

Sulfate attack. Internally, sulfate attack is comparatively rare, though it can occur where cement and gypsum are in contact with each other and remain wet for long periods. Unfortunately, there is a temptation for some builders and, perhaps, DIY enthusiasts to add gypsum plaster to a cement/sand mix to accelerate its set. However, as previously explained in the sub-section on gypsum plasters, this risks future problems of sulfate attack.

A similar problem occurs where, on walls intended for cement/sand, angle beads are bedded in gypsum plaster because it sets quickly. In damp conditions, sulfate attack can occur and leads to expansion of the plaster and its eventual failure.

Sulfate attack can also occur in cement-based undercoat plasters if they contain ash (a sulfurous material) and if they remain wet for long periods. The plaster will expand slightly, forcing away the finish coat – whether it is gypsum-based or lime-based (see photograph below).

Chapter 6 discusses sulfate attack in more detail.

Staining and damp penetration. Galvanised or stainless steel angle beads are normally installed to form corners and angles in the plasterwork. Unfortunately, the galvanising coat can be scraped off the beads when plasterers run their sharp trowels along the bead's edge and the unprotected steel can also be exposed when the beads are cut to length. Rust stains can then occur in rooms of high humidity, e.g. bathrooms and kitchens, because of condensation, and rust on angle beads can also be a result of damp penetration of the wall. The problem is more likely to occur where gypsum plasters are used because cement-based plasters and lime are alkaline materials which prevent rust from occurring.

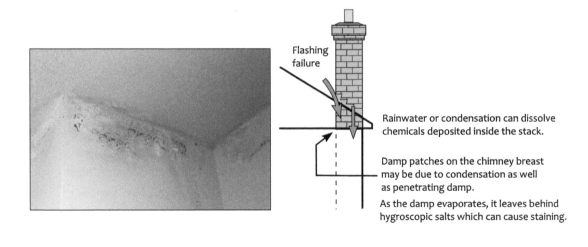

Flashing failure

Rainwater or condensation can dissolve chemicals deposited inside the stack.

Damp patches on the chimney breast may be due to condensation as well as penetrating damp.

As the damp evaporates, it leaves behind hygroscopic salts which can cause staining.

Staining of plaster is a common problem on chimney breasts. It is often assumed that this is due to water running down the stack itself, or possibly due to flashing failure where the stack runs through the roof. Although this can be the cause of staining, there are other possibilities. Ammonia and sulfur dioxides (from solid fuel or gas or oil) form salt deposits within the stack or flue. These salts are hygroscopic and are often accompanied by rusty-looking stains. Condensation (or rain water) occurring in a flue can absorb these soluble salts and deposit them on the plaster surface when the water eventually evaporates through the flue wall.

Long-term damp within a wall, caused by rising damp or damp penetration will result in plaster damage and staining of the surface. These effects were caused by long-term saturation and both rooms are at ground level.

Other surface defects

This sub-section includes a number of common problems which, though unsightly, will not necessarily affect the durability of the plaster.

Popping and pitting. These occur in plasters which contain lime. They are caused by small particles of unslaked lime which slowly continue to slake after the plaster has been applied. As the plaster slakes, it expands. The small particles form small conical holes about 50mm in diameter. At the base of the cone, a small white lump of slaked lime can often be seen.

Lime bloom. This can occur in plasters and renders which contain lime. Lime bloom is a fine film or mist of calcium carbonate which forms on the face of the work. It is caused by hydrated or slaked lime dissolving in water and rising to the face of the plasterwork on drying. There it reacts with carbon dioxide to form calcium carbonate. It can also occur in cement/sand mixes which do not contain lime as calcium hydroxide is formed during the hydration process.

Pattern staining. Pattern staining, where the faint outline of the joists or battens can be seen on the surface of plasterboard, is caused by dust in the air which is deposited to a varying degree on the materials it comes into contact with. Cold surfaces are more likely to attract dust than warm ones. Thus, a plaster ceiling fixed to timber joists is more likely to find dust settling on the underside of the plaster in between the joists than directly under the joists themselves. The plaster immediately below the joists is better insulated and therefore attracts less dust. This problem often occurs beneath uninsulated pitched and flat roofs

Mould growth. In conditions of high humidity, mould growth can occur on plastered walls and ceilings. Gypsum plasters are particularly at risk. The problem is explained in more detail in Chapter 16. It is most likely to occur on cold walls and ceilings in poorly heated buildings and/or in rooms where ventilation is limited (in bathrooms, kitchens, behind wardrobes, etc.). Well-heated and well-ventilated buildings are not usually at risk.

Mould growth, but not the underlying problem of its cause – condensation – can be prevented to some extent by adding a small proportion of lime to a gypsum finishing plaster. This is because mould growth is less likely in an alkaline environment. However, adding lime will result in a slightly softer surface finish.

Grinning. This usually occurs where thin or insufficient plaster coats are applied. The joints of the block or brick wall or the scratch marks of the previous coat appear to 'grin' through the top coat. (See Chapter 11 for an example.)

Lack of drying time between coats.
Gypsum plasters normally set within a few hours. If a finishing (or setting) coat is applied before the base coat is firm, it is very difficult to achieve a smooth, even finish.

Paint defects. Insufficient thickness of finish plaster can also result in problems with painted finishes. The finish coat, where it is thin, is unlikely to provide even suction. This affects the painted finish because there is an uneven build-up of paint film. Trying to use single coats of emulsion on plaster finishes of the correct thickness can result in a similar appearance. Most paint manufacturers recommend a 'mist coat' followed by two full coats.

The setting coat was applied too soon (within a couple of hours). Extensive sanding was required as well as lining paper to provide a suitable finish for painting.

Problems can also occur where plasters which are still wet are painted with impervious paints. The trapped moisture may cause subsequent blistering or bubbling of the paint work and, possibly, breakdown of the plaster itself.

Plaster failure following DPC installation and problems of dampness

See the sub-section on problems with replacement DPCs in Chapter 16.

Failure of lath and plaster

Timber lath ceilings (and partitions) can fail for a number of reasons. A common problem occurs where the laths are fixed too close together or are too far apart (the optimum is about 10mm and never less than 6mm). This prevents the first coat of plaster from squeezing between the laths and forming a good key. Another problem is fixing laths to wide joists. This also prevents the plaster from forming a key. Counter-battening prevents this problem from occurring, but its use is rare in all but the best-quality work.

Other problems with lath and plaster include:
- *excess weight from items stored directly on the lath in a roof space – leads to bulging or collapse*
- *rusting of nails – usually caused by damp penetration or old age*

- *dampness causes rotting of lath – in ceilings this can lead to bulging or collapse*
- *vibration from traffic and human activity breaks the mechanical key between lath and plaster – in ceilings, this can result in bulging or detachment*
- *the use of oak laths (rare) – tannic acid in green oak will corrode ferrous fixings and stain plaster.*

If laths are too close together, the pricking up coat cannot squeeze between them.

If laths are too far apart, it places greater stress on the pricking-up coat.

Where joists are wide, there is no bond for the plaster.

General bulging can be caused by excess load on laths (storage in roof), detached coats, rusting nails and rotten laths. If the joists are undersized, excess 'spring' in the floor can dislodge plaster.

Lath too wide

Lath too close together

Lath too far apart

Plasterboard

Generally

The use of plasterboard first became common in the 1930s when it slowly, but steadily, replaced plaster and timber lath in ceiling construction. It was also used extensively during the Second World War to help patch bomb-damaged buildings. Since then, its popularity has grown steadily and in modern buildings it is also used to line internal walls and in partition systems.

Wallboard with square and tapered edges

Typical thickness–9.5, 12.5, 15, 19, 25mm
Typical width–600, 900, 1200mm
Typical length–1800, 2400, 3600mm

Plaster setting coat

Plasterboard lath is a board designed to receive plaster. The boards are quite small (1200 by 400 typically) for ease of handling.

Laths, unlike other plasterboards, do not require reinforcement at the joints.

Their small size means that stresses from joist shrinkage can be spread across the ceiling, thus reducing the risk of cracking.

Plasterboard is available in a range of sizes and thicknesses. Plasterboards comprise a gypsum plaster core covered on both sides (and along the long edges) with a heavy paper. The long edges can either be square, rounded or tapered. Tapered edges are designed for taping and direct decoration and there is no need to provide a plaster skim to these boards.

As a dry lining, plasterboard can be fixed in a number of ways. Probably the most common is to apply dabs of adhesive to a block or brick background. Another option which is becoming common is to fix the boards to a system of steel channels.

Plasterboard is also used as a finish to metal and timber stud partitions and is invaluable for loft conversions and the like. Originally, plasterboard was nailed to timber supports – nowadays, it is usually screw-fixed.

Plasterboard can also be used below existing lath and plaster ceilings, for example, to improve fire protection or to hide minor defects.

Specialist boards

A number of specialist boards are available. These include plasterboards backed with a metallised polyester film to provide water vapour resistance, and plasterboards bonded to expanded polystyrene or polyurethane for better thermal insulation.

Finishing the boards

Where boards have been fixed, they are usually taped and jointed, as shown in the photographs below. Additional compound is feathered across the tape, thus disguising the joint.

(Photos courtesy of British Gypsum.)

When the joint treatment has dried, a slurry of special sealer or similar product should be applied to the board surface using a roller. This provides an even texture and suction for the painted finish. It also provides a sealed surface so that any wallpaper can be 'wet stripped' before repapering. A second coat of sealer can be applied over the first to provide a vapour control layer.

An alternative approach often used for ceilings or for stud partitions is to apply one or more coats of plaster to the boards. Where a setting coat alone is applied to the boards, they can be finished by filling in the joints between the boards with plaster (a grade suitable for boards) and then pressing joint tape or a cotton scrim into the plaster. The tape or scrim should then be covered again with a thin layer of plaster. When the joint treatment has stiffened, the plaster should be laid on the boards with firm pressure, to give an overall thickness of about 2mm. Each board manufacturer provides specific advice on the procedure.

If a thicker ceiling finish is required (for improved fire resistance or to hide any imperfections in the ceiling line), the boards can be given a floating coat of plaster (about 5mm thick) followed by a skim coat (2mm).

Defects

Plasterboard is an inert material and should be normally maintenance-free. It can be affected by dampness, impact damage or by incorrect loading. Shelves, for example, on dry-lined walls should be fixed to the wall behind and not just to the plasterboard, though light loads, such as ornaments or pictures, can normally be supported using special fixings designed for plasterboard.

Some defects are largely cosmetic; the boards on the next page were not taped before being plastered.

Most defects result from incorrect nailing, inadequate background support or incorrect preparation prior to applying a plaster finish. The diagrams on the next page show typical problems which can occur if correct installation procedure is not followed.

A continuous band around the perimeter (and around all openings) minimises air movement and improves thermal insulation.

On this staircase wall, the dabs are too far apart.

NB: If thermal board is used (with a lining of insulation), mechanical fixings (i.e. screws and plugs – usually two per board), are usually required to ensure the plasterboard protects the flammable lining in a fire.

The dabs should cover at least 20 per cent of the board area. Three rows are normally recommended for 12.5mm boards 1200mm wide.

(Diagram courtesy of British Gypsum.)

If boards are forced into an inadequate gap, the boards may bow.

If framing is more than 6mm out of alignment, it is difficult to bring the board into firm contact with framing members.

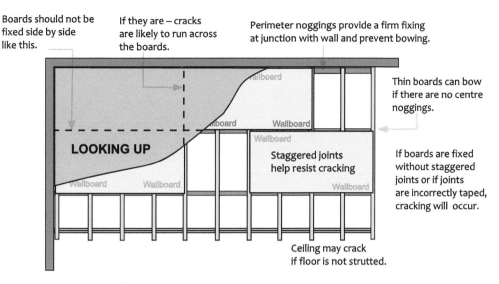

Boards should not be fixed side by side like this.

If they are – cracks are likely to run across the boards.

Perimeter noggings provide a firm fixing at junction with wall and prevent bowing.

Thin boards can bow if there are no centre noggings.

LOOKING UP

Staggered joints help resist cracking

If boards are fixed without staggered joints or if joints are incorrectly taped, cracking will occur.

Ceiling may crack if floor is not strutted.

(Diagram courtesy of British Gypsum.)

Boards should be held firm against framing members when nailed. Nails should not project beyond board face.

If boards are not fixed firmly against framing members, subsequent pressure on board may cause 'popping'.

Paper lining should not be punctured by 'over driving' nails.

'Over driving' nails will tear paper lining and reduce the board's attachment to the timber.

Careless hammering can be hidden by ceiling finish. It is sometimes less easy to disguise on self finished boards.

Nails not central in the framing will not be secure; board can pull away from framing member.

If drying timbers twist or warp, they can loosen the nail or pull it through the board.

About to 'pop' (above) and 'popped' (below).

Battens stored externally may contain high levels of moisture.

When the battens shrink, they can cause popping.

Ridging can occur if battens shrink. As the boards are squeezed together, the compressive forces form a ridge along the joint.

Ridging can also occur if battens twist.

Plasterboard in damp conditions

In damp conditions, plasterboard will quickly deteriorate. The plaster can swell and the paper linings can become detached from the inner plaster core. Perhaps the most common cause of dampness is applying plasterboard to solid walls which are prone to penetrating damp. In many cases, uninformed builders and DIY enthusiasts see this as a solution rather than a potential problem. Plasterboard used on ceilings can also fail if it is subject to constant wetting from leaks. Ceilings below bathrooms (especially showers) and pipe ducts lined with plasterboard are particularly at risk.

Vapour checks (vapour control layers)

Even if solid walls are dry, there is a risk of dampness, caused not by penetrating damp, but by interstitial condensation. Where external solid walls are dry-lined, the risk of interstitial condensation requires that a vapour check (sometimes referred to as a vapour barrier or vapour control layer) be installed on the warm (i.e. internal) face of any insulation. The purpose of the vapour check is to prevent warm moist air from inside a building coming into contact with the insulation or the cold wall behind it and condensing. The wall is cold because of the insulation. (See Chapter 16 for a detailed discussion.) This is explained in the diagram below.

Insulated plasterboard bonded to the wall.

Damp patches may appear on the face of the plasterboard. Condensation can also run down the wall and collect at skirting level.

It can be confused with penetrating damp and, at ground floor level, rising damp.

Adding a vapour control layer prevents the moist air from reaching the cold surface of the wall.

Asbestos boarding and fibreboard

The use of asbestos products has been closely controlled for many years. However, there are still many houses where asbestos boarding has been used to dry-line walls or finish ceilings. It was a very common lining material for framed buildings and for internal partitions in the post-Second World War reconstruction of Britain, particularly in non-traditional system housing. Some grades of asbestos boarding were suitable for painting, others received a plaster finish.

Where asbestos boarding is suspected, the key point to note is that the use of specialist advisers and contactors is mandatory under current health and safety legislation in terms of any procedures for its identification and its subsequent removal. (See Chapter 18 for a detailed discussion of asbestos problems.)

Asbestos boards were usually finished by painting (no plaster skim).

Timber cover strip to hide joints

Example from 1930s

Ovolo Moulded Cover Strip
No. 4982.
Size, $2 \times \frac{1}{4}$ in.
Price, Double or Single   **6/3** per 100 ft.

Thin boards can twist and bow, especially if the joist centres are 'stretched'. There are post-war examples of joist centres as wide as 900mm.

Many houses dating from the 1940s and 1950s have ceilings (and walls in many non-traditional houses) lined with fibreboard or hardboard. These can be quite durable, but will weaken and disintegrate if they get wet: they can also crack through impact damage or through incorrect fixing of shelving, etc. They also tend to naturally sag and bow, causing an uneven wall or ceiling finish. Fibreboard and hardboard, unlike asbestos or plasterboard, will burn and are, therefore, also potential fire risks.

This advert is from the late 1930s when various building boards made by companies such as Essex, Celotex, Sundeala and Masonite were growing in popularity.

Internal walls and partitions 13

Introduction

This chapter briefly describes the construction of load-bearing and non-load-bearing walls (partitions) and explains the most common defects which can occur. Many of the most common defects are, in fact, caused by incorrect plastering or incorrect fixing of plasterboard (see Chapter 12 for details).

Load-bearing or non-load-bearing walls

Internal walls and partitions divide space within a building and they can be load-bearing or non-load-bearing. Load-bearing partitions usually provide support to upper-floor joists and, in buildings with 'cut', or traditional roofs, support for part of the roof structure. They may also provide restraint to the external walls. Non-load-bearing partitions divide space and should not carry any loads from floors or roof structure. Depending on their construction, they may also provide some restraint to external walls.

Load-bearing and non-load-bearing partitions can be made from brick, concrete blocks, clay blocks and timber studding. Non-load-bearing partitions can also be made from compressed straw boards and proprietary plasterboard systems, neither material being suitable for load-bearing.

In practice, it can be quite difficult to determine whether or not an existing partition is load-bearing. Although the construction of the partition (where it is known) may give some clues, caution is required before reaching any firm conclusions. Timber studding is a common form of load-bearing partition, particularly in Georgian and Victorian construction where it was often used to form internal walls prior to the mass production of common bricks. In some Georgian and Victorian houses, timber partitions were trussed; in other words, they were constructed to span unaided across an open space, possibly, at the same time, providing support to the joists immediately beneath them.

The age of a building will give some clue as to the nature of its partitions. Modern houses, with roof structures constructed with trussed rafters, are unlikely to have load-bearing partitions on the upper floor as modern trussed rafters are normally designed to span from external wall to external wall. Modern houses are, however, likely to have one or more load-bearing walls on the ground floor (and the first floor, if the house is of three storeys) to support the floor joists. Only in the very smallest of houses will the upper-floor joists span from external wall to external wall without intermediate support.

Houses built before the 1960s are likely to have traditional or 'cut' roofs. In modest houses, an internal load-bearing wall was provided to support the upper-floor ceiling joists. In larger houses, where there are strutted purlins, the wall will also be supporting part of the roof structure.

Construction until 1950s

Double or purlin roof

purlin

strut

Construction from 1960s onwards
Trussed rafter roof spans full width.
First floor partitions carry no load
from roof structure.

Non-load-bearing
wall

Load-bearing
wall

Traditional roofs will have struts bearing on the internal
load-bearing wall. Where there are no struts the internal
wall will still be carrying the load of the ceiling joists.

Ground-floor walls support first floor only. They may
have foundations or be supported on thickening in slab.
Post-1990 floors are often suspended concrete.

To identify whether or not a partition is load-bearing requires, therefore, a broad knowledge of building construction together with an understanding of the development of construction methods. A roof inspection will fairly quickly identify the nature of its structure. However, to assess whether or not partitions are supporting floor joists is not quite so easy and may require lifting of floor boards. Where joists run across internal walls and bear upon them directly, it is likely that the walls are load-bearing though it is possible that the joists are merely providing the wall with lateral restraint. If joists meet, and lap, on internal walls there is no doubt that they are carrying load.

Left: Once fully plaster-boarded, it will be difficult to determine the exact construction of the wall. Right: An example of this problem can be seen with the partition to the right of the staircase. At the first-floor level, it is of plasterboard over a steel frame; at the ground-floor level, the plasterboard is fixed with dabs to a load-bearing concrete block wall. The wall sounds hollow over its full height; as the floor above is chipboarded, only its thickness gives a clue as to construction.

Construction summary

Brick and block partitions

Load-bearing partitions in modern housing are usually formed in blockwork and are most likely to be found at ground-floor level where, besides dividing space, they support the upper-floor joists. They will usually be built on a foundation though, in some buildings, they may be built on a thickening in the concrete slab. The partitions can be built from dense or lightweight blocks and are usually 100mm thick. In very small houses, where the joists can span from external wall to external wall, load-bearing partitions are unnecessary. Load-bearing partitions should always be fixed securely to the external walls as they provide mutual restraint. 'Toothing-in' is the best structural method though precautions will be necessary to ensure a cold bridge is not created (see next section on defects). An alternative is to bond the two walls together with wall ties, stainless steel mesh or 'crocodile' ties.

The photographs are of a typical Victorian house. The spine (central) wall is carrying part of the upper floor, ceiling joists and roof. This internal wall has a foundation, although it is often smaller and shallower than the foundations under the external walls

Non-load-bearing partitions formed from lightweight blocks are rare in modern construction, but were quite common from the 1930s to the 1960s. The blocks can be made from a variety of lightweight aggregates, most of which are no longer available. These include breeze, coke and clinker.

The blocks are usually 50–75mm thick and are often built directly off concrete and timber floors. The thinner blocks often require storey-height door linings to maintain stability. A storey-height lining has jambs (sides) which run up beyond the head of the door and are fixed to the joists above. Often a glazed storey-light is formed above the door opening. Although these blocks do not support loads, they may provide restraint to the external walls and certainly depend on the external walls for their own stability.

Brick partitions can be load-bearing or non-load-bearing. Despite the economic advantages of blocks, brick partitions were quite common until the 1960s. They are usually a half-brick thick (112mm), though in larger Victorian houses one brick-thick load-bearing partitions sometimes formed the spine wall which often supported part of the roof structure.

Hollow pot partitions are no longer manufactured, but their use was quite common in the 1930s to 1950s. They were made from concrete or clay and were usually non-load-bearing. The thinner examples required storey-height door linings to prevent deformation.

Left: A timber stud partition can be seen in the foreground; in the distance and to the right can be seen concrete block partitions. Right: Sound insulation is now a Building Regulation requirement in partitions between bedrooms or rooms containing a water closet and other rooms. Earlier houses are unlikely to meet this requirement.

Timber partitions

Timber partitions have been used in housing for several hundred years. Since the beginning of the Georgian period in the early eighteenth century, imported softwood has been the most common framing material. Timber partitions can be load-bearing or non-load-bearing, the former usually constructed from timbers of a larger cross-section (100 x 50mm is typical). Timber studding comprises a series of vertical softwood studs, at centres ranging from 350 to 600mm. The studs are fixed to softwood horizontal bottom and top plates and will normally be stiffened with softwood centre noggings to prevent deflection and twisting. Early stud partitions were faced with timber laths covered with three coats of lime plaster.

Left: In this Victorian building, partial demolition has revealed timber stud partitions to the upper floors. The upper partitions are trussed as they provide support to the roof. Right: Partial removal of lath and plasterwork has exposed the timber studs beneath.

In Victorian and Edwardian housing, the studs may have an infill of brick-on-edge, i.e. bricks laid on their stretcher face, to provide improved sound insulation. This was often the construction of the partition beside the entrance hall and vestibule.

Top and bottom rails   Noggings (stiffness)   Strips of fir lath
                                                        nailed to studding

Two or three
coats of lime
plaster

75 or 100 x 50mm studs   Brick-on-edge infill (mostly
                              ground floor partitions only)
Thicker studs around opening

In modern construction, studding is normally covered with plasterboard or, possibly, with a metal lath finished with in-situ gypsum plaster. Timber stud partitions are used in new timber frame buildings and sometimes in other new housing, though their use is more common in rehabilitation work.

As previously mentioned, in large Victorian houses trussed partitions were sometimes used when a load-bearing partition was required over a void. This might be necessary where rooms in an upper floor were situated over a larger room on a lower floor. Trussed partitions usually comprised two or more angled compression members to spread the load from above into the supporting walls and a tension member (often an iron or steel rod) to prevent the lower beam from deflecting. Trussed partitions are very effective, but problems may arise if subsequent building alterations cut through the compression or tension members.

Compression member
                                            Floor
Timber studs   Timber
                                                        beam
Opening cut for
new door will cause
twisting of partition
and sagging of floor
below
                                                        Floor

                                                        Timber
                                                        beam
Wrought Iron or steel rod in tension

Void underneath trussed partition – no
need for supporting wall below

Proprietary partitions

In modern housing, non-load-bearing partitions are likely to be formed from laminated plasterboard, cellular core plasterboard or steel studding – the latter finished with a 12.5 or 15mm plasterboard covering. However, laminated and cellular core partitions have fallen out of favour in recent years due to their relatively poor sound insulation.

Left: A cellular core partition. Right: A metal stud partition. A quilt (positioned in-between the studs) or, alternatively, thick plasterboard improves the sound insulation.

Other partition types

In the 1950s and 1960s, 'Stramit' partitions were quite common. 'Stramit' partitions were made from large slabs of compressed strawboard covered with a thick paper facing. Storey-height panels approximately 50mm thick and 1200mm wide were fixed to a timber or metal framing. The side joints were covered with a jute scrim and finished with a skim of gypsum plaster.

Summary of construction errors

This chapter describes the main generic causes of defects in partitions. It should be recognised, however, that there are also a number of common construction errors related to specific partition types.

Stud partitions

- *Studs at stretched centres or studs of inadequate size. As a rule of thumb, 100 x 50mm timber studs are suitable for load-bearing partitions, 75 x 50mm for non-load-bearing.*
- *Inadequate fixings, especially to external walls.*
- *Lack of noggings to stiffen the partition.*
- *Lack of additional noggings to support shelving loads.*
- *Poor sound insulation due to inadequate thickness of plasterboard finish and/or insulation.*
- *Poor fire protection due to undersized timbers, lack of noggings or inadequate thickness of plasterboard.*
- *Inadequate nailing of plasterboard to studs.*

Badly fitting studs can work loose.

Studs can deflect if undersized or at the wrong centres.

End stud should be securely fixed to supporting wall.

Light loads can be supported by plasterboard.

Heavy loads fixed direct to boards can pull them away from studding/framing.

Without centre noggings, studs can twist.

100 x 50 studs are suitable for load-bearing partitions; 75 x 50 for non-load-bearing partitions.

Left: A load-bearing partition in a timber-framed house. The partition contains a raking brace because it is helping to keep the timber frame rigid. Right: A non load-bearing partition.

Block partitions

- *Poor fixing to external walls; partitions can be bonded to pockets in the internal leaf or secured by mesh/wall ties.*
- *Incorrect choice of mortar; a dense (cement rich) mortar may shrink with consequential loss of bond and cracking.*
- *Inadequate thickness of blocks; 75mm blocks are generally suitable for non-load-bearing partitions; 100mm for load-bearing partitions.*
- *Use of wet blocks and subsequent problems of shrinkage.*
- *Blocks laid on ground floor slabs without DPC below.*
- *Lack of bond with other partitions.*
- *Use of thin blocks without storey-height linings.*

Plasterboard systems

- *Inadequate fixings at top, sides and bottom.*
- *Use of incorrect door linings.*
- *Hanging heavy doors on thin partitions.*

- *Reduced fire protection and sound insulation caused by addition of electrical sockets or pipework penetrating the partition.*
- *Excess loading caused by heavy shelves, etc.*

Defects in partitions

There are a number of common defects in partitions. These can occur through:
- *imposed loading*
- *lack of support*
- *poor connections to external walls and other partitions*
- *inadequate thickness*
- *poor sound insulation*
- *poor fire protection*
- *thermal/moisture movement, condensation and damp penetration*
- *poor initial construction (see above)*
- *overloading (from shelving, etc.).*

Imposed loads

In modern construction, one of the most common forms of failure is caused by erecting non-load-bearing partitions on the upper floor before tiling the roof. The weight of the coverings (3–4 tonnes) will inevitably cause slight deformation in the trusses. If a truss compression member is over, or near, a partition, it acts as a strut, transferring load from the roof into the partition. This may cause buckling in the partition and may also cause deformation in the floor immediately under the partition. Partitions which are not in close proximity to the compression members may cause upward deflection in the bottom chord of the truss.

Problems of a not dissimilar nature occur in traditional construction. Settlement of foundations (often of inadequate depth) below internal walls will manifest itself in sloping floors and out of square door openings. Where roofs are recovered with heavier materials, the same effect will occur. If the wall is central under the ridge, the force from the struts will be downwards. If the wall is offset slightly, the uneven forces may cause lateral movement as well.

Roof recovered with heavy materials.

Joists bearing on partition develop pronounced sag.

Partitions bearing on sagging floor will have openings out of square and/or picture rails out of level.

Half Brick thick int. walls

Differential settlement will depend on size and depth of original foundation.

In modern and traditional roofs, excess loads in the roof space can also cause buckling in non-load-bearing partitions. A cold water storage tank, for example, can impose a considerable load and, in a modern trussed roof, may also cause damage to the trusses unless correctly supported.

Incorrectly installed water tank will cause deformation in bottom chord of truss.

Lightweight partitions under the tank will bow or may cause deflection in floor below.

The load from tanks should be supported as shown in the left-hand graphic; i.e. using bearers near the node points of the trusses, and sitting on at least four trussed rafters.

Centre trusses cut away for clarity, and braces omitted.

Changes of use which require the addition of extra partitions (e.g. converting large houses into flats) may also cause deflection in a floor, and subsequent overloading of partitions on the storey below. (This is covered in more detail in Chapter 8.)

Lack of support

The problem of internal load-bearing walls without adequate foundations has already been mentioned. Non-load-bearing walls can be built directly on a concrete or timber floor. Additional floor joists are necessary if the partitions, whether of blockwork or studding, run parallel to the floor joists. Brickwork or blockwork supported by floor boarding alone will result in board deflection. In addition, it is virtually impossible to remove the boards at a later date.

Two closely spaced joists have been positioned to support a partition. The floorboards are fixed above the joists and beneath the partition (which runs off to the right).

Lack of side fixings

Most partitions, load-bearing or non-load-bearing, are quite thin. A secure connection to the external walls and other partitions is vital because they may be providing mutual support. The restraint from the internal walls is sometimes vital to ensure the structural integrity of the external wall. This is particularly important where the external walls receive little or no restraint from the floor joists. (This type of failure is covered in more detail in Chapters 4 and 8.)

Brick and block partitions should be bonded or 'toothed-in' to the wall; an alternative is to use wall ties or expanded metal mesh or 'crocodile' ties. Timber partitions can be plugged and screwed to the external walls. In earlier construction, partitions were often nailed to timber pads built into the brickwork. This method was common before the introduction of cavity walls and is not good practice; dampness in the solid wall can lead to rot in the pads and, possibly, the partition itself.

Some forms of roof construction may impose a lateral thrust on the external walls which can overcome the bond between partitions and wall. The right-hand diagram (at the top of the next page) shows how this can occur. A vertical crack (often wider at the top) indicates a problem of lateral movement.

Where partitions are not necessary to the structural integrity of the external wall, a good connection is still important because thin partitions rely on this connection for their horizontal stability. Thin partitions, whether built from bricks, blocks, studding or plasterboard, can twist or bend if this support is not provided.

Cracks between partition and wall.

Both load-bearing and non load-bearing partitions can help restrain external walls.

In some houses, the joists offer no restraint to the external walls. Altering partitions during conversion work can have a disastrous effect upon the integrity of the external wall.

In three- and four-storey houses, the risk of movement is much higher.

Roof loading pushes flank wall out.

Cracking between cross-wall and flank wall – crack wider at the top.

A typical Victorian end-of-terrace butterfly roof bearing onto load bearing walls.

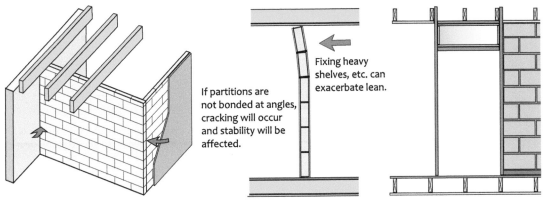

If partitions are not bonded at angles, cracking will occur and stability will be affected.

Fixing heavy shelves, etc. can exacerbate lean.

A long, even crack, where partition meets external wall suggests poor bond in original construction.

50mm and, possibly, 75mm block partitions may lean or buckle if not wedged or fixed to joists above.

Storey-height linings provide additional stiffness.

In these two photos, you can see some of the stresses cause by lateral movement of external walls. In both examples, movement of the left-hand wall (an end-of-terrace flank wall) has caused cracking at the junction with the internal wall.

Dampness, condensation and thermal/moisture movement

Penetrating and rising damp can affect partitions in a number of ways. Where partitions are built into solid walls or into defective cavity walls, damp may penetrate the edge of the partition. In studding and proprietary plasterboard partitions, this can cause rot in the framing materials and deterioration of the plasterboard. Rising damp can affect partitions at ground-floor level. This can be due to failure (or non-existence) of DPCs. Where partitions are built directly on concrete ground floors with a membrane below the slab, a separate DPC is still necessary. This ensures that any construction water, or rainwater trapped in the slab, cannot migrate up the internal wall.

Condensation can occur on the face of internal block partitions, particularly if they are built from dense concrete. It occurs in two common situations. The first is where partitions are built next to open or partly ventilated spaces; a typical situation is a wall separating a garage from the main body of the house. The cold wall will allow condensation to form on its inner face. The second problem can occur in modern construction where dense blocks are built into the inner leaf of a well-insulated cavity wall. The dense blocks form a cold bridge which allows condensation to form on the face of the partition adjacent to the external wall.

Finally, where blocks are left out on site uncovered and subsequently used in partitions, they can shrink as the building slowly dries out. As the blocks shrink, fine cracks occur along the bed and perpend mortar joints.

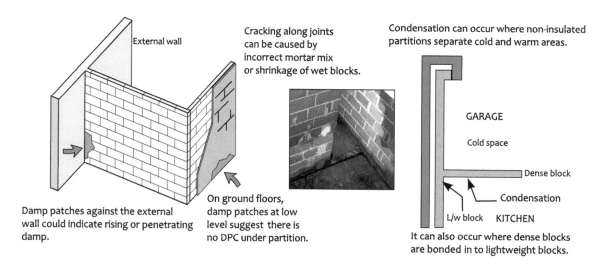

External wall

Cracking along joints can be caused by incorrect mortar mix or shrinkage of wet blocks.

Condensation can occur where non-insulated partitions separate cold and warm areas.

GARAGE

Cold space

Dense block

Condensation

L/w block   KITCHEN

Damp patches against the external wall could indicate rising or penetrating damp.

On ground floors, damp patches at low level suggest there is no DPC under partition.

It can also occur where dense blocks are bonded in to lightweight blocks.

In timber stud partitions, cracking may occur if wet timbers are used. Subsequent drying out of the timber causes cracking and usually manifests itself as fine cracks along the plasterboard joints.

In long lengths of blockwork (usually 6m or more), cracking can be caused by thermal and moisture movement. However, walls of this length are comparatively rare in housing.

Greater than 6 metres

On lengths of 6 metres or more, cracks may appear if movement joints are not provided. Depending on the strength of the mortar, these can appear as vertical lines through the blocks or cracked joints.

If partitions are altered through improvement works or alterations, blocks of the same type should be used. Different blocks have different characteristics and cracking at the interface of the two materials is a possibility.

In most houses, block walls are less than 6 metres long; movement joints are not usually required .

Other cracking

In new houses, where partitions abut ceilings, cracks often appear within a few months of construction. This is usually caused by shrinkage of new timber joists as their moisture content slowly settles into equilibrium with their surroundings. A more serious problem occurs in older housing (as shown in the photograph on the next page). This cracking appeared after a long hot summer and is probably caused by minor differential settlement or temporary changes in moisture content of the roofing or ceiling timbers.

Cracks like these may sometimes close slightly when weather conditions change. They are rarely of structural significance and are best dealt with by raking out and filling prior to decoration. Modern silicon-based flexible fillers are probably more durable than traditional plaster-based ones.

Cracking can also be caused by excess vibration (e.g. from traffic or machinery) and heavy doors which are continuously slammed.

Overloading

Heavy shelving should not be fixed to lightweight or thin partitions. There are two potential problems caused by the weight of the shelving and/or the loads supported: the failure of the fixings and buckling of the partitions.

External joinery and painted finishes 14

Introduction

This chapter identifies the nature of defects which can occur to external joinery: doors and windows. It also looks at the options and common defects for painted finishes for external timber joinery.

'Joinery' is a term which refers to components and elements which are manufactured away from a building site and are inserted or fixed into buildings in order to enable particular functions. In the case of external doors and windows, this will include control over:

- *entrance and egress – security being an essential consideration, as well as concerns about egress in emergencies, such as fires*
- *natural illumination of the interior*
- *heat loss*
- *environmental pollution: noise, air pollution, and so on*
- *opportunities for ventilation.*

Joinery components such as external doors and windows are also an important element in the aesthetics of houses. Like other aspects of houses, they are required to be durable. Given that they are operational, moving elements of a house, their durability is, to a degree, as much a factor of their maintainability as is the material from which they are made.

Two late Victorian houses – one with the original softwood joinery, the other with replacement UPVC joinery. The visual impact can be significant; so can the short- and long-term costs of replacement, which has been big business over the last 30 years, particularly since the economic availability of double glazing. The long-term impact of replacement is not clear-cut. Currently, UPVC is the material of consumer choice; this seems to be linked to the low on-going maintenance requirement in the short to medium term. However, there are longer-term costs which have to be accounted for.

Materials

There are a range of materials that external joinery can be made from: steel, aluminium, UPVC and timber. The two materials which dominate the external joinery market in Britain are UPVC and timber.

UPVC

Over the past few decades, UPVC (un-plasticised polyvinyl chloride) has become the predominant material used for external joinery. UPVC is a chemical compound of carbon, chlorine and hydrogen. 'Un-plasticised' refers to the fact that the PVC has not been softened by the addition of chemicals, known as plasticisers, which are used to create softer and more flexible PVC materials, such as are used for the insulation of electrical cables, and so on. Essentially salt and petrol or natural gas are used to create polyvinyl. A polymerisation process changes this to polyvinyl chloride. However, pure UPVC is not suitable for joinery purposes and to ensure its durability and suitability as a joinery material, it requires the addition of stabilisers and other additives: these provide resistance to the effects of heat, and particularly, ultra-violet radiation in sunlight.

Typical UPVC windows: mitred and heat-welded, these weather-stripped windows dominate both the new build and replacement market. Left: UPVC sliding sashes. Right: A UPVC hinged casement window.

UPVC joinery has been around for several decades, but really took off in the late 1980s housing booms. Although originally used in commercial application and for window replacement, UPVC now accounts for perhaps as much as 85 per cent of the new window market. UPVC joinery consists of a variety of extruded plastic sections which are heat-welded together to form frames and opening casements or sashes. In specific circumstances, the extruded profiles can be reinforced. For example, in some 1920s and 1930s houses with bay windows, the original timber window frames were designed to be structural and replacement plastic windows in such circumstances would definitely require reinforcement. Additionally, concern about the security in early UPVC joinery, which had relatively thin and simple plastic profiles, encouraged the near universal inclusion of metal reinforcement. Improvements in extrusion technology have reduced this requirement. This, together with the increasing thickness of double-glazed sealed units, has required more complex extruded sections than in the past. In addition to the need for more complex profiles, the glazing itself provides additional stiffness to the frames once this has been added.

The advantages of UPVC joinery are:

- *It requires low on-going maintenance (generally cleaning and lubrication only) – this is a major reason why they are popular.*
- *It is not affected by moisture movement – this reduces the need for on-going easing and adjustment – once correctly installed, they are relatively 'worry-free' in the short to medium term.*
- *It has reasonable thermal insulation qualities; this can be improved with additional thermal breaks and added insulation.*

Typical UPVC profiles. Left: The complexity of the extrusion. The compartmentalisation adds to the stiffness of the profile. Note that these sections are also reinforced with galvanised metal stiffeners. Middle: A detail of the mitred and welded joint. Note the weather stripping on the edge of this opening casement; this will be compressed as the casement closes up against the fixed frame.

Right: A typical friction hinge used to secure the casement into the window frame. These need to be substantial to support the significant load of the glazed window. They are called 'friction' hinges because they have sufficient friction on opening to avoid the casement being subject to significant gusts of wind and smashing. This sophisticated hinge also enables near 180 degree opening enabling the occupant to clean the windows from the inside of the house. However, there have been a number of instances where difficulty in sourcing exact replacement ironmongery, including hinges, has led to premature replacement of the entire window.

However, for UPVC windows and doors:

- *Repairs and adjustments are very difficult to undertake – often, damaged or failing units will require complete replacement, i.e. on-going, piecemeal repairs are very difficult. Some argue that manufacturers' claims of 'maintenance-free' would be better described as 'un-maintainable'.*
- *There is uncertain long-term durability; compared to the known hundreds of years of potential life of timber, UPVC's life is uncertain and likely to be significantly shorter.*
- *There are environmental concerns at a number of life cycle stages – extraction, processing and at the end of life.*
- *Regular cleaning of UPVC and lubricating hinges, locks and seals is vital to ensure longevity – these recommended tasks are frequently ignored.*

Like all building materials, UPVC is affected by a range of environmental factors, in particular ultra-violet light radiation (UV). The impact on UPVC is that it becomes increasingly brittle by exposure – time and the strength of exposure are both factors. Cracking and surface blemishes can become quite significant, particularly on south- and west-facing façades. The surface and structural impact of UV degradation can be exacerbated by environmental pollution though there have been some significant advances in the UPVC manufacturing processes, e.g. the employment, in particular, of stabilisers. A number of paint manufacturers produce a paint for UPVC to enhance protection against UV and to improve surfaces affected by UV and pollution.

All UPVC manufacturers recommend regular cleaning for all UPVC surfaces. It is also important to pay particular attention to the glazing gaskets and weather stripping too – these are also affected by UV exposure.

Perhaps the most significant defect with UPVC is the difficulty in undertaking piecing-in repairs, hardware replacement (compatible hinges, locks, etc.) and some adjustments. In the medium term, this is likely to be the biggest cause of premature failure of UPVC joinery.

Timber

Prior to the introduction and promotion of UPVC, the traditional material for joinery was timber, indeed timber joinery now only represents a small proportion of the joinery selected for new construction. However, millions of timber windows and doors still remain in existing houses, and timber remains an important material of choice in the replacement market, particularly in conservation.

Four examples of timber windows – from left to right:
1. *Typical eighteenth-century sliding sash window. Single-glazed with small panes of glass supported by glazing bars.*
2. *Typical nineteenth-century sliding sash window, behind a masonry reveal. Single glazing, but with large panes as glass technology advances.*
3. *An unusual bespoke twentieth-century hardwood casement (i.e. a hinged and outward opening window), double-glazed and with a translucent low-solid wood stain finish.*
4. *Typical twenty-first-century commercial softwood double-glazed casement window, with a high-solid wood stain. This provides a greater degree of body or build to the finish, which is essential for avoiding excessive moisture absorption for softwood.*

Although some joinery makes use of either indigenous or imported hardwoods, it is imported softwood which dominates timber window and door production. The two main commercial species used for joinery in the UK are pine (Pinus sylvestris) colloquially known as 'redwood' and spruce (picea abies) known as 'whitewood'. Whitewood is marginally less durable, but because of its cell structure will absorb less moisture and is arguably better for external uses. However, it is not nearly as widely used for joinery as is redwood. Other species are used, hemlock and Douglas fir, for example, but these tend to be used in bespoke production rather than in mass joinery production – both are stronger, more durable and more expensive than the principal commercial species.

Chapter 15 explains some of the issues with the nature of timber which impact on its durability. Essentially, the growing tree has two distinct types of wood: the sapwood is the first few annual growth rings found under the bark, and beyond this, right up to the centre of the tree, is the heartwood. Heartwood was once sapwood, but as the tree grows, an additional layer is added each year under the bark. As this growth continues, the inner layers of sapwood stop providing active conduction and become the location for the deposit for naturally occurring preservatives within the tree. The heartwood is denser, more durable timber. By contrast, the sapwood is the active part of the tree where sugars (the product of photosynthesis) and water are transported around the growing tree. The lower density and increased sugar content make the sapwood significantly more vulnerable to defects related to organisms which use timber as a food source, principally fungi and wood-boring insects.

The issue with much commercially produced softwood timber since 1945, and particularly since the 1960s, is that, because of increasing global demand for timber, the typical age of trees felled commercially has dropped markedly; the younger the tree, the larger the proportion of sapwood compared to heartwood. Additionally, the older the tree, the denser the heartwood is. All of this means that, in order to minimise risk of defects caused by fungi and insects, effectively all commercially produced softwood windows will be chemically preserved to guarantee their longevity in this respect.

While biochemical approaches are now viewed as an important guarantee of durability, it is essential, also, that other preventative approaches are adopted to minimise risk. This includes minimising the risks associated with the use of highly poisonous preservative chemicals.

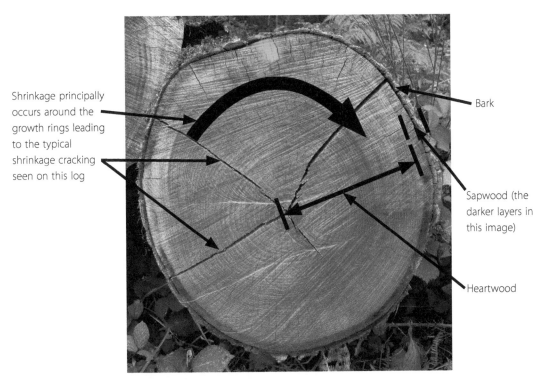

Shrinkage principally occurs around the growth rings leading to the typical shrinkage cracking seen on this log

Bark

Sapwood (the darker layers in this image)

Heartwood

This shows a typical section through a log. Just inside the outer protective layer – the bark – each year an additional growth ring is added. The first few layers of these are the sapwood. This section is concerned with transport of starches and water around the tree. This section of the tree has a greater vulnerability to the principal timber defects – fungi and wood-boring insects. Beyond the sapwood is the heartwood – the more durable and denser timber. Shrinkage and greatest distortion occur around the annual growth rings.

Understanding the life cycle of the fungi and insects which attack timber is also key to their control. Both organisms like moist timber, so restricting the moisture content of timber in joinery to less than 20 per cent is essential.

There are additional reasons why management of the moisture content of timber joinery is essential. The dimensional stability of timber is also significantly affected by moisture content. When a tree is growing, it will have a substantial amount of moisture within the body of the tree. The most economic method of 'converting' a tree to timber is by sawing. Once converted, a process of managed evaporation – known as seasoning – carefully reduces the moisture content of the timber. The careful removal of moisture increases the stiffness and strength of the timber. However, it also causes the timber to shrink. Longitudinal shrinkage, along the grain of the timber, is negligible. The most significant shrinkage of timber occurs around the annual growth ring. The majority of commercial joinery timber is sawn 'through and through'; this produces timbers which can be subject to dimensional instability and shrinkage distortion. 'Quarter sawn' timber is a far more expensive (and wasteful) way of sawing the tree, but will produce timber with greater dimensional stability.

Ideally, timber used in construction needs to be seasoned to a moisture content similar to the ambient moisture conditions where that timber will be located. Typically, the moisture content of internal joinery timber will need to be 5–9 per cent, at the lower end, if in centrally heated spaces. Seasoning was traditionally undertaken by stacking the sawn timbers so that air could circulate to all the cut faces enabling evaporation. Air seasoning could take many months, even years for large section timbers; attaining low moisture contents, such as those required in centrally heated houses, using this method, was difficult. Nowadays, seasoning processes have been significantly speeded up through the use of powered timber drying kilns which can season timbers down to a specific moisture content in a relatively short time period, i.e. a period suited to the commercial production cycle.

Timber in a sawmill that has been converted and is 'in-stick' – with small section sticks inserted between the planks in the sawn log to allow air seasoning through air circulation.

To a limited extent, the seasoning of timber is reversible; if it is exposed to moist conditions, then it will act rather like a sponge – it is a hygroscopic material – and it will take up or give off moisture to match those ambient moisture conditions. This can be highly problematic, as shrinking and swelling timber can significantly affect the serviceability of timber joinery, i.e. swelling up making it difficult to open and close, or shrinking and enabling excessive uncontrolled ventilation.

The end grain of timber is particularly susceptible to the uptake of liquid water and significant dimensional change. These vulnerable areas occur at the horizontal ends of frame and casement assemblies and at all the areas where there are joints. Restricting the exposure of timber to moisture is essential if dimensional stability and the risk of fungal and insect attack are to be avoided and timber joinery is to be durable and serviceable in the long term. For timber joinery, this is neither straightforward nor a one-off task and is normally achieved by the use of paints and other protective coats.

Paints and protective coatings for timber joinery

Generally

Ensuring that all timber surfaces are provided with an effective and durable coating is essential if the defects associated with excessive moisture content are to be avoided. Painting obviously provides a decorative purpose, it also avoids exposure of timber to ultra-violet (UV) light – this is one of the key mechanisms in degradation of external building materials and a cause of defects in all building materials. However, the primary purpose of painting is to avoid timber getting wet for the reasons discussed earlier, i.e. avoiding dimensional instability and movement and reducing the risk of fungi and wood-boring insect infestations.

Traditionally, oil-based paints have provided both the decorative and protective coatings to timber joinery. Nowadays, paint technology has developed considerably and there is an increased range of options.

All paints consist of three elements:

1   **Pigments**, *which impart colour and opacity, i.e. the prevention of light penetrating through to the substrate. This is useful in protecting from UV radiation and hiding previous colours. In the past, naturally occurring and metal-based pigments were common; nowadays synthetic pigments offer both economy and reduced health risks.*

2   **Binders**, *these are resins which are the film and body-forming component of paint, binding the pigments together and giving the paint 'body'. In the past, milk, bones, lead and various oils were commonly used. More recently, synthetic resins such as vinyl, polyurethane and alkyds predominate.*

3   **Dilutent**, *the base liquid into which the pigments and binders are suspended. In the past, natural and man-made solvents – turpentine, white spirit, alcohol, etc. – were common. Nowadays, water is increasingly common as a paint dilutent as there has been increasing regulation concerning the environmental and human health impacts of Volatile Organic Compounds (VOC) such as white spirit and other previously mentioned solvents. Once applied, the dilutents evaporate, leaving the pigments and binders as a surface coating.*

Paints

In the past, lead-based paints were the most frequently used for external timber joinery. The most common binder was lead carbonate mixed with linseed oil – this was highly toxic, but did provide a durable, flexible and dense body to the paint. Lead paint has been effectively banned for all but very few conservation applications since the 1960s.

Until recently, alkyd paints dominated oil-based paints for joinery. These required a minimum of three coats to provide sufficient protection for bare timber. A primer coat to stop subsequent coats being absorbed and help fill the timber grain; an undercoat to further fill the timber grain, provide opacity and a flat and uniform surface; this was followed by the top coats – two were commonly specified – this was a hard-wearing colour-fast gloss paint. The build-up of coats necessary with alkyd paints provided cost-effective waterproof protection, opacity and good ultra-violet (UV) light repellence.

However, these multi-coat paints could be complicated to apply, require regular maintenance (a 3–5-year repainting cycle was common) and are relatively brittle. Although the provision of a waterproof barrier might sound ideal, because this barrier is quite brittle, it can cause significant problems. In reality, however well painted a window is, as cyclical changes in humidity occur, some moisture will find its way into the timber. If it slightly swells or shrinks, then alkyd paints can only withstand a very limited degree of movement without cracking. Once the paint covering is cracked, this will allow moisture to enter the timber, accelerating the moisture movement of timber. The waterproof nature of the surface coating will mean that moisture within the timber will have restricted opportunity for evaporation. Further cracking of the paint may occur, which then enables further moisture to enter the timber and further movement, and so on.

In the past 10 years, EU regulations concerned with air quality and the health of construction workers have required the development of low VOC paints. Essentially, this means that many external timber paints are now water-based. There is evidence that these paints have better characteristics than earlier paints. They are more flexible than their oil-based alternatives, allowing for inevitable timber movement without excessive cracking or flaking creating the opportunity for further movement. They are also more vapour-permeable than previous paints, allowing better transmission of moisture vapour out of the timber through the body of the paint once applied, while preventing liquid water from entering. Both these qualities diminish the risk of moisture-related defects for the joinery and are of significant importance to the durability and frequency of maintenance that is required for timber joinery and for its finishes too. These paints are often marketed as 'microporous'.

Cracking caused by moisture movement in timber. The putty retaining the glass has hardened and a hairline crack has appeared between the putty and the timber. This has enabled liquid water, running down the glazing, to enter the crack causing the timber to swell. The crack has enlarged – allowing additional water to enter. The end grain timber at the joint has eventually absorbed some of this water, swelling and cracking the paint at the joint too. This provides a further opportunity for more water to get into the timber and further cracking of the paint finish, and so on.

Wood stains

These remain less common in existing buildings, but are relatively common in new houses. Many of these are marketed as microporous. They are of two types:

1 *'Low solid' stains, which consist of a very thin paint with a very small proportion of binder. Here the pigment is drawn into the surface of the timber and there is little surface-build, leaving the grain pattern of the timber intact. These, generally, have very high porosity, but as there is little surface build, they have reduced water-shedding qualities.*

2 *'High solid' wood stains are significantly more popular and are, in effect, a form of translucent paint. They contain less pigment but, as they have a surface-build, they offer better resistance to moisture penetration for timber joinery than low solid stains. They also have less moisture vapour permeability than the low solid stains – meaning that there is a compromise between the need for surface-build to avoid liquid water entering timber joinery and the vapour permeability of the finish, where water vapour can easily evaporate through the finish.*

One advantage of both low and high solid wood stain products is that, generally, a single product applied in two coats can offer several years of service – five to ten years is frequently claimed for high solid stains. Many wood stains are solvent-based and there is some evidence that these are more suitable for application on timber joinery, but there are increasing numbers which are water-based.

Defects of painted timber joinery

In order to avoid excessive absorption of moisture, it is important that timber is provided with its protective coating prior to any opportunity to exposure to saturation. In the past, much timber joinery was supplied to builders' merchants or sites unpainted or with only a factory-applied primer coat – though a primer coat did provide some limited protection, it still required subsequent finishing coats to be applied rapidly in order to ensure adequate protection. Increasingly, it is recognised that complete painting of new timber joinery prior to installation on site is essential. There is clear evidence to suggest that, if this is the case, then the durability of timber joinery is significantly enhanced – particularly, if care has been taken in paint selection.

The use of preformed insulated plastic sub-frames has supported the increasingly adopted strategy of supplying and fixing completely finished and glazed joinery to site. These sub-frames provide an insulated cavity closer with integral DPC to the openings in cavity walls. The frame itself is UPVC and provides a precise template for the construction of the opening in the cavity wall. They also provide a secure, accurate and permanent fixing for completely finished joinery – doors or windows. These can be UPVC or timber, but in the case of timber such an approach protects the pre-finished quality of the product.

Specific defects in paints and protective coatings

Blistering and flaking

This can occur because excessive moisture in the timber was present immediately prior to painting. Ideally, the moisture content of timber for external joinery prior to painting should be between 10 and 12 per cent. In practice, this can be hard to achieve. A long run of relatively dry and warm months would provide suitable ambient conditions and, traditionally, late summer was considered to be ideal. However, this is subject to the vagaries of the British summer weather. Dry and constant conditions are the ideal time for decoration of external timber.

Cracking and flaking can also very commonly occur where liquid water has been absorbed under the paint in service, i.e. through a crack in the paint finish caused by minor moisture movement at a joint.

Blistering and flaking occur because the surface is insufficiently vapour-permeable to allow the trapped moisture – as a vapour or a liquid – to dissipate and, as a result, the paint/coating is forced off of the timber. Cracking or flaking tend to be a vicious circle with the cracking and flaking caused by moisture movement of the timber enabling more liquid moisture to enter the timber followed by more movement and more cracking, and so on.

Dark-coloured, south-facing timber joinery is likely to be subject to significant solar radiation UV and direct heat. Not only will this cause problems with UV, which will degrade the paint surface, if the paint does not have sufficient flexibility to withstand stretching, cracking will occur. Dark colours will concentrate and heighten temperatures. Direct and excessive heat will cause moisture within the timber to evaporate and, sometimes, quite significant movement of the timber. The vapour will need a route for escape and, if the paint does not have sufficient permeability, paint cracking will occur.

Adhesion between coats and compatibility

Poor adhesion between coats and the incompatibility of paints used in separate coats can exacerbate flaking and cracking. Preparation and sanding between coats are critical to avoid such defects.

Ensuring compatibility between specific paints is essential. While microporous finishes do have their advantages, their effectiveness is compromised if they are used over existing non-microporous finishes, such as existing alkyd finishes. There is also some evidence that, with some of the high performance finishes, the quality of the application becomes a significant issue. While this is true of all finishes to some extent, with high performance finishes, the quality of workmanship, particularly with the reduced number of coats, is an even more critical factor in their long-term durability.

Flaking caused by a lack of adhesion between coats – there is no effective key and, over time, the top coat has lost adhesion with the undercoat. In this case, this is probably a lead-based paint.

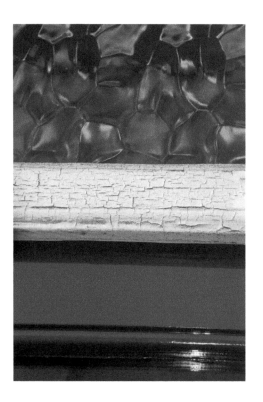

Resin and knot seepage

While many manufacturers avoid timber with large knots as part of their selection process, the softwood species commonly used for joinery will have knots. Knots will exude resin under paint finishes and this can cause staining and localised discoloration. It can also cause blistering and direct resin seepage through the painted finish. The effective use of knotting (a shellac-based coating for all knots) prior to priming bare timber is usually effective in dealing with knot seepage.

Some timbers are particularly resinous, e.g. Douglas fir, and can have substantial resin pockets. These are not always very obvious to the eye during manufacturing, but will exude resin for many years potentially until the pocket is empty – this can be very disruptive. Resin exudation is exacerbated by direct sunlight and heat. Painted timber joinery in full sun, for instance, can be particularly prone to such problems. Careful identification of resin pockets at the pre-manufacturing timber selection stage is essential. Knotting will help seal relatively small resin pockets and some specifiers recommend an aluminium primer as well.

This south-facing softwood door was not treated with knotting (a spirit-based coating for knots in bare wood). This would have stopped the resin residue in the knot from leaching out of the knot and subsequently disrupting the finished paint surface.

Inappropriate maintenance

Obviously, the failure to effectively maintain painted or protective coatings on timber joinery is a significant cause of defects. Because timber can undergo significant moisture movement, there is a need to ensure that any protective paint finishes are well maintained. The on-going maintenance of painted finishes is a significantly important preventative measure. Once timber is exposed to changes in moisture content, it will be harder to guarantee that subsequent paint maintenance will be effective.

Too much paint and a failure to ensure that timber joinery has undergone appropriate and timely easing, adjustment and repair prior to maintenance painting, can mean that the excessive build-up of paint will render the joinery inoperable or difficult to operate, i.e. hard to open or close or paint-sealed shut. Thus, the failure to paint can lead to excessive moisture movement and operative failure, but excessive layers of paint without suitable pre-adjustment can also lead to functional defects.

This is not an issue for UPVC joinery, as this has virtually no moisture-movement characteristics. However, this advantage does emphasise one of the significant disadvantages associated with joinery in UPVC – there is little if any adjustment and minor repair possible with UPVC. This is a significant factor in its long-term durability.

This softwood door and its decoration have been neglected for several years. The silvering and cracking of the exposed softwood are effects of UV exposure and substantial exposure to wetting and drying. Repainting softwood joinery in this condition requires significant surface preparation prior to refinishing. It is likely that repainting will never be durable, given the nature of the deterioration encountered. Regular maintenance attention and repainting will prolong the life of timber joinery (in some cases for hundreds of years), but the process requires vigilance and is time-consuming and labour-intensive.

Glazing in external joinery

The glazing of joinery has been a significant cause of building defects in new build until relatively recently. The practice of site finishing, and in particular site glazing of windows, was identified as the principal culprit. The use of pre-formed cavity closers mentioned above has gone a long way to resolve this. Before discussing the potential defects associated with double glazing (now the standard glazing in all new construction), single glazing, a common practice until the 1990s and still practised in many conservation situations, will be discussed.

Single glazing

Single glazing (a 3mm, or more commonly, a 4mm single sheet of glass) is fixed into an opening sash or casement using sprigs, which are headless nails which mechanically fix the glass into the glazing rebate (onto a back bed of linseed oil putty). Once the sprigs are fixed, an additional bead of putty is added. This is used to create a smooth bevelled bead which covers the tops of the sprigs, completes the space between the glass and the edge of the rebate and provides a waterproof seal.

As long as the linseed putty remains relatively soft, a good waterproof joint can be retained. If the timber rebate is bare wood, then the linseed oil in the putty will be absorbed by the timber and the putty becomes brittle over time and starts to fail as a waterproof seal. Linseed oil putty needs to be painted to ensure its durability too. This cannot happen to newly applied putty for at least a month until an effective skin has formed on the putty. It is advisable to periodically replace linseed oil putty, particularly the lower horizontal sections, to ensure that the waterproof seal is maintained at this critical point immediately above joints in the timber window (an area of significant vulnerability in terms of water ingress). (See the photograph on p. 370.)

Linseed putty and the sprigs that retain glass in casements will always require periodic replacement. In order to avoid this necessity, timber glazing beads were sometimes used as the means for retaining glass. Glazing beads are located externally on the casement, i.e. the window is glazed from outside. Beads consist of small-section machined timbers used to retain the glass in the rebate. The glass itself requires a back bed (sometimes putty, sometimes a mastic or silicone bead) to cushion the glass into the back of the rebate.

Beads were cut to form neat mitred joints at each of the corners of the rebate. To effectively retain the glass, the beads were pinned into the rebate. This use of a mitred joint was suitable at the top corners of a glazed frame; however, on the lowest horizontal rebates significant amounts of liquid water drain to these points. Here, a mitred joint on the beads actually channels water into the most vulnerable area of the joint, i.e. between the horizontal rail and the vertical upright timbers. This particularly occurred if the beads shrank over time or were poorly maintained. In addition to this moisture-ingress opportunity, most beads were finished flush with or slightly inside the face of the frame to which they were applied. At the lower horizontal level, this could actually encourage water to be retained in the rebate.

It is better practice to have all the lowest glass retaining beads running the full width of the rebate; with the vertical side beads cut onto the horizontal bead and to have them weathered, i.e. shaped with an angle that actually throws water off away from the face of the frame. This is also assisted by having this horizontal bead extend out beyond the face of the joinery. This projection should incorporate a drip to ensure that there is no chance of water tracking back into the joinery. There is some evidence that the nails used to secure beads into the frame are potentially a source of water problems, though this is considered to be marginal.

Single-glazed Douglas fir double doors. The side bead finishes inside the glazing rebate and its lower end is cut over the lowest bead. Here, the lowest bead is a different profile and extends right across the width of the glazing rebate, minimising the opportunities for rain to penetrate the joinery above the joints in the lock rail. The lower bead extends out beyond the face of the door and has a drip arrangement intended to throw water off of it and away from the door itself.

Multiple glazing

Multiple glazing improves the energy performance of external joinery by trapping air, or specific gases, in the space between the multiple panes of glass. It has become the standard approach for external joinery. Multiple layers of glass interrupt the conduction path and the rate of heat flow is reduced. Double-glazed windows are the most commonly encountered with two separated panes of glass, but triple glazing, with three panes of glass, is now becoming more common.

In order to maintain the insulating qualities of multiple glazing units, it is essential that the air trapped between the panes is dry and that the unit has a completely airtight seal – hence the term 'sealed unit'. In the past, some seals have been poorly formed. This has enabled moist air to leak into the sealed unit, effectively negating almost all of its insulating qualities. This has been a very common failure and has been a significant cause for concern, provoking significant changes in multiple glazing manufacturing in the last decade. Typically, the units mist up as moisture condenses between the two panes. Attention to the detailing of the seal and the spacer bar, which permanently separates the panes, has reduced the incidence of failure. In the past, the spacer bar was made of an aluminium alloy. This was found to create problems of cold bridging. This localised drop in temperature also caused condensation at this vulnerable point and the liquid water so created was found to corrode the seals on the double-glazed unit from the inside – leading to failure and increased condensation potential. Nowadays, spacer bars are generally made from plastic materials, which avoid significant cold bridging, and the units have desiccants within them that ensure that any residual moisture in the units is removed effectively over the long term.

The seals and spacer bars have also been found to be degraded by exposure to UV. The retaining beads need to be deep enough to ensure that they fully cover the seal and spacer area; in practice, this is about 20mm. The depth of the rebate has been increased on both timber and UPVC windows in order to ensure that it is deep enough to provide sufficient cover for the spacer bars. It is also increasing in plastic windows because the greater the area of glass there is, the lower the U-value is. This is because the sealed unit has a lower U-value than the plastic frame.

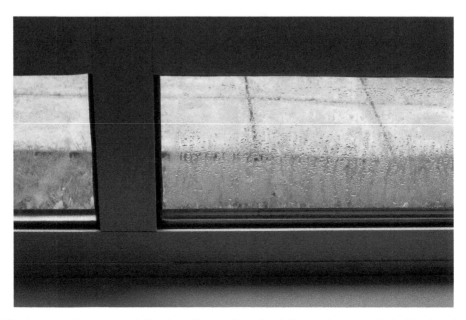

'Misting' in the space between sealed units – this condensation indicates that water is leaking into the sealed unit; more critically, much of the insulation performance has been lost. This has provoked improvements in sealed unit manufacturing and significant changes to the detailing of glazing in all forms of joinery. It has also led to much new joinery being completely glazed and finished before being fixed on site.

Double glazing in timber windows

Linseed oil putty is not appropriate for the glazing of windows and doors with multiple glazed sealed units. These will all be beaded and the glass sits in the rebates against a neoprene, or similar, glazing gasket.

For timber joinery, it is most common for window beads to be located externally. However, the joint between the bead and the glass can be vulnerable to liquid water ingress and, nowadays, most timber beads incorporate neoprene glazing gaskets located in a groove within the bead. Compressing this, as the beads are fitted and fixed, enables effective water drainage between the glass and the beads.

Ensuring that the lowest horizontal glazing bead (where most of the rain drains to) is effective at shedding water has proven to be essential. This is not just a matter of providing the compressible neoprene glazing gasket, it requires careful and specific detailing for this particular bead. If this is not the case, then the bottom edge of the sealed unit can potentially sit in water for significant periods of time. The effective detailing of this lowest horizontal rebate and bead involves allowing the bead to be ventilated, i.e. it has gaps on its underside enabling trapped water to drain out and to enable any residual moisture to be rapidly evaporated. It is also important to ensure that the sealed unit itself does not sit directly on the rebate, but sits on top of spacers, avoiding saturation of the vulnerable seal. Now, unlike in the past, the rebate itself is rarely at right angles to the face of the casement and is angled so as to encourage water to drain out of the rebate.

For timber windows, this bead is so vulnerable that a number of manufacturers are now using an aluminium bead for the lowest rebate to provide assured durability. The bead itself has no nails or other fixings through to the rebate (a potential source of water ingress). It is clipped to an integral glazing gasket which itself provides the required ventilation and drainage for the bottom of the sealed unit in the rebate. Spacers are also used in the side rebates to allow for the thermal expansion of the glass itself.

Internal glazing gasket: supports the double-glazed sealed unit against the glazing rebate. The gasket is located in a groove in the opening casement.

Double-glazed sealed unit: This unit totals 28mm (4:20:4). The edge seals are vulnerable if they get and stay wet for long periods.

External glazing gasket Located in a special bead, it is compressed against the double-glazed sealed unit, retaining it and providing a seal against precipitation. The lower horizontal timbers on the window are the most vulnerable to rain. To avoid saturation, the bead is of aluminium. It extends out beyond the window surface – throwing water away from the timber window. At the same time this bead enables significant ventilation of the glazing rebate – just in case any water ingress does occur. Unlike the timber beads at the sides and top of the glazing, which are of a different profile and are fixed in position by nailing, this bead is clipped onto a special plastic bracket – this avoids any nails piercing the bead, a potential path for moisture ingress.

Compressible neoprene draught stripping: This is located in a groove in the opening casement. It is compressed as the casement is closed against the rebate in the window frame. Weather stripping is now standard in windows and allows for fairly open joints between the frame and casement which encourages effective ventilation and drainage in the window.

Wedge-shaped drainage spacer: This supports the double-glazed sealed unit. These spacers stop the unit from sitting in a wet rebate, they also enable moisture to be drained and ventilated out of the rebate under the aluminium bead. The rebate itself is angled so that any moisture ingress is drained away.

A cross-section through the bottom rail and sill of a quality modern softwood casement window

Double glazing in UPVC windows

Most UPVC windows are now glazed with internal rather than external beadings. Though this is partly related to concerns with damage to the sealed unit because of water leakage, it has become common practice because of concerns with security. The beads on UPVC windows are not mechanically fixed, as they are in most timber windows. The beads on UPVC clip into the frame. They are relatively easy to prise off and remove (an easy move for burglars!). Locating them internally eliminates this and it makes re-glazing more easily undertaken at height.

This arrangement also means that the detailing of the edges of the door or window retaining the glass requires effective glazing gaskets. In particular, the horizontal member supporting the bottom of the glazed sealed unit requires careful detailing for all the reasons mentioned above with regard to the lowest bead in timber windows. The lowest horizontal member requires weathering (i.e. it is angled to shed water), and will also incorporate a neoprene gasket, additionally, the frame will have to incorporate the opportunity for both ventilation and drainage. It must do all this without creating a cold bridge. Careful detailing and the effective use of UV-resistant gaskets are still required to avoid glazing problems in the long term.

Weather stripping of joinery

Where there are opening casements, sashes or doors, weather stripping is now standard to all new joinery of whatever material. It can also be retro-fitted to existing joinery. Given the increasing concern with energy efficiency, weather stripping can effectively control unregulated ventilation and it can also enable improved thermal comfort for house occupants.

Most weather stripping consists of plastic or neoprene seals located into integral grooves in the rebates of the fixed frame of the joinery. Sometimes they are attached to the opening element; sometimes to both. When the joinery is closed, the weather strip is compressed and, when this occurs, the weather stripping controls excessive air infiltration. To be effective over the long term, weather strips need to be reversible, i.e. they need to have a long-term flexibility and, once the joinery is opened, they resume their original profile, only to compress again and again as the window or door is opened and shut over its service life.

Over time, weather strips become brittle and lose their flexibility. Others become degraded and ineffective if they are painted. Some, in the past, have impeded the effective drainage of liquid water and this is clearly undesirable, particularly in timber joinery and where this impedes drainage from glazing. When problems occur, replacement of the weather strips to the whole unit is probably required.

Timber pests 15

Introduction

Timber has been used successfully in construction for many hundreds of years. Although the seventeenth century saw the increasing popularity, and then dominance, of stone and brick for external walls, timber has remained a significant material for house construction. Timber is valued for its functional and its aesthetic qualities. It is also a relatively durable material. Unfortunately, this quality of durability may be seriously affected if it is attacked by what are known as timber pests. In the context of buildings, the term timber pests refers to those organisms which attack timber for food. These organisms can be divided into two groups:

1 *fungi.*
2 *wood-boring insects.*

The risk of damage by timber pests for any given piece of timber in a building will be related to a number of factors including:

- *the natural durability of the species of timber.*
- *the 'food value' of the timber.*
- *adverse environmental conditions – particularly the moisture content of the timber.*
- *the protection afforded through any treatment by chemical preservatives.*

The existence of timber frame buildings dating from the medieval and Elizabethan periods is a testament to the natural durability of some timber species. But even as far back as the Elizabethan period there were problems in obtaining sufficient economical supplies of appropriately durable timber for building. This was largely due to demands from other uses, such as ships, but made worse by the poor management of timber supplies. During the Georgian and Victorian periods there was a vast increase in the use of imported softwood. This coincided with changes in design and building use which resulted in these timbers being utilised in particularly vulnerable positions (e.g. suspended timber ground floors). The problem of quality being reduced as demand increased has been repeated in more recent times with the use of smaller sections of timber from relatively immature trees.

Some timbers are naturally more resistant to fungal attack than others. This durability is essentially determined by the presence in the tree of naturally occurring substances which are toxic to fungi. Natural durability is primarily determined by the type, amount and distribution of these substances. Because this development of an in-built resistance has evolved with varying degrees of success between species, some timbers are more durable than others. In terms of timber use in buildings, knowledge of natural durability will have an effect on a number of decisions, including where the timber might be used and the necessity for preservative treatment. Specifying an appropriately durable timber and reducing the likelihood of adverse, particularly damp, conditions by attention to appropriate design and construction details can reduce the risk of attack by timber pests. Although preservative chemicals can reduce the risk of fungal or insect attack, there are financial and environmental costs arising from their use.

Wood is the raw material; 'timber' is the term for wood that has been converted for use. Wood consists of millions of cells, whose function is to provide the physical support to the tree as well as to store and circulate food. The cells consist of cellulose fibres. There is a 'crust' around these cellulose cells formed by a material called hemicellulose. A material known as lignin is the matrix between the cellulose and hemicellulose. Lignin helps to determine hardness and resistance to fungal decay. The physical nature and structure of the cells are factors in durability.

The seasonal cycle of growth of tree cells produces the rings that can be seen when the tree is cut. Sapwood is that part of the tree that carries nutrients from the ground to the branches and leaves and, in the other direction, the products of photosynthesis from the leaves to the roots. As a tree ages, the older cells die. The portion of the trunk that consists of these dead cells is known as heartwood. Obviously, with age, the proportion of heartwood to sapwood increases. The proportion of sapwood to heartwood will also vary between tree species. Heartwood is in effect dead sapwood; it does not carry nutrients but it functions as a depository for waste matter and as a skeletal support to the trunk. The heartwood can often be distinguished from sapwood because of a distinct colour difference – it is usually darker due to the presence of toxic organic substances. The colour distinction is often less clear once the timber has been incorporated in a building. In living trees the combination of natural toxins and the fact that the cells are full of water means that they are relatively well protected against attack by fungi. Where attack does take place, it is usually associated with damaged or dead areas. In some instances, heartwood can be relatively vulnerable due to its lower moisture content. However, when a tree is felled, and then converted into timber for use in a building, the moisture content will be reduced (through natural means and by drying processes). As the timber dries, the moisture content of the sapwood will fall within the optimum range that will encourage attack by fungi (and wood-boring insects).

Timber should not be used in a building until its moisture content is below the minimum required for fungal attack. In some cases this may not happen or, due to a defect, the moisture content may rise. In such situations the sapwood is vulnerable to decay because it contains the nutrients that timber pests feed on. Because it will contain natural toxins and does not carry nutrients, the heartwood of timber in a building has a greater resistance to insect and fungal attack than sapwood.

Softwood approximately 350 mmm in diameter and 40 years old. The heartwood can be clearly seen (as a lighter colour in the centre) as can the annual growth rings.

Many older properties will contain timbers with a higher percentage of heartwood than is the case in newer properties. In the recent past, commercial pressures on forestry production have meant that timber for building use tends to have a high proportion of sapwood and early wood because it is taken from younger, faster-growing trees. This goes some way to explain why, for instance, relatively new timber windows, notably from the 1960s onwards, may prove to be less durable than those from the Victorian or Georgian periods.

Timber is classified as either softwood or hardwood. The terms hardwood and softwood can be confusing as they refer to botanical differences between broad-leaved trees (hardwood) and coniferous trees (softwood) rather than the actual hardness of the material. Balsawood is a much quoted example of timber which is categorised as a hardwood but is physically soft. Conversely, yew is a very hard softwood. In reality, though, the majority of hardwood timber used in buildings is harder than the majority of softwoods. The terms may also mislead in assumptions of durability because, for example, some softwoods are more resistant to fungal decay than some hardwoods.

Many older buildings will have been constructed using hardwoods, though as mentioned previously, softwoods became the most common type of timber during the Georgian period.

Fungal attack

Fungi are important organisms. They are present in great numbers in many situations and they play a number of roles. They are responsible for the breaking down of organic matter, they can cause disease in both animals and humans, they are the basis of fermentation, and they help to produce chemicals and medicines (such as antibiotics). They are important factors in enhancing the growth of living trees but are also responsible for the decay of timber 'in service'.

In very simple terms, fungi are a group of plants which do not possess chlorophyll and do not, therefore, have the ability to produce their own food through the process of photosynthesis. Because of this, they feed by attacking a cellulose-based food source such as wood. They do so by secreting acids and enzymes into the wood, which break it down – the soluble end product then is used as food by the fungi. This process has evolved so that dead trees and plants which fall to the ground can be broken down into their essential components and re-integrated into the ecological chain. If fungi did not exist, the Earth would be covered in dead trees.

Fungi are an important part of the ecological chain.

Fungal attack to a suspended timber floor (photo courtesy of C. Edmans).

Fungi which attack timber in buildings will usually develop and thrive in certain environmental conditions. It has been observed that fungi will thrive in those parts of a building which go some way to replicate the conditions of the forest floor. The key factor is the moisture content of the timber but the amount of light, air movement and humidity are also important. If timber is attacked by fungi, then the key to effective remedial treatment is to alter the environmental conditions and deal with the fault which is allowing or causing the high moisture content of the timber.

There are generally recognised to be six groups of fungi growing on timber:

1   *brown rots*

2   *white rots*

3   *soft rots*

4   *stains*

5   *moulds*

6   *plaster fungi.*

Of these, the ones which will actually break down timber belong to the groups known as brown rots and white rots. Fungi belonging to the other groups will not usually cause serious damage to timber, often because they cannot break down the cell walls. Moulds and stain fungi are not a serious problem, though they can increase a timber's porosity and they are able to neutralise some of the timber's natural fungicides. Soft rots can cause some surface decay, particularly in very wet timbers, but the growth is usually not very extensive in buildings (it can be more significant with timber in contact with the soil – garden fence posts, for example). However, because their formation is associated with damp conditions, the presence of these 'non-destructive' fungi will indicate that conditions are suitable for the more serious rots to occur.

Brown rots

These can cause severe decay in timber. The timber cracks along and across the grain in a manner that forms cubes during the decay process. This cuboidal cracking is a primary identifier of a brown rot attack. Brown rots can cause decay and a consequent loss of strength in the timber quite quickly. As a consequence of the attack, the wood becomes darker in colour and often becomes so dry that it can be crumbled easily between the fingers.

Cuboidal cracking is caused by brown rots (note also flight holes from wood-boring insects in bottom image). (Bottom photo courtesy of Johan Mattsson.)

White rots

With white rots the timber develops a rather fibrous consistency which is sometimes described as being similar to lint. As a consequence of the attack, wood generally becomes lighter in colour – it is some cases, almost as though it has been bleached. The cracking that appears tends to be along the grain, rather than along and across (i.e. cuboidal cracking) as is the case with brown rot. As a generalisation, white rots tend to prefer a higher moisture in timber than is the case with brown rots.

White Rot damage to external cladding (photo courtesy of Johan Mattsson).

Life cycle

The life cycle of wood-rotting fungi is shown in the diagram below.

The cycle begins when the spores (seeds) are dispersed from the fruiting body. These spores are very small, but numerous and light, and can be carried great distances by air currents. A fruiting body can produce literally millions of spores over a short period of time. Some of these spores will inevitably alight on timber. If the conditions are suitable, the spores will germinate and develop microscopic root-like filaments, known as hyphae, which penetrate the wood. A massing of hyphae, which usually renders them visible, is known as mycelium. Fungi use enzymes and acids to dissolve the timber cells. The nutritious products of this breakdown are absorbed and used as food. When a fungus is well established, and if the conditions are suitable, it may complete its life cycle by producing a fruiting body. The fruiting body is known as a sporophore. It is the most easily identifiable part of the fungus and will produce the seeds/spores which can be distributed in order to start the cycle again.

The process described above may be interrupted if environmental conditions change.

The timber floor in this building was replaced following an attack by wood rotting fungi (photo courtesy of Robert Shadbolt).

Essential conditions for wood decay by fungi

Fungi will only be able to survive and to thrive in certain conditions; these are discussed briefly below.

Food. Susceptibility to fungal decay varies between the different species of timber. In general, the sapwood will be attacked because it contains the nutrients that the fungi needs (though in certain timbers the heartwood can be susceptible). Modern construction, as we have already observed, tends to use timber from relatively young trees which therefore tend to have a high proportion of sapwood.

Moisture. Moisture, sufficient to allow spores to germinate and the fungus to develop, must be present for decay to occur. In a building which is correctly designed, constructed and maintained, the moisture content of the timber should remain low enough to prevent fungal attack. The moisture content of the timber is usually the most critical factor in the establishment and continuation of decay – without a high moisture content in the timber, fungal decay will not start, and if excess moisture is removed growth will stop.

Moisture content may also have an influence on the type of fungi that develops. There is some debate about what the minimum moisture for fungal growth is. Although some sources have a lower figure, a minimum moisture content of around 22 per cent (for dry rot) is often referred to. In practice, this figure is commonly reduced to 20 per cent to allow for margins of error in assessment. The optimum moisture content range for the growth of fungi will be higher than this figure.

Temperature. Fungal growth can be killed off by low and high temperatures and its growth will be slowed down as temperatures are lowered beyond a certain point. Some fungi can become dormant at lower temperatures but may start to grow again if the temperature rises. The optimum temperature for growth will depend on the type of fungus.

Air. Fungi require air, or more specifically, oxygen, for growth. If air is excluded, then decay will not occur. For example, timber lying underwater may have no oxygen in its cells – because they are full of water – and therefore the wood will not decay.

If one or more of the above conditions are absent or unfavourable, then fungal decay will not occur. Even when decay has become established, a change of conditions may kill off the fungus, alter its pattern and rate of development, or cause it to enter a dormant state. If the fungi remains in a dormant state, it will eventually die, but if the change of condition is only temporary, the fungus may resume its evolution.

Daylight. This apparently has a complex relationship with fungal growth. Daylight is not generally required, indeed, for most of the time it will be detrimental to growth. However, it is believed that at certain stages of fungus evolution, particularly in relation to the formation of the fruiting body, exposure to daylight may actually accelerate growth.

The feet of these rafters have partially collapsed following an attack by wood rotting fungi. (note: the white/yellow material is thermal insulation). The source of the dampness was a fault in the roof covering aggravated by condensation (this was in a kitchen).

Effect of fungal attack on wood

The most important effect of an attack is the loss of strength. As decay proceeds, the wood becomes weakened, producing splits and cracks, and at a more advanced stage the wood may crumble at the touch. Other effects include a loss of weight and a change in colour. A further effect may be the production of a condition where the wood is more easily attacked by wood-boring insects because it has been 'softened up' by the fungi.

Types of fungal attack

There are a number of fungi which will attack and break down timber but in the context of buildings they are divided into two groups known as wet rot and dry rot. Dry rot is a specific type of brown rot. All other rots which attack timber – whether they are brown or white rots – are referred to as wet rots.

Dry rot

Dry rot is often referred to by its Latin name Serpula Lacrymans. It is the term given to a particular identifiable brown rot. The popular name refers to its effect on wood – which becomes dry and crumbly – rather than the conditions that create it. Dry rot spores are probably present in most buildings, though some buildings are more prone to attack than others. This is because of the presence of excessive moisture, perhaps due to defects, but it can also be that, once a successful attack occurs, there are more spores around, increasing the likelihood of further successful dry rot germination. Figures for the level of moisture content which make timber susceptible to dry rot vary. It has been suggested that it can occur in timber with a moisture content of about 17 per cent, but figures of 20 per cent or 22 per cent are more usually referred to.

Dry rot prefers damp still air. An environment where timber is in contact with damp brickwork, and where ventilation is poor, is ideal for its growth. Therefore sub-floor voids or cellars with no ventilation are examples of ideal locations for its propagation.

Because of its sensitivity to environmental change, the dry rot fungus is only rarely found outside buildings. Reports of specimens being found in the pine forests of central Europe or the Himalayas seem to result in expeditions being mounted with an enthusiasm similar to that involved in searching for the Yeti.

How the fungus spreads

The mycelium of dry rot can spread extensively on the surface of affected wood, and as the growth progresses, it can develop various visual characteristics, often in response to differing environmental conditions, such as air temperature, air speed, light, etc. For example, it may look like what can be best described as a mass of cotton wool, bright lemon patches may be seen and other patches of colour, particularly lilac, may occur. In other circumstances a thin rather 'leathery' skin may appear. Dry rot produces conducting strands which develop within the mycelium. These strands (rhizomorphs) supply water and nutrients to the active growth area. The strands can move not only over wood but also over the surface of inert materials, such as masonry, and they are able to penetrate some plaster and mortar joints, which effectively allows an outbreak to spread from building to building via a masonry party wall. This is most likely to happen in older properties where the alkaline nature of lime-based mortars and plastering has reduced over time (and where the plaster may have lost its key). However, replastering or repointing, for example, may help to restrain growth because the mortar will be relatively highly alkaline. Obviously, the strands can obtain no sustenance from masonry or plaster – though the fungi may utilise, and indeed depend upon, some of the chemicals present in these inert materials in order to assist in its development. The important point is that the dry rot strands can travel over or through the inert materials, in search of a suitable food source. As they do this, they will, in the right conditions, continue to supply food to the growing area over a considerable period of time.

Many older buildings have timbers 'buried' in the brickwork behind plaster and this can provide a potential food source for the fungi (if the timber is sufficiently 'wet').

It is sometimes stated that dry rot can create its own condition for growth by producing water and therefore wetting timbers sufficiently to make them vulnerable to attack. It would appear that while the breakdown of timber can produce significant amounts of moisture, this will occur over a period of time and will often tend to be lost by evaporation and other mechanisms.

Identification of dry rot

There are some distinguishing features of dry rot attack which can help with identification.
* *The wood becomes dry and crumbly. It becomes light in weight and has a dull brown colour.*
* *Although both dry rot and the brown rot species of wet rot produce a cuboidal cracking effect in the timber under attack, in the case of dry rot, the cracking is often more pronounced. In practice, however, it is often difficult to make this distinction.*

Cuboidal cracking caused by dry rot (photo courtesy of Rentokil).

Dry rot damage (photo courtesy of Johan Mattsson).

- *The strands are grey or white and 2–8mm thick.*

Mycelia strand of dry rot fungus (photo courtesy of Johan Mattsson).

- *Depending on environmental conditions, the mycelium may develop some different physical attributes, for example, as noted previously, it may have a leathery, pearly white skin or it may look like 'cottonwool' or silky sheets.*

These photos show different stages in the development of the mycelium of dry rot.
Left: photo courtesy of Rentokil. Middle: photo courtesy of F. Blampied, Bristol City Council. Right: photo courtesy of F. Blampied, Bristol City Council.

- *The spores are often the first indication of dry rot. Individually they are microscopic but 'en masse' they appear as a rusty red coloured dust.*
- *The fruiting body is usually only found indoors and is quite distinctive. It is usually either pancake or bracket-like in shape. The fungus has a cratered surface, sometimes with folds, and is a reddish brown with grey-white margins.*

The reddish dust-like particles on top of the jars (left) and the pans (right) are dry rot spores. (photos courtesy of Ole Martin Stensli).

Dry rot fruiting body. Note the multitude of red spores which dominate its surface (photo courtesy of Johan Mattsson).

Dry rot fruiting body, in this case growing on the underside of damp chipboard (photo courtesy of Fiona Burrows).

Dry rot fruiting body and mycelium (photo courtesy of Johan Mattsson).

In practice, identification of the fungus can be difficult, and confusion can arise. A common confusion is to think that cuboidal cracking only occurs with dry rot whereas in fact it is, as observed previously, a symptom of brown rot.

At least one specialist uses specially trained dogs ('rot hounds') to seek out (living) dry rot by scent.

Dry rot – note the fruiting body in the top left and the mycelium in the top right of the image (photo courtesy of F. Blampied, Bristol City Council).

Dry rot showing fruiting body (top left-hand corner) and hyphal threads (as mycelium) (photo courtesy of F. Blampied, Bristol City Council).

The images below show a dry rot outbreak to a house following damage caused by vandals to sanitary fittings and water pipes. Note the mycelium in its various stages, for instance, in the right-hand photograph where it looks like cotton wool and this is growing over a 'leathery skin'.

(Photo courtesy of John Thatcher.)

Wet rot

As mentioned previously, there are a number of fungi which can be identified by the term 'wet rot', but it is normal to consider them under this single heading.

Table 15.1 summarises the identifying characteristics of the common wood-rotting fungi. It is produced by the Building Research Establishment (see BRE Digest 345).

Identification of wet rot

There are a number of wet rot fungi and their characteristics differ as, to a certain extent, do the optimum conditions for growth. The mycelium produced may differ between the species, as may the fruiting body and the strands. Because wet rots can be either brown rots or white rots, some of the wet rots produce a cuboidal cracking reminiscent of dry rot attack, some may cause a darkening of the wood, while others may bleach it.

Table 15.1 Identifying characteristics of the common wood-rotting fungi

FUNGUS	MOST COMMON SITUATION	FRUIT-BODY	STRANDS	MYCELLUM	DECAY TYPE	TO DISTINGUISH FROM DRY ROT
Serpula lacrymans (dry rot)	Timber in contact with wet brickwork	Rusty red folded surface: flat or bracket-shaped. Spore dust produced	White to grey, robust; brittle when dry	White to grey silky sheets, often with patches of lemon or lilac	Brown rot, commonly with large cuboidal cracking	—
BROWN ROTS						
Coniophora puteana	Very damp situations	Olive green–brown, surface irregular lumps. Flat on substrate	Dark brown, yellowish when young	Rare. Cream to brownish; off white under impervious floor coverings	Brown rot, commonly leaves a skin of sound timber	Brown strands
Coniophora marmorata	Floors, cellars	Pinkish-brown, smooth to lumpy. Flat on substrate	As *C. puteana*	As *C. puteana*	As *C. puteana*	Brown strands
Fibroporia vallantii Poria placenta Amylorporia xantha	Areas of higher temperature	White to cream to yellow; surface covered in minute spores	White to off-white (only well developed in F. vallantii); flexible when dry	White to off-white, woolly or fern-like	Brown rot; usually paler than dry rot	Flexible strands; fruit-body has spores
Dacrymyces stillatus	Exterior joinery	Small orange fleshy lumps	None	None visible	Brown rot	Fruit-bodies
Lentinus lepideus	Skirting boards	Pale beige mushroom with gills; commonly 'staghorns' produced	None	White with pinkish brown patches; distinct sweet organic smell	Brown rot	Mycelium and smell
Ozonium stage of *Coprinus*	Laths	None in this growth form	Ochre, thin	Ochre; can be very resilient to the touch	Brown rot; laths crumble	Colour of mycelium and strands
Paxillus panuoides	Very wet situations	Yellow to amber with gills	Yellow to amber	Yellow to amber, woolly	Brown rot; yellow coloration initially	Colour of fungal growths
Ptychogaster rubescens	Timber in contact with wet brickwork	Pinkish-brown powdery cushions	White, not common; may be brittle when dry	White, woolly	Brown rot; may leave skin of sound timber	Fruit-body
WHITE ROTS						
Donkioporia expansa	Ends of large oak beams embedded in walls	Brown or buff, covered in minute spores	None	Not always present; yellow to red-brown felted growth	White rot	Decay type, fruit body
Asvomycete decay e.g. *D. concentrica*	Hardwood floors particularly sports halls	Hard black rounded, concentric rings inside	None	None visible	White rot plus superficial date zone	Decay type
Asterostroma spp	Skirting boards	Cream to beige to light-tan, smooth, flat on substrate	White or light tan, sometimes across masonry for long distances	White, cream or buff sheets	White rot; often little decay	Decay type; strands flexible when dry

FUNGUS	MOST COMMON SITUATION	FRUIT-BODY	STRANDS	MYCELLUM	DECAY TYPE	TO DISTINGUISH FROM DRY ROT
Phellinus contiguus	Exterior joinery	Ochre to dark brown, covered in minute pores	None	Tawny brown tufts	White rot	Decay type; colour of fungus growth
Pleurotus ostreatus	Chipboard	Grey or fawn on top, white gills underneath	None	Whitish woolly mat; strong mushroom smell	White rot; chips tend to separate	Decay type; fruit-body

A wet rot attack to the frame and threshold of an external door.

Wet rots tend to thrive in timber with a moisture content higher than the optimum for dry rot. Wet rot is often found where wood is repeatedly wetted, for instance, as a result of faulty plumbing or leaking gutters, or where there is poor protective detailing. For the reasons mentioned above, a description of every wet rot fungi is inappropriate in this publication. The brief description below illustrates the point by comparing the characteristics of just two types.

A common wet rot is cellar rot (Coniophora Puteana). It is a brown rot which attacks both hardwoods and softwoods. The wood becomes darker and suffers from cuboidal cracking. The fruiting body is thin and flat and is olive-coloured. The fruiting body is rarely found in buildings. The strands are yellowish-coloured during their early development but become darker, reaching a blackish colour with age. Because of the cuboidal cracking, this fungi is sometimes confused with dry rot and undoubtedly often treated as such.

Coniophora puteana (photo courtesy of N.J. Donnithorne).

Phellinus Contiguus is a white rot that is commonly found on external joinery. Its mycelium looks like brown tufts. The fruiting body is rare and it is an elongated shape which is coloured dark brown. As it is a white rot, the wood becomes bleached and has a stringy, fibrous appearance.

Coniophora puteana (photo courtesy of Johan Mattsson).

The remedial treatment of dry and wet rot

It is important that the cause as well as the symptoms of a rot outbreak is adequately dealt with. The first step in remedial treatment for rot is to determine the cause of the outbreak. This will involve finding and eliminating the source of dampness and rectifying those faults which have contributed to the problem. Fungal decay happens because conditions are created which allow for the germination and evolution of an attack and the key factor in creating and maintaining these conditions is timber with a high moisture content – without which, there will be no rot. Even when wood is infected, if its moisture content is reduced to an appropriate level, the fungus will stop developing and then die off. It is important to remember that the removal of damp conditions is the key treatment in dealing with fungal decay. Therefore, the cause of the excessive moisture should be identified and rectified. Measures should also be undertaken to dry out the structure and fabric as quickly as possible. Clearly timber which has been unacceptably weakened structurally should be replaced. In reality, it is common practice for all infected wood to be removed plus, in the case of dry rot, seemingly unaffected timber, say, 300–450mm beyond, to allow for the possibility of less obvious infection. There is a debate about the necessity for this approach, because drying out and/or selective 'preservative' treatment will kill the fungi (though infected timber will remain a risk while it dries out).

There may of course be situations where it is not possible to cost-effectively ensure that dampness does not reoccur. Timber which is removed can be replaced by timber which has been pre-treated with preservatives, or alternatively material which will not rot can be substituted (for example, concrete or steel). It may be necessary to treat remaining timbers in situ with preservatives, particularly where damp conditions may remain for some time, or where isolation from, say, damp masonry is difficult.

Because the dry rot fungus can spread rapidly throughout a building, including through mortar joints and behind plaster, it is best if the full extent of the outbreak can be established in order to ensure that all vulnerable or infested timbers within and near the infected area are located. The investigation should take account of timbers such as noggins, levelling pieces, etc., which may have been incorporated in the masonry. Also, because the dry rot fungus can penetrate mortar joints and plaster, the separation of vulnerable timbers from 'infected' masonry needs to be considered.

In many cases this investigative work can be a costly and destructive exercise. Additionally, where historic buildings are involved, there is a specific concern about the damage to historic fabric where large areas of timber or plaster are removed for precautionary reasons. Decisions on the scale and scope of investigation and treatment are essentially a professional judgement made in the light of the particular circumstances. Excessive destruction of materials or the unnecessary use of chemicals can be avoided, if the causes and risks associated with dry rot are better understood. The balance between a rightful appreciation of the damage that can be caused by dry rot, and damage which is often caused by the methods which are sometimes adopted for its eradication, should be considered.

A detailed description of remedial treatment is inappropriate here (there are a number of publications produced by the Building Research Establishment, to which further reference can be made). It should, however, be noted that concern over the health of humans, and indeed the wider environment, in relation to the toxicity of timber preservatives has raised debate about what might be called 'traditional' approaches to the eradication of rot in timbers. Although the use of in-situ timber preservatives on remaining timbers may be necessary in certain circumstances, the decision to use them – and where used, the amount and type of chemical specified – should be carefully considered. If they are used, the extent and intensity of treatment should be commensurate with the particular situation. For example, the practice of drilling holes in masonry and saturating the irrigated walls with fungicide has been discredited. This is because not only was it ineffective in transporting the fungicide through the wall but it also introduced water (as the carrier for the fungicide) into the masonry, i.e. it was a supposed solution that actually increased the supply of one of the main elements essential for the development of fungi, while saturating the wall with a toxic substance. It is also clear that many other treatment specifications still include the use of chemicals for unjustifiable precautionary reasons.

While dry rot can certainly cause severe damage, it is a relatively vulnerable organism which can be destroyed by environmental change, particularly in relation to moisture content. It should be remembered that once dry rot is identified, the primary treatment, without which all others will fail, is the drying out of the structure and the prevention of further damp conditions. While this may be difficult to achieve in some circumstances, the remedy is in fact no more than to return the building to a fit state.

Insect attack

Insect damage to timber can occur in standing trees, freshly felled logs and other unseasoned timber, as well as in seasoned timber, both in storage and in use. Usually those insects that attack unseasoned timber die out as the cut timber dries. It is possible for live insects to be incorporated in a building because they have infested the timber before it has been used. In most cases they cannot re-infest the timber in the condition it is found in buildings and therefore no specific treatment is necessary. Other insects do not feed on wood but bore into timber for shelter. Wood wasps are an example of insects in this latter category.

Damage to timber caused by wood-boring insects. Left and middle: Flight holes produced as the adult beetle emerges. Right: Flight holes plus some of the internal damage (note that this also shows evidence of fungal attack – see reference to death-watch beetle below).

This section is concerned with insects that infest timber after it is incorporated into a building. The insects which cause problems for timber used in buildings are often referred to as 'woodworm', but they are not worms and, in fact, most can fly; the insects most commonly encountered in timber (and which cause damage) are beetles.

As with wood-rotting fungi, the insects in question use wood as a food source as well as a habitat, and can cause damage to building timbers by eating away at it.

Life cycle

The general description of the life cycle is useful for a better understanding of the problem and the remedial treatments.

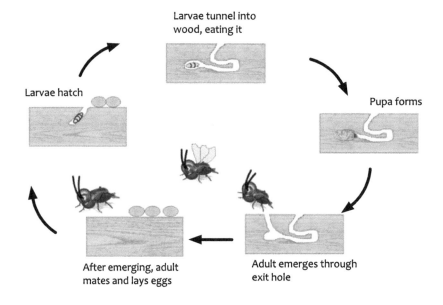

Eggs

These are laid by female insects in cracks, splits, rough surfaces or the holes created by previous insect attacks. Under the right circumstances the eggs will hatch and emerge as larvae.

Adult house longhorn beetle laying her eggs (photo courtesy of Johann Mattsson).

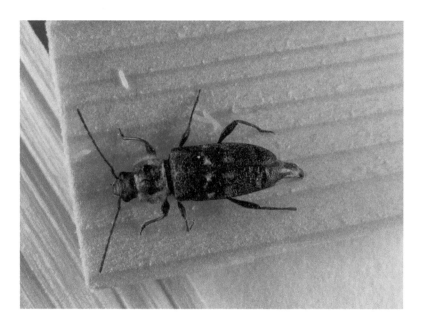

Larvae

It is during this stage that the most damage is done, as the larvae burrow into the wood to obtain food and shelter. Unfortunately there is usually no obvious external sign of activity as the damage is taking place within the timber. The length of this period will vary between species, and for each species there will be variance, perhaps between 2 and 10 years, which may be affected by the conditions within the timber.

Larvae of house longhorn beetle (photo courtesy of Johann Mattsson).

Pupa and adult

During this stage the insect ceases to feed and a process of metamorphosis occurs, at the completion of which the adult insect emerges from the pupal skin and bores its way out of the wood, forming an exit hole. The size and shape of the exit hole are a useful indicator of the type of insect involved, as to a lesser extent is the bore dust (known as frass).

Adult death-watch beetle – note nearby exit holes.

Exit holes formed by adult beetle (in this case both death-watch and common furniture beetle) emerging from the timber.

The insects are generally capable of flight but some are more likely to fly than others. The adults mate quite soon after emerging and the female will then look for a suitable place to lay her eggs. Often the most suitable place will be timber from which she has recently emerged. This means that timber can be continuously re-infected as new insects invade the timber year after year. Soon after laying the eggs, the adult will die. The period from emergence to death is generally only a matter of weeks.

Conditions favourable to insect attack

The likelihood of attack will depend on a variety of factors, including the species of timber, its condition and the temperature and humidity of its environment. It is noticeable, for instance, that flight holes in the floorboards to suspended timber floors often exist predominantly on the underside; that is facing into the damper void rather than the drier room. Although the insects generally like damp timber, excessive moisture is by no means a critical factor. However, low moisture content in timber can kill larvae by causing them to desiccate (which may be why the introduction of central heating systems into existing property is thought to have an effect on some insect infestations). Ideal requirements will vary between the different insects and between the different life cycles of the same insect. If conditions are good, the life cycle will be shorter. This might explain why life cycles are often longer in buildings than they are in controlled laboratory situations.

Sensitivity to environmental conditions as well as the variable food value within timbers (or parts of the same timber) may mean that attacks can be extremely localised.

As the insects are seeking food as well as shelter, they will mainly attack sapwood but they may move around any part of the timber while seeking food supplements such as starch or proteins. A much-quoted example is the way in which older plywood is readily attacked by the common furniture beetle because it was bonded with animal glue, whereas modern plywoods are rarely attacked.

The flight holes shown in these two photographs illustrate that it is the sapwood that is attacked rather than the heartwood.

Table 15.2 (from BRE: Remedial Treatment of Wood Rot and Insect Attack in Buildings) gives indicators which will help towards recognition.

Table 15.2 Recognition of damage and remedial treatment for insect attack

TYPE OF BORER	HABITAT	RECOGNITION OF DAMAGE		
		EMERGENCE HOLE SHAPE & SIZE (mm)	TUNNELS	BORE DUST
DAMAGE FOR WHICH REMEDIAL TREATMENT IS USUALLY NEEDED				
Common furniture beetle	Sapwood of softwoods European hardwoods	Circular 1–2	Numerous, close	Cream, and granular, lemon-shaped pellets (x 10 lens)
Ptilinus beetle	Limited number of European hardwoods	Circular 1–2	Numerous, close	Pink or cream talc-like, not easily dislodged from tunnels
House longhorn beetle	Sapwood of softwoods	Oval 6–10, often ragged	Numerous, often coalesce to powdery mass beneath the surface	Veneer cream powder, chips and cylindrical pellets
Lyctus powderpost beetle	Sapwood of coarse-pored hardwoods	Circular 1.5	Numerous, close	Cream, talc-like
Deathwatch beetle	Sapwood and heartwood of decayed hardwoods, occasionally softwoods	Circular 3	Numerous, close, eventually forming a honeycomb appearance	Brown, disc-shaped pellets
DAMAGE FOR WHICH REMEDIAL TREATMENT IS NECESSARY ONLY TO CONTROL ASSOCIATED WOOD ROT				
Wood-boring weevil	Any, if damp and decayed	Ragged 1	Numerous, close, breaking through to surface in places	Brown, fine, lemon-shaped pellets (x 10 lens)
Wharf borer beetle	Any, if damp and decayed	Oval 6	Numerous, close, often coalescing to form cavities	Dark brown, mud-like substance; bundles of coarse wood fibres
Leaf cutter bee/ solitary wasp	Any, if badly decayed	Circular 6	Sparse network	Brown chips, metallic-like fragments, fly wings, barrel-shaped cocoons of leaves
DAMAGE FOR WHICH NO REMEDIAL TREATMENT IS NEEDED				
Pinhole borer beetles	Any in log form	Circular 1–2	Across grain, darkly stained	None
Bark borer beetle	Bark of softwoods in bark, few in sapwood	Circular 1.5–2, some bark and few in wood	Network between bark and few in sapwood	Cream and brown round pellets underlying wood
Wood wasp	Sapwood and heartwood of softwoods	Circular 4–7	Few, widely spaced	Coarse, powdery
Forest longhorn beetles	Any	Oval 6–10 on bark edges only; may be larger in some hardwoods	Few, widely spread; sections on sawn surfaces, oval 6–10 mm	None or, rarely, small piles of coarse fibres
Marine borers	Any	None, but tunnel sections can be exposed by sawing	Circular up to 15mm	None, but tunnels may have white chalky lining

Examples of insects that damage timber

For the purposes of this book it is not necessary to consider in detail all of the insects which attack timber, but it is useful to consider one or two of the more common species. Some insects will only be able to attack part of the timber's cellular structure and others will be able to attack all parts. For example, powder post beetles cannot break down hemicellulose or cellulose and pinhole borers cannot break down cellulose, but others can attack cellulose, hemicellulose and lignin. Examples of this latter, more destructive group, include the following.

Common furniture beetle

This has the Latin name Anobium Punctatum and is the most common form of insect to attack timber in this country. Despite its English name it does not of course attack just furniture (though it is often transported from building to building by furniture which has been infested) and it will attack softwoods and hardwoods in many situations, from a damp under-floor area to the roof. The beetle attacks mainly sapwood; heartwood is only usually attacked where it has suffered from fungal decay. If the sapwood is only a small part of the timber section, as it will tend to be in older buildings (with original timbers), then the amount of damage caused by this insect may be relatively small.

Adult common furniture beetle (photo courtesy of N.J. Donnithorne).

Damage by common furniture beetle can be identified by the round emergence holes which are about 1–2mm in diameter. The adult insect is about 3–5mm long. The female will lay about 30 eggs in a convenient place in the timber. In practice, this often means cracks, rough surfaces or even an old exit hole. The larval stage will typically last between 2 and 5 years. The adult beetle that normally emerges between May and August can sometimes be seen in favoured places such as roof spaces.

Death-watch beetle

This beetle tends to be well known because it is associated with attacks on the type of hardwood, particularly oak or elm, which may be present in more prestigious or historic buildings. Other hardwoods including walnut, chestnut, elder and beech, are also at risk from attack, as are softwoods which are in the vicinity of the hardwood.

Adult death-watch beetle (photo courtesy of Rentokil).

The death-watch beetle will usually only attack damp, decayed timber. Infestation is often therefore restricted to, for instance, timbers or parts of timbers built into damp walls. Widespread infestation may occur following the type of diffuse dampness associated with condensation.

Examples of this may include the badly heated roof areas of medieval halls or churches or the underside of lead roofs. Although the damage caused by this insect may be limited in scope, it is often structurally significant. This is because it attacks heartwood as well as sapwood and because in-built timbers are often carrying loads. Its name, incidentally, is supposed to derive from the fact that the beetles sometimes make a knocking sound, during courtship. It is possible that, in the still of the night, those watching over the dying could hear the noise.

The female will usually lay between 40 and 60 eggs and the larval stage can last 10 years or more. Damage can be identified by the round 3mm holes left by the emerging adult. The adults are 6–9mm long. As with the common furniture beetle, the insect moves towards the outside of the timber to undergo the pupal stage, and this often occurs in July or August, but unlike the common furniture beetle it usually does not emerge from the timber for another 9 months or so.

House longhorn beetle

House longhorn beetles attack mainly the sapwood of softwood and are an important insect menace because of the relatively severe damage which can be caused in a short period of time. They are also interesting in that they are only common in certain geographical areas around London and the Home Counties. The reason that they are only found in this area is thought to be connected with climate. This geographical factor is acknowledged by the Building Regulations which stipulate that buildings in the prescribed areas should be treated against infestation. The Building Research Establishment asks that suspected outbreaks should be reported to them in order that records of the spread of the attacks can be kept.

Male house longhorn beetle (photo courtesy of Johan Mattsson).

Flight holes of house longhorn beetle. Note how the frass (wood dust from the boring activity) can be clearly seen in the middle right of the image (photo courtesy of Johan Mattsson).

The larvae can grow up to about 24mm in length and this, combined with a long larval period, means that the beetle can inflict considerable damage. An attack can be recognised by the relatively large oval-shaped emergence holes which are about 6–10mm in diameter. Also the larvae burrow quite near the surface of the timber, leaving a thin intact skin and this can produce a rippling/blistering effect on the surface. About 140–200 eggs are laid and the larval stage usually lasts between 3 and 6 years, with the beetle typically emerging in the period July to September.

Eradication of beetle attack

Insect attack may die out naturally due to a number of reasons. These might include desiccation of the larvae in timber where the moisture content has been lowered, or attack by parasites, or because the nutrients in the timber have become exhausted.

A detailed description of remedial treatment is inappropriate in this publication though the following brief points are worthy of note: insecticides are usually applied as a liquid or as a paste. In certain circumstances smoke, gas or heat treatments are used. Liquids can be either emulsion or solvent-based.

The purpose of anti-insect treatment is to both kill the insects already in the timber and to prevent further attack. The success of a treatment will depend on a number of factors including: the cross-sectional area and porosity of the timber, the type and formulation of the insecticide, and the manner of its application. Treatments applied to the surface of the timber will only penetrate to a certain depth and, because of this, not all of the larvae will be killed immediately. The intention is that those larvae within the area of timber penetrated will be killed immediately – as will the eggs and adults on the surface – and a residual layer of preservative will be left behind to kill newly evolved adults as they tunnel out of the wood and any eggs laid at a later date.

The insect should first be identified in order to facilitate judgements about the possible extent of damage to the timber. If the attack is one which is associated with fungal attack, this, and the associated damp problem, need to be addressed. Attempts should be made to estimate the extent of the damage and whether or not the attack is still active. Establishing either of these factors is often difficult in practice because the damage is essentially taking place in the body of the timber and cannot be accurately assessed without some level of destructive investigation. It is not always easy to determine whether an infestation is active, or has died out naturally, or indeed has been treated in the recent past. However, if the flight holes have sharp edges and clean fresh wood can be seen, then recent or current activity can be assumed. If structural damage has been caused, then obviously repair or replacement of timbers will be necessary.

Appropriate insecticides can be applied in situ, if an insect attack is identified as active. As with fungal treatments the focus, extent and intensity of treatment should be appropriate to the particular situation. Pressures on surveyors may mean that timber is treated even when an attack is no longer active. Fixed identification tabs following treatment can be useful in preventing unnecessary double treatments. As with the chemicals used for treatment of fungal attack, care must be taken to address health and safety issues and to guard against general environmental damage. Of particular relevance are occupiers and operatives as well as other creatures, such as bats. Bats are a protected species under the Wildlife and Countryside Act 1981. They often nest in roofs and are vulnerable to the toxic chemicals used in pesticides. With both fungal and insect treatment, the Control of Substances Hazardous to Health (COSHH) Regulations and the Control of Pesticides Regulations will apply.

In recent years some work has been carried out into the use of light traps. The specific targets were death-watch beetles which were found to be causing damage to historic buildings. The beetles were attracted to a light source and then trapped.

Preservation of new timbers

Because of the problems described above, one might assume that the sensible approach with new timbers that are to be incorporated into a building is to apply preservatives which will prevent attack by fungi or insects. The only statutory requirement for treatment is where the

Building Regulations require it in those geographical areas liable to attack by house longhorn beetle. Therefore, not all timbers incorporated in a building will be routinely treated. There are a number of reasons for this, including concerns about the environmental consequences of preservative use, but the main one is cost. Other reasons include the possibility of corrosion to metal fixings if water-borne chemicals are used, and the fact that some chemicals can affect glued timber joints.

The factors which might be taken into account in deciding whether to preserve or not include:

- *the required life of the building related to the natural durability of the various timbers within it*
- *the type and position of the timber component and, therefore, the likelihood of attack – example, roof tiling/slating battens are more vulnerable than internal door linings to attack by wood-rotting fungi.*
- *the environmental implications of chemical treatments.*

Although many specifications simply include an instruction to use preserved timber, the subject is more complex than this would imply. There are a variety of preservative chemicals. There are also a variety of processes which can be used to apply the preservative. These different methods offer different levels of effectiveness and must be matched with the type of timber, its position in the building and, therefore, its vulnerability. These factors must then be related to the type and toxicity of preservative.

Damp: Rising damp, penetrating damp and condensation 16

Introduction

This chapter explains the main causes of penetrating and rising damp and condensation. It complements some of the detail in other chapters where dampness is inevitably discussed (for example, in Chapter 7, dealing with ground floors, and Chapters 9 and 10, dealing with roofs). This chapter defines dampness, explains how dampness can occur, how it can be identified and how it can be measured. Incorrect diagnosis is a common problem caused, in many cases, by insufficient thought and over-reliance on inconclusive instrument checks. Incorrect diagnosis at best leads to wasted effort and wasted resources; at worst, it can lead to 'repair' works which can actually exacerbate the initial problem. The chapter includes some diagnostic check lists which are intended to help the reader approach a suspected problem of dampness in a systematic way.

The problem of dampness

The presence of excess water in buildings probably gives rise to more defects than any other cause. Dampness can:
*   *be a health hazard*
*   *reduce the strength of building materials such as chipboard and plasterboard*
*   *cause movement in building elements, e.g. block floors*
*   *lead to outbreaks of dry and wet rot (see Chapter 15 on timber pests)*
*   *cause chemical reactions in building components (see Chapter 16 on walls)*
*   *reduce the effectiveness of insulation*
*   *damage decorations.*

Unfortunately problems of dampness can be quite complex; a sensible starting point is to consider a definition of dampness and to examine how water is held in materials.

Dampness and moisture content

The word 'damp' is potentially confusing because most building materials (with the exception of plastics and metals) are porous and will always contain some moisture. Assuming there are no problems of condensation, spillage, or damp penetration through the fabric, the moisture they contain will vary according to the relative humidity of their surroundings. This is why it is so important to ventilate timber ground floors. Inadequate ventilation, particularly if the ground below the void is wet, will allow the relative humidity of the air below the floor to rise. This occurs because the air steadily absorbs water. At the same time the moisture content of the floor structure will slowly settle into equilibrium with the damp air, and, as it does so, its moisture content will also rise. If the air below the floor has a relative humidity of 85 per cent or more (see later for a definition of relative humidity), the timber's moisture content will rise to a point where biological decay is likely in most untreated softwoods; this is about 20 per cent.

The relationship between relative humidity and moisture content of softwood is shown in the chart above.

If the relative humidity of the air below the floor is above 85% the timber floor will eventually reach an equilibrium moisture content of about 20%.

An example of a partially blocked vent to a suspended timber ground floor.

The moisture content of the timber floor will increase until it is in equilibrium with the air below.

Blocked vent reduces flow of AIR.

Moisture evaporating from the ground raises relative humidity of the air below the floor.

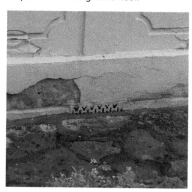

Ventilation of the under-floor void should ensure that the relative humidity of the air remains below 75 per cent. Timber in equilibrium with a relative humidity of 75 per cent has a moisture content of about 17–18 per cent.

Moisture content of timber

Timber with a moisture content of less than 17–18 per cent is safe from decay. At 18–20 per cent the risk of decay is more likely and, above 20 per cent, it is inevitable (assuming the dampness is chronic and the presence of spores, see Chapter 15). The moisture content can be calculated as shown below:

$$\frac{\text{Wet weight of material} - \text{dry weight of the material} \times 100}{\text{dry weight of the material}}$$

So, for example, if a sample of timber weighs 150g and if the same sample weighs 130g when dried, its moisture content will be:

$$\frac{150 - 130}{130} \times 100 = 15\%$$

The method is often referred to as oven drying. The sample of timber is weighed, dried in the oven, and then weighed again. However, this method is of little practical use and site testing usually requires the use of a moisture meter.

It should also be remembered that moisture content will vary with the seasons; roof timbers in a well-ventilated roof, for example, can vary from 15 per cent in a wet January to 5 per cent in a dry July.

Moisture meters

When materials absorb moisture, they can conduct electricity. By measuring a material's ability to conduct electricity, a measure of the amount of water in it can be assessed. Thus it should be remembered that most moisture meters measure conductivity not moisture content. It should also be remembered that the ability of a material to conduct electricity is not in direct proportion to the amount of water it holds. Very low levels of moisture cannot be detected. This is because low levels of water cannot dissolve the soluble chemicals required to form the solution which actually conducts the electricity.

If there is sufficient 'free' water, soluble salts will dissolve. When soluble salts are dissolved in water, each molecule splits into two parts, one with a negative charge and one with a positive charge. These charged particles (known as ions) are attracted to the negative and positive electrodes of a conductance meter (the most common form of moisture meter – see below) where they give up their charges. This results in a flow of electricity, the level of which is measured by the moisture meter.

Conductance meters

These instruments, which are by far the most common form of meter, usually have two metal probes (the electrodes) which are firmly pressed into the material being tested. The electrical resistance between the two probes can then be measured. The meters (depending on type) can provide analogue or digital readings. The Protimeter 'Mini' meter, for example, has coloured LED lights (green, yellow and red) representing 'safe', 'borderline' and 'decay inevitable' situations. More sophisticated meters provide specific percentage measurements and can be used with attachments to provide relative humidity and surface temperature readings to help identify condensation.

Protimeter Mini, Survey Master and Timber Master.

Most moisture meters are specifically calibrated for timber and cannot give accurate readings (i.e. percentage moisture content) for other materials. Even with timber there are differences between timber species, and between hardwoods and softwoods. However, the differences are not particularly significant and measurements can be adjusted if an accurate moisture content is required for a known species. There are, however, one or two pitfalls which should be recognised by anyone carrying out testing to assess moisture content. For example, readings from an electric moisture meter can be affected (readings will be too high) if the timber has been treated with water-based preservatives. The temperature of the wood will also affect readings.

A definition of damp

In a well-ventilated room with a fairly constant relative humidity of about 30 per cent, a softwood timber skirting board may contain as little as 6 per cent moisture (see chart on p. 404). In a kitchen with a constant relative humidity of about 70 per cent, the moisture content of the skirting will be approximately 16 per cent. In both cases the timber contains moisture but, because its moisture content fluctuates according to the relative humidity, and because a relative humidity of 70 per cent will not encourage decay in timber, it can be referred to as 'air dry'. However, if the relative humidity reaches 85 per cent and stays at, or above, this level for a length of time, conditions are created which can cause decay in the timber, damage to decorations, the development of moulds and the appearance of mites. Thus, when a timber skirting is in equilibrium with this high relative humidity (over 85 per cent), the material can be regarded as 'damp'. Its moisture content will be about 19 per cent or even more. Thus dampness can occur through persistent levels of high humidity where the timber in contact with the moist air reaches a point of equilibrium. Similarly timber can be regarded as 'air dry' (even if it contains some moisture) if it is in equilibrium with a 'normal' atmosphere which is typically 30–70 per cent relative humidity. At these humidities its moisture content will be 7–18 per cent.

The problems shown in the two images above are caused by long-term persistent dampness.

Materials other than timber

Of course, the external wall of a house contains a number of elements in addition to timber skirtings. These elements vary considerably in their acceptable percentage moisture content. Timber, as demonstrated above, can vary from 4 per cent to 18 per cent, depending on the relative humidity, and still be regarded as 'air dry'. Brickwork (depending on specific type), however, is wet at 5 per cent and saturated at a moisture content of 10 per cent; a more acceptable level would be between 1.5 per cent and 2.5 per cent. Some plasters are dry at 1 per cent and saturated at 5 per cent. Like timber, all porous materials will settle into equilibrium with the air surrounding them. The higher the relative humidity, the greater their moisture content. If a moisture meter is used to measure dampness in these materials, it must be remembered that the moisture meter is calibrated to measure moisture content in timber. A reading on any other surface will only give a comparative reading, not an accurate indication of moisture content.

Measuring the moisture content of concrete, plaster and brickwork, etc. can be carried out using the oven drying method mentioned earlier. Another option is to use carbide testing (see below). Whichever technique is used, it must be recognised that establishing moisture content provides little help unless the particular characteristics of the material are known. In addition, it should be recognised that spot readings on their own may show high levels of moisture, but will give little indication as to their cause. One of the first stages in any diagnosis, therefore, should be to establish contours, or patterns, of dampness across the face of a wall. The section on diagnosis explains this in more detail.

Carbide testing

For masonry products, bricks, blocks, mortars, etc., carbide testing can be used to assess moisture content. Its use should be regarded with caution because, as already mentioned, these materials will differ in the amount of moisture they can contain and still be regarded as dry. By drilling into a wall with a large slow bit (to minimise heating and, therefore, drying) a sample of suspect material can be collected. Part of the sample is weighed and tipped into a cylinder. A specific amount of calcium carbide is added and the container sealed. Vigorous shaking of the cylinder will then mix the two materials. Moisture in the sample reacts with the calcium carbide to form acetylene gas. As the gas forms, the pressure in the sealed cylinder increases. A pressure gauge within the cylinder is calibrated to show the moisture content of the sample.

Although these instruments will fairly accurately measure moisture content, they do not help diagnosis of dampness unless several readings are taken across a wall. This is obviously a fairly destructive exercise.

The oven drying method of calculating moisture content and the carbide tests show that if two materials of differing densities contain the same amount of water, they will have different moisture contents. Even similar materials can be different, e.g. weak and strong renders. Therefore, to use this method to interpret whether a material is wet or not would require a detailed knowledge of a whole range of building materials. Gypsum plaster with a moisture content of 2 per cent would in fact be very wet, lime mortar at 4 per cent would be fairly dry. In practice, this is not a practical option and most surveyors tend to rely on conductance meters.

Interpreting a conductance meter

Conductance meters can be used to help determine levels of dampness in materials other than timber, though, as indicated above, they should be used with caution. A simple laboratory-based experiment illustrates this.

A sample of facing brick and a sample of cement/sand plaster were oven-dried for 36 hours and then weighed. They were then soaked in water for 48 hours. Surplus water was removed and they were weighed again. This allowed the moisture content to be calculated in their saturated state. Over the next few days the samples were allowed to dry in a well-ventilated room of average humidity. At regular periods, while they were drying, the samples were wrapped in polythene for several hours to ensure water was evenly distributed within each sample. After unwrapping, the samples were weighed again, and measured with a moisture meter. The graphs below show the relationship between their actual moisture content and the relative readings on the moisture meter.

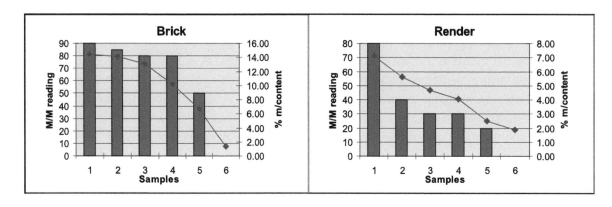

It can be seen from the above graphs that the relative readings (the vertical bars) roughly follow the curves of the actual moisture content (the black line). In ambient conditions (18°C with an RH of 50 per cent), both samples had a moisture content slightly below 2 per cent, and in these conditions the moisture meter showed them both to be dry. Similarly, when both materials were saturated (7 per cent render, 14 per cent brick), the meter readings were both very high. The meter was also fairly accurate in providing a comparative assessment of dampness. However, it should be clear from the above results that the relative readings provided by the meter should not be confused with actual moisture content.

Conductance meters are therefore ideal for establishing contours or patterns of dampness and for establishing whether or not materials are dry. Establishing contours of dampness should therefore be an operation which should be one of the first stages in diagnosing dampness.

Measuring dampness within a wall

Carbide testers, by their very nature, measure the dampness within a wall rather than at its surface. Conductance meters, however, measure dampness at, or just below, the surface. Surface dampness can be caused by condensation. However, a range of deeper probes can be used if the internal part of a wall needs to be measured. Deeper testing is often necessary if inaccurate surface readings are to be avoided. The deep probes have insulated sides to make sure that readings nearer the surface are ignored.

Salts

Another problem which affects the accuracy of conductance meters is related to the salt content of the material being tested. Salts, such as nitrates and chlorides, are commonly deposited on the inner surface of walls as groundwater rises up a wall and evaporates. Other salts deposited on the surface or near the surface of the plaster can come from chemicals naturally present in the walling materials which are dissolved by dampness and deposited as salts on, or near, the surface of the wall. These salts readily conduct electricity and, when tested with a conductance meter, can indicate levels of moisture far higher than the actual one. In addition, nitrates and chlorides are hygroscopic; in other words, they absorb moisture. Thus, a high moisture reading may not indicate problems of ongoing rising or penetrating damp but may be measuring water from the air absorbed by the hygroscopic materials left by previous dampness. Fortunately, there are instruments which can detect these salts either through site testing or laboratory analysis. The nature and detection of these salts are covered in more detail later in the chapter.

Sources of damp

There are four main sources of dampness in housing: rising damp, penetrating damp, water from leaks (pipes bedded in screeds, etc.), and condensation. In new buildings, an additional problem is trapped construction water. This next section explains the most common ways in which penetrating damp can occur.

Structural movement and consequential cracking can form obvious paths for water to

penetrate a wall. Similarly, frost attack, sulfate attack and surface erosion caused by atmospheric impurities can reduce a wall's resistance to damp penetration. These phenomena, and their potential consequences in terms of chemical attack and wall deterioration, have been considered in earlier chapters.

Penetrating damp

Penetrating dampness occurs in a variety of ways; the main 'mechanisms' are shown below.

Mechanisms for water ingress include:

- gravity
- capillary action
- surface tension
- kinetic energy (splashing)
- wind force
- differential air pressure (inside and out).

Water is also absorbed, i.e. taken directly from the air. The amount of water absorbed depends on the ambient humidity and the nature of the material.

Solid external walls

Until the 1920s the majority of buildings in the UK were constructed from solid, i.e. non-cavity masonry. The materials used, mainly brick and stone, varied widely in their resistance to moisture absorption. Some materials, such as engineering brickwork, salt-glazed brickwork and granite are virtually moisture-resistant. Others, such as limestone, sandstone and most clay-facing bricks are more permeable. Contrary to what one might expect, it is often walls constructed from the denser materials that are most at risk from damp penetration. The diagram below shows the effect of rainwater on two solid walls. The moderately porous wall will absorb a proportion of the rainwater running down its face. As long as the wall is of adequate thickness (usually 1 brick) and contains no significant cracks, it will resist damp penetration in all but the worst of exposures. In wet conditions the wall will absorb substantial amounts of rainwater but when the weather changes the absorbed water will slowly evaporate from the wall. Minor imperfections or cracks in the wall will not necessarily lead to damp penetration due to the limited amount of water running down the wall's face.

Walls built from denser materials such as granite or engineering brick lack the ability to absorb any rainwater running down the wall's face. Minor cracks in the walling material or in the mortar itself are therefore more critical and can allow moisture ingress. In addition, the dense, smooth nature of some of these materials precludes a good bond with the mortar. The interface between walling material and mortar can therefore provide a capillary route for damp penetration. In some properties the effect of this can be seen when observing the internal plasterwork; 'lines' of dampness in the plaster correspond with the position of the mortar joints.

Vernacular traditions, such as the slate cladding shown in the photograph on the previous page, can be effective against penetrating damp and can often be seen in areas exposed to high levels of wind-driven rain. They may also be used as a retrofit solution as shown in the photograph below.

External cladding is being added to the external wall in order to combat penetrating damp.

Causes of dampness

General defects

The diagram below shows some of the common routes for damp penetration in houses with solid and, in some cases, cavity walls. The photos on p. 412 show a few more examples.

- Inadequate chimney/parapet flashings
- Chimneys not capped where flues redundant; cracked flaunching or cappings.
- Copings without drips or not bedded on DPCs
- Dropped arches or soldier courses
- Cracked render
- Overflowing hopper heads
- Defective pointing, render or poor quality bricks
- Rusted downpipes or broken joints
- Warning pipes 'running'
- Movement cracks
- Blocked gulleys can result in rainwater splashing against wall
- Badly fitting windows
- Sills without drips
- Poorly designed threshold

Cracks in masonry   Mortar flashing   Slipped tile   Patched downpipes

Overflowing warning pipes   Gutters full of vegetation   Sills without drips

Render over DPC (and cracked)   Patched lead work; inadequate steps   Decayed pointing   Leaking gutter union

Mortar joints

The purpose of the mortar joint is to:
- *provide an even bed to distribute the loads of the wall*
- *bond the bricks or stones together*
- *seal the joint against rain penetration.*

An effective seal depends on a good bond between the mortar and the bricks or stones. In addition, to be effective over a long period, the mortar must be able to accommodate thermal and moisture movement without cracking.

It is generally accepted that the quality of the mortar and the nature of the pointing are of more significance than the bricks themselves in combating moisture penetration. Strong mortars, in other words, those with a high cement content, are likely to crack through drying shrinkage and they are less capable of withstanding the thermal and moisture movement stresses the mortar will have to cope with during its life. This is particularly important where the walling material is concrete blockwork or artificial cast stone. Both these materials have relatively high thermal movement which places considerable stresses on the mortar. Similarly, where mortar joints are very thin, they may not have the capacity to tolerate movement without cracking.

Weaker mortars are more flexible, less prone to drying shrinkage and generally more durable. They 'breathe' more readily and will encourage evaporation.

Unfortunately, many older houses, originally built with lime mortar, have been repointed with strong cement mortars. This is often counter-productive; for reasons already outlined, they may encourage penetrating damp. At the same time their dense nature restricts evaporation. Minor problems of penetrating or rising damp become more significant as the reduced rate of evaporation encourages water to dry out on the inner face of the wall. At the same time, rising dampness may move further up the wall as its ability to evaporate is impeded. Changing the 'equilibrium' of a building's external fabric is a common cause of damp problems.

Where old solid walls are damp, many builders will suggest external rendering as a solution. However, it should be recognised that renders, like mortars, are more effective if they are made from relatively weak, porous materials (see Chapter 11). Strong renders are more likely to crack (drying shrinkage) and will impede evaporation. This can actually increase the likelihood of penetrating damp.

The nature of the pointing finish will also affect a wall's ability to withstand rain penetration. Generally, experience shows that tooled joints are more effective. Tooling the joint helps seal the mortar against the brick, and compresses the mortar slightly, thus increasing its impermeability. Some joints, such as the recessed joint, are very vulnerable. Whichever joint is used, the mortar must be laid as a continuous bed and perpend joints should be well filled.

Bucket handle – a good bed and the face of the mortar is compressed by tooling.

Flush pointing – a good bed but the face mortar is not compressed because the joint is not tooled.

Tooled joints, full mortar beds and well filled perp joints help resist damp penetration.

Recessed – only a partial bed and poor at shedding water. Brick edges vulnerable to frost attack; algae growth common.

Struck joints – good at shedding water but very skilled work. The re-pointing on the right is vulnerable to physical damage and frost attack.

String courses, parapets and subsills

These are all vulnerable areas. If copings are not weathered (sloped), they will be slow to shed water. Cracked, pervious or loose pointing in between the copings will allow vertical penetration if the copings are not bedded on a DPC. Copings are obviously in a very exposed situation and, over the years, erosion and chemical attack can cause deterioration in the stone which reduces the profile of the drips. The drips prevent water running back under the coping and into the wall. In some rehabilitation projects, paving slabs are used as a substitute for proper copings. These are often bedded level and, by their very nature, do not contain drips.

In many parts of the country, string courses are cut from fairly soft stone. These are often fairly porous materials which readily absorb water. If the top is not weathered or not protected by a lead or copper flashing, problems of damp penetration can occur. String courses are often subject to frost attack; as the ice forms, parts of the stonework break away. Not only does this further reduce the ability of the string course to resist water penetration, it also places pedestrians below in very real jeopardy.

Defective stone or concrete sub-sills can also lead to problems of damp penetration. Problems include insufficient fall, gaps at junction with window, deterioration of drips (often blocked through painting) and lack of adequate projection. The latter often occurs where buildings are rendered at some point after their original construction.

Chimneys

Chimneys are very vulnerable to damp penetration, particularly if they are no longer in use. Correct diagnosis of the source of the moisture can be difficult as chimneys are of relatively complex construction. Damp penetration can occur from:
- *rainwater running down through the pots – a particular problem on low chimneys or low pitched roofs.*
- *failure of the flashings where the chimney runs through the roof.*
- *lack of DPCs in stack itself.*
- *sulfate attack in the brickwork.*
- *inadequate detailing around the top of the stack.*

On buildings with fairly steeply pitched roofs, problems of damp penetration may not manifest themselves in the rooms below. More at risk are buildings with shallow roofs particularly where the chimneys are situated above external walls at eaves level.

Mortar flashings such as shown to the chimney stack as follows can also cause damp penetration problems when used at the junctions of the roof and the party wall.

Cracked flaunching or decaying pointing can lead to damp penetration. They can also lead to pot or stack instability.

The joint between the roof covering and the stack is vulnerable to rain penetration.

Minor dampness will evaporate

AIR

AIR

In a high roof with a well ventilated roof void minor damp penetration of the stack is not likely to manifest itself in the rooms below.

Mortar flashings often crack and allow water ingress.

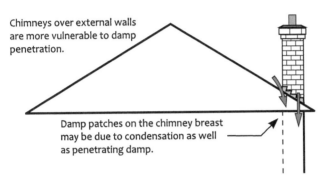

Chimneys over external walls are more vulnerable to damp penetration.

Chimneys in this position should include a horizontal DPC just above, or just below, the roof line.

Penetration can occur through defective flashings, through penetration of the brickwork and rainwater down the pots.

Damp patches on the chimney breast may be due to condensation as well as penetrating damp.

Note: Problems of sulfate attack or defective rendering can increase the risk of damp penetration.

Cracks to a mortar flashing at junction of roof and adjoining party wall.

Basements

In modern construction, basements are fairly rare though, until the 1930s, they were quite common. The examples below show typical late Victorian construction – all designed for areas where the ground was well drained and where the water table was well below the foundations. The dampness will be a combination of penetrating and rising damp.

Asphalt tanking

Asphalt becomes brittle after many years and may crack.

Two brick wall

Terrazzo screed

Asphalt tanking

Differential movement cracks asphalt

⟹ *Routes for damp penetration*

Basement with vented (and maybe drained) cavity

Air brick

One brick wall

Half brick wall

Air brick

Timber floor

DPC

Concrete or brick footing

Cavities are sometimes drained at the base (with gulleys)

Cavity construction – but non ventilated

Rusting iron ties and/or ground pressure.

Build-up of water

No tanking present

One-and-a-half brick wall

Most basement construction methods depend on well drained ground and a low water table. In this situation the building techniques protect against capillary action. Protecting existing basements against direct water pressure requires very expensive tanking works.

Other problems

There are a number of other problems which can cause penetrating damp. In many cases these are fairly easy to diagnose because the dampness will be localised and this will help to pinpoint the external defect. Problems include:

- **Leaking gutters and downpipes.** *Cast-iron downpipes often rust and develop leaks at the back, i.e. against the wall (this part of the downpipe is the hardest to paint). Gutters can fail through lack of correct support which allows backfalls, damaged joints and lack of regular clearing.*
- **Overflowing warning pipes** *which do not throw water clear of the wall. On some buildings lead overflows have even been cut off for their scrap value!*
- **Poorly fitting windows.**
- **Door frames with poorly designed thresholds.**
- **Cracked render.**

Aid to diagnosis

The chart below, which should be read in conjunction with the charts on pp. 427 and 448 explains the basic steps in trying to identify and pinpoint problems of penetrating damp. Mistakes in diagnosis are common, and where the problem is not obvious, remedial action should not be considered until all the steps have been carried out. Specialist advice is often necessary though it should be remembered that many of the so-called 'Damp Specialists' have a vested interest in promoting their own products.

Diagnosing penetrating dampness

Stage 1   **Pinpoint damp contours with moisture meter**   ⟹   **Key features**

Broken joint on downpipe

Defective eaves detail

25% wood scale

80 (R/scale)

Defective sill detail

60 (R/scale)

- Often in localised areas.
- Readings show sharp change from wet to dry.
- Patches of efflorescence (crystallisation of sulfates and carbonates present in building materials).
- Mould growth unusual.
- Timber in area of damp has high moisture content.

External inspection may reveal obvious defects including cracked render or brickwork, sills without drips, damaged downpipes etc.

Stage 2   ⟸   ⟸   ↵

- Deep wall probes indicate high readings in centre of wall.
- Salt analysis shows zero level of nitrates and chlorides (eliminating rising damp).
- Measure wall temperature, air temperature and establish humidity levels to eliminate condensation.

Cavity walls

The introduction of cavity walls in the 1920s was intended, in part, to improve the resistance of a wall to damp penetration. Unfortunately, failure of cavity walls, in terms of damp resistance, is a not uncommon problem. The defects, in the main, relate to poor design and poor workmanship rather than inadequate materials. A cavity wall may have a saturated external leaf after prolonged rainfall. Water should, in theory, run down the inner face of the brick and disperse back into the ground or substructure.

Unfortunately, there are a number of ways in which water can bridge the cavity and cause problems of dampness on the inner leaf of the wall. Careless bricklaying is probably the most common cause. Mortar droppings in the cavity can cause a build-up on wall ties or on floor joists which project beyond the internal leaf. It only needs a relatively small amount to bridge the cavity. Wall ties, if laid upside down or laid sloping towards the inner leaf, can also allow water to cross the cavity.

Cavity walls 1920s to 1960s.

Water running down cavity can escape through weepholes at base of wall.

Cavity tray

DPC

Cavity tray

Vertical DPCs at the sides of openings prevent water crossing to the inner leaf. Cavity trays over windows and doors direct any water outwards. Both were sometimes omitted in early cavity walls.

View inside cavity

Inner leaf   Closer

In exposed areas heavy rain can result in water running down the inner face of the outer leaf. In sheltered areas the outer leaf can become damp. Cavities must be kept clear of any debris in order to keep the inner leaf dry.

Mortar on ties bridges cavity.

Uneven courses result in ties sloping towards internal leaf.

Top

Some ties, if laid upside down, can actually channel water across the cavity.

These joists should not project beyond the inner leaf – the ledge forms a bridge across the cavity.

This photograph was taken using an endoscope. It shows a wall tie which is covered in mortar.

Cavity insulation

Most forms of cavity insulation (complete cavity fill) are treated to make them water-repellent. Unfortunately, if the manufacturers' recommendations regarding the sequence of construction are not followed, situations can occur which permit damp penetration. Where cavity batts are used, there is the risk that mortar can collect on the top of each batt and create a bridge. This is a particular problem where batts are inserted down into the cavity rather than being positioned against the wall as it is being built. Evidence suggests that new houses with filled cavities suffer more failures than older houses with 'retro' filled cavities – this suggests that construction errors might be the cause.

There is, however, a particular retrofit insulation material which has caused rain penetration problems. The material is urea-formaldehyde (UF) foam. UF foam was a popular material in the 1970s and 1980s, but two particular problems emerged:

1   *UF insulation was inserted in existing properties by pumping it into the wall in order to fill the cavity. The foam would then harden and set. Earlier forms of the foam tended to shrink as it set, and this would sometimes cause cracks which could allow moisture to pass from the outer skin to the inner skin.*

2   *In the 1980s, problems arose because it was found that a vapour produced by UF foam during the curing process could cause health problems. The material was banned in some countries.*

In the UK, its use should be related to the exposure of the building (to try and avoid rain penetration). Its use is therefore subject to control by the Building Regulations, which in effect state that it can be used in low exposure situations but only in masonry walls, i.e. it cannot be used in timber-framed buildings because of the possibility of formaldehyde vapour entering the interior of the building.

The damp patch (which can be seen just above the framed picture on the wall) appeared after the cavity wall was filled with UF foam in this 1970s house.

When cavities are filled, the insulation which fills the cavity will restrict evaporation. If an outer leaf becomes saturated through prolonged rainfall, it will dry more slowly as evaporation into the cavity is restricted. At the same time a wall which remains wet is at greater risk from frost attack. In addition, any mortar droppings in the cavity will prevent complete cavity fill and create a cold bridge. This can result in localised spots of condensation on the inner face of the wall.

Water can pond in depression caused by mortar or bridge cavity by capillary action.

Batts should never be pushed down into cavities. They will dislodge mortar forming a potential bridge for damp.

Damp patches at first floor ceiling level

If batts stop at ceiling level a cavity tray must be provided. Otherwise water can run across top of batts.

The laminations in batts must always be vertical. Off-cuts must not be used horizontally as infill pieces.

At gable ends, batts should be taken right up to the verge unless a cavity tray is provided.

Note that current Building Regulations do not permit the use of cavity fill in exposed situations unless some form of impervious cladding such as weather-boarding protects the masonry. This obviously reflects concerns about long-term resistance to damp penetration.

Where cavity boards are used (cavity boards do not fill the cavity), the special ties, which include large washers to hold the insulation in place, must not be omitted. These can allow the boards to fall against the outside leaf and form a path for water penetration. They can also form ledges which may collect mortar droppings.

Cavity boards must be held securely against the inside leaf. If they touch the outer leaf, a path may be created for water penetration.

Cavities must be kept clean – if mortar touches an insulation board, water can run down its face, into the blockwork.

Parapets

The problems of parapets in solid walls have been mentioned earlier in the chapter. Cavity walls are equally at risk. Damp stains below the parapet at ceiling level may be caused by failure of the roof covering, be it pitched or flat. They can also be caused by water penetrating the coping stones. Coping stones without drips or which are not bedded on a DPC, supported over the cavity, can permit water to run down the outer face of the inner leaf. Similarly, if a cavity tray is omitted at roof level, water can penetrate the inner leaf and run down into the main building.

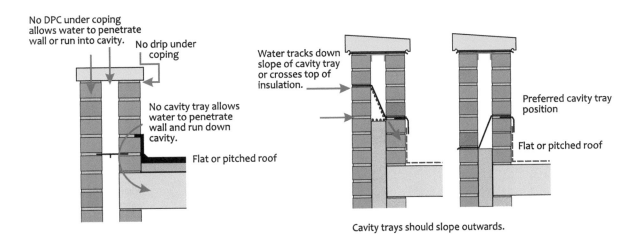

No DPC under coping allows water to penetrate wall or run into cavity.

No drip under coping

No cavity tray allows water to penetrate wall and run down cavity.

Flat or pitched roof

Water tracks down slope of cavity tray or crosses top of insulation.

Preferred cavity tray position

Flat or pitched roof

Cavity trays should slope outwards.

Cavity trays

Problems of missing cavity trays or mortar build-up on trays have already been mentioned. There are two common situations, apart from over windows and in parapets, where correct installation of cavity trays is important. These both occur where structures abut a cavity wall. Problems may occur through incorrect installation but are more likely to be caused by subsequent building alterations or additions.

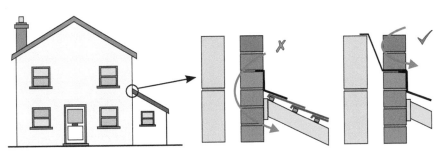

If there is no cavity tray (in a cavity wall), water can penetrate the new addition. When an extension is added, a tray is sometimes omitted because it is difficult to fit.

Cavity trays and flashings are required where, for example, a roof meets a cavity wall.

Windows and doors

In cavity walls the cavity is normally closed to form openings for windows and doors. At the jambs it is normal practice to return the blockwork to the brickwork and insert a vertical DPC between the two materials to prevent damp penetration. If built correctly, problems of damp penetration will not occur. However, the construction around windows and doors can be quite complex. Mistakes in construction and detailing occur quite commonly. There are three vulnerable areas: the head, the jambs and the sill.

Rising damp

This section explains the phenomenon of rising damp and describes its most common causes. It also identifies the various steps which may be required in order to confirm (or otherwise) a diagnosis. Rising damp is often confused with penetrating damp and condensation, and, as already explained at the beginning of the chapter, incorrect diagnosis often leads to remedies which are wholly inappropriate to the cause.

How rising damp occurs

In liquids, surface tension is a result of cohesive intermolecular forces. These cohesive forces are the reason why rain falls as small droplets. The molecules at the surface of the liquid are attracted inwards; this attraction results in the liquid trying to minimise its surface area and forming the shape of a sphere. If the liquid is in contact with another material, there will also be some adhesion between the molecules in the liquid and the material itself. So, there are two forces: the cohesion (surface tension) within the liquid, and its natural inclination to adhere to another material (wetting). If the solid material is a surface such as plaster, brick or block (known as wettable surfaces), the adhesion forces will be greater than the cohesion forces and the water will tend to spread outwards and form a thin surface layer. However, if the adhesion forces are less significant than the cohesive forces, as occurs when water lies on a waxed surface, the liquid will remain in droplets. This explains, for example, why a newly waxed car is covered in tiny droplets after rainfall.

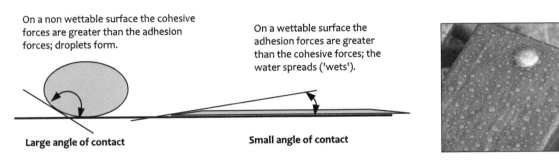

On a non wettable surface the cohesive forces are greater than the adhesion forces; droplets form.

Large angle of contact

On a wettable surface the adhesion forces are greater than the cohesive forces; the water spreads ('wets').

Small angle of contact

This effect is seen after waxing a car or varnishing garden furniture; the surface becomes non-wettable and, after rainfall, is covered with small droplets of water.

Most building materials are porous; in other words, they contain tiny spaces or pores. The pores can occur naturally and may also be created by the shrinkage of materials such as render and mortar. Water's natural inclination to adhere to wettable surfaces will cause it to spread into these tiny spaces, and the smaller the diameter of the pore, the greater the attraction (capillary action). This occurs because in narrow pores the adhesive forces (wetting) at the interface of the pore sides and the water are greater than the cohesive forces (surface tension). As the pore gets wider, the changing ratio between the cohesive forces and the adhesive forces results in the water only rising to lower levels.

If the pores are coated with a non-wettable solution (the basis of chemical DPC injection), the adhesive forces are reduced and the water will settle below its natural level.

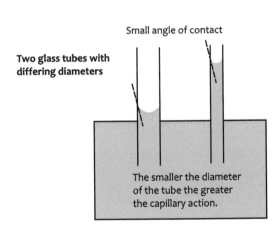

Two glass tubes with differing diameters

Small angle of contact

The smaller the diameter of the tube the greater the capillary action.

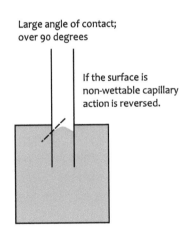

Large angle of contact; over 90 degrees

If the surface is non-wettable capillary action is reversed.

Rising dampness in a building is caused, therefore, by water in the sub-soil rising up the pores or capillaries of the materials in the wall. Since the early 1900s most houses have been constructed with damp-proof courses.

The images that follow show a selection of damp-proof courses (DPCs).

This image shows a vitrified clay DPC in a early twentieth-century house.

A slate DPC in a house dating from the early 1920s.

A bitumen felt DPC.

When functioning correctly, DPCs provide an effective barrier to rising damp. However, DPCs can fail for a number of reasons. These include deterioration of the original material or physical damage caused by building movement. Early DPCs included tar and sand, lead, copper, bitumen felt, lead cored felt, asphalt, slate and engineering bricks. Some of these materials are quite brittle and even minor building movement can cause damage. However, in many cases, rising damp is caused by bridging of the DPC or changes in ground water levels, rather than failure of DPC material. The diagrams below show a number of ways in which DPC bridging can occur.

There are a number of ways in which ground water levels can change. These include natural or man-induced changes to the water table, leaking drains, blocked land drainage systems, leaking water mains and springs. If rising dampness occurs fairly suddenly in a wall that has hitherto been dry, changes in water levels are more likely to be the cause than any breakdown of an existing DPC. Changes in water levels may ultimately cause the expansion of certain sub-soils with obvious implications for the building structure.

The rising damp is caused by a recently installed screed bridging the DPC.

In the two examples shown above the render has been taken below the DPC and so bridging it – a problem made worse by the render touching the external ground finish thereby increasing the likelihood of rising damp problems.

The height of rising damp

The height to which damp rises depends on a number of factors. These include the following.

The porous nature of the material. Research has shown that pores of about 0.01mm provide the main routes for rising damp and that, where these are present, damp can rise about 1.2–1.5 metres. Occasionally, it may be higher. Theoretically, finer pores should allow damp to rise to even greater heights but, in practice, finer pores can actually resist the flow of water.

The amount of water present in the ground. The height to which damp rises will, to some extent, be influenced by the amount of water in the ground. This, in turn, will be affected by the moisture content of the soil and the level of the water table; both of these will depend on the season and the amount of recent rainfall. During wet, humid conditions, for example, evaporation from the wall will be reduced and this will encourage the damp to rise further up the wall.

Limits on evaporation. The impervious nature of wall finishes such as internal plastering, external rendering or painted finish will limit evaporation from a wall. A reduction in evaporation will lead to an increased height of rising damp. Thick walls can also limit evaporation because the surface area is comparatively low in relation to the volume of the wall. Conversely, thin walls, where the surface area is comparatively high in relation to volume, are less likely to have high levels of rising damp.

Internal finishes can impede evaporation and push the rising damp further up the wall. In this case, the damp is appearing just above the level of the timber panelling.

The level of heating inside the building. A well-heated, dry environment will encourage evaporation and thus limit the height of rising damp.

Chemicals in the ground and walls. The ions of dissolved mineral salts affect patterns of rising damp in a number of ways. Sodium, calcium and potassium salts can be present in the ground or in the masonry. In solution they are drawn up into the wall through capillary action. Where evaporation occurs, these salts may crystallise (and cause efflorescence on the surface) and block the capillaries through which the water is evaporating. This can reduce evaporation and drive the water further up the wall. Perhaps the worst culprit is calcium sulfate. This chemical is not readily soluble and is most likely to crystallise.

The pattern shown by this tidemark is typical of that left by rising damp (albeit usually more clearly visible internally).

The white pattern staining shows the salts (known as efflorescence) deposited as moisture from rising damp evaporates in this case to the external face of the wall (the salts in a vertical pattern are caused by damp from a leaking rainwater gutter).

Chemical analysis

Moisture meters are helpful in providing a first step towards the diagnosis of rising damp. However, as described earlier, it should be remembered that they have limitations when used on materials other than wood and are easily confused by the presence of salts. The various stages in achieving a diagnosis are included in a chart later in this chapter. Definitive evidence can be provided by chemical analysis.

Sub-soil naturally contains nitrates, from decaying plant matter and fertilisers, and chlorides. These chemicals are very soluble and are drawn up into the wall with the water. When the water evaporates from the wall, these chemicals are left behind. They form heavy concentrations on the plaster surface and wallpaper. Nitrates and chlorides are both hygroscopic, in other words, they absorb moisture from the air. As they absorb water, they continually re-dissolve and this prevents any crystallisation. Research has shown that these hygroscopic salts are most abundant at the peak of rising damp because this is where most evaporation takes place. In nitrates and chlorides hygroscopic action commences at very low relative humidities: 55 per cent or less.

Dampness lower down the wall is usually the result of capillary action and analysis has shown that the amount of hygroscopic salts is usually quite low at the base of the wall. Rising dampness is often accompanied by efflorescence. This is caused by a number of sulfates and carbonates which are always present in building materials. These materials crystallise quite readily because they are not that soluble. They are not hygroscopic and merely indicate that moisture is evaporating from a structure. They can sometimes block pores in brickwork, thus preventing evaporation. The effect of this may be to drive damp higher up a wall.

A scraping of wallpaper and finish coat plaster can be sent for chemical analysis or tested 'on site'. The degree of contamination by chlorides and nitrates will give an indication of how long the damp has been rising. Condensation or penetrating damp should show zero levels of nitrates and chlorides.

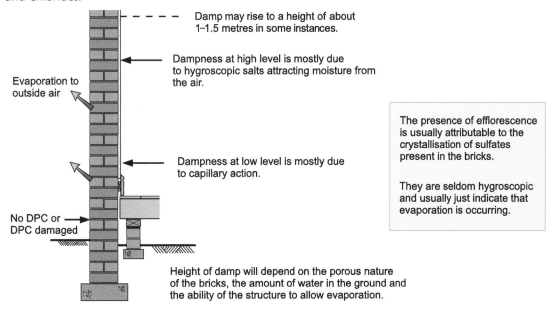

Damp may rise to a height of about 1–1.5 metres in some instances.

Dampness at high level is mostly due to hygroscopic salts attracting moisture from the air.

Evaporation to outside air

Dampness at low level is mostly due to capillary action.

No DPC or DPC damaged

Height of damp will depend on the porous nature of the bricks, the amount of water in the ground and the ability of the structure to allow evaporation.

The presence of efflorescence is usually attributable to the crystallisation of sulfates present in the bricks.

They are seldom hygroscopic and usually just indicate that evaporation is occurring.

Diagnosis

The diagnosis of rising damp needs careful and systematic thought because it can easily be confused with penetrating dampness and condensation. The BRE have suggested that only 10 per cent of the dampness problems it investigates are attributable to rising damp. Unfortunately, there are a number of companies specialising in DPC replacement who obviously have a commercial interest in finding problems of rising damp. Their diagnosis needs to be treated with caution. Although there are several reputable companies working in this field, it may be wise to seek independent advice. Further 'encouragement' to find problems of rising damp is provided by banks and building societies who often request a damp report as a condition of a mortgage advance.

The diagnostic stages and key features of rising damp are shown in the diagram.

Rising dampness

Stage 1   Pinpoint damp contours with moisture meter

Key features

- Limit is usually 1–1.5m above ground level.

- Readings will drop quickly above peak of dampness.

- Stains and/or tide mark on wall paper/plaster.

- High % moisture content of timber skirtings.

- Areas of dampness appear to get wetter in humid conditions (hygroscopic salts).

2.0m
1.5m 0
1.0m 70
0.5m 80
      90

25% moisture content in skirting board

Examples of 'relative scale' readings

External inspection may indicate missing or bridged DPC

Stage 2

- Deep wall probes indicate high readings in centre of wall; eliminates condensation.

- Salt analysis shows high levels of nitrates and chlorides.

- Establish internal surface temperature and relative humidity to determine whether condensation is a possibility.

Tide marks are common in rising damp but do not confuse them with leaks/spills.

Typical patterns of dampness

The contours of dampness shown in the previous diagram, i.e. high readings (on the relative scale) at the base of the wall and at 0.50 metres, slightly lower readings at 1 metre, and virtually zero levels at 1.50 metres, are typical of rising damp. There are, however, other common patterns which can confuse an inexperienced surveyor. The diagram shows some typical patterns and suggests likely causes.

	A	B	C	D
1.75	20	10	00	100
1.50	30	80	10	100
1.25	40	70	60	100
1.00	50	00	50	100
0.75	60	00	40	100
0.50	70	00	20	100
0.25	80	00	20	100
0.00				

A Typical of condensation. The lower part of the wall is often colder; hence the higher readings. Condensation running down the wall also accounts for higher readings.

B Typical of a failed replacement DPC. Replastering works ensure that the base of the wall remains dry. The dampness manifests itself above the limit of the replaced plasterwork. There may also be contamination by salts.

C Typical of long-term, but controlled rising damp. The low readings at the base indicate minor problems of rising damp. The high readings further up the wall suggest long-term contamination by salts.

D Consistent maximum readings suggest that wall is lined with metal foil. This is probably a past remedial measure against dampness.

Problems with replacement DPCs

Replacement DPCs or new DPCs (where none existed before) can be of various types. In the past, electro-osmotic systems and atmospheric tubes have been popular. Electro-osmotic systems can normally be identified by a horizontal continuous copper strip or thick wire running along the base of the wall (they may be hidden by render). Atmospheric tubes are normally made from ceramics or clay and are about 50mm in diameter. They are usually fixed at centres between 200mm and 400mm and penetrate approximately $^2/_3$ of the wall. In many cases a plastic vent will cover the end of the tube.

Nowadays the two options recommended by the BRE are physical DPCs and chemical DPCs. The former are more expensive and can only be laid in horizontal course, i.e. in brickwork or coursed stonework. They are not suitable, therefore, for rubble walls. In practice, chemical DPCs are the cheapest and most common form of remedial treatment. There are a number of patented methods but most work on the same principle, i.e. they line the pores with a non-wettable surface to reduce capillary action. Their effectiveness depends on their successful penetration of the wall and a number of concerns have been raised about how successfully this can be achieved. There are other specialist repairs but these are beyond the scope of this text.

Whichever system is used, it must be recognised that providing a DPC is only part of the solution. The hygroscopic nitrates and chlorides present in the wall structure and plaster need to be dealt with as well. Most of the hygroscopic salts will be present in the plaster and this should be renewed (usually to a height of 300mm above the level of rising damp). The new plaster must prevent any remaining salts in the wall from migrating to the plaster surface. This usually requires the use of cement-based plasters as most gypsum plasters cannot prevent the passage of hygroscopic salts and quickly break down in wet conditions. Thus, apparent failure of a new, replacement DPC may, in fact, be caused by failure to renew the plasterwork or incorrect choice of plastering materials.

Whichever plaster system is used, it should be recognised that limiting evaporation from the wall, e.g. by using an external render system, may encourage any remaining dampness in the wall, or dampness caused by an ineffective DPC to rise even higher.

The triangle is the cap for an evaporative ceramic tube. The holes below it indicate that a chemical DPC has also been used. It is not clear if one has failed – nor if so, which one.

Condensation

Incidence

Many people associate damp patches on walls with penetrating or rising damp. In fact, most dampness is probably caused by condensation. Unfortunately, for some of those involved in the inspection and maintenance of buildings, condensation is a baffling problem. It is compounded by the fact that, in some cases, condensation may be exacerbated by problems of rising or penetrating damp. This chapter explains the phenomenon of condensation and describes how, when and where it can occur. Failure to understand its nature can lead to incorrect diagnosis of damp problems, resulting, in many cases, in wasted effort and wasted resources.

Dealing with condensation is often much harder than its diagnosis. In practice, condensation can be prevented by a careful consideration of a number of factors. These include moisture generation, ventilation, heating and insulation. These factors are discussed in more detail towards the end of the chapter.

Why does condensation occur?

Air contains water vapour, the warmer the air, the more vapour it can hold. That is, the amount of moisture that a volume of air can hold will vary according to its temperature. If moist air comes into contact with a cold surface, the air is cooled; if the air is cooled below a particular temperature, the water vapour will condense on the cold surface. Therefore, whether or not condensation occurs depends on the amount of water vapour in the air and the temperature of the surfaces in contact with the air. The temperature at which water vapour will condense is known as its dew point. It is also referred to as the saturation point because at this temperature a given volume of air cannot hold any more moisture. The moisture content of the air is usually referred to as a percentage of the amount of moisture it can hold at a given temperature, i.e. its relative humidity (RH).

Everybody is familiar with condensation in one form or another. If you have a cold glass of wine or beer on a summer's day you may have noticed that the outside of the glass is covered in droplets of moisture. This is condensation. When the element in a car's heated rear windscreen is switched on, it is to combat the problem of condensation caused by the warmer moisture-laden air hitting the colder windscreen. The warm(ed) glass prevents moist air condensing on its surface.

The vapour trails left by planes in the sky are usually caused by condensation as water vapour from the jet engines hit the cold air.

Condensation on a single-glazed kitchen window.

The pictures show three examples of condensation. Left: Condensation on a kitchen wall; at the time the photograph was taken the kitchen was unheated and had no ventilation other than a small window with a top-opening light. The house was situated on top of a hill and the walls were solid brickwork, one brick thick. Middle: Condensation inside a double glazed unit (caused by failure of the seals). Right: Condensation on an upstairs ceiling caused by carelessly laid insulation. The presence of black spot mould (as in the left- and right-hand images) is a good indicator of condensation.

Recognition

Condensation can be confused with penetrating damp and rising damp. Distinguishing the three is not always simple and may require instrument checks to help an accurate diagnosis. In some cases all three problems may be occurring at once. Condensation can be particularly difficult to diagnose because, despite evidence of its occurrence, it may not be taking place when a property is inspected. In practice, however, condensation manifests itself in a number of ways:

- *the wall has a 'misty' surface*
- *stains or streaks of water running down a wall, particularly in bathrooms, kitchens and below windows*
- *damp patches with no definite edges*
- *dampness behind wall cupboards or inside wardrobes against external walls; areas where air circulation is restricted*
- *patches of mould growth.*

In practice, diagnosis cannot just depend on a visual inspection of the dampness; humidity, insulation, ventilation, and heating patterns also need to be taken into account. These are explored in more detail in later sections.

Why is condensation such a significant problem?

One of the ways in which condensation can be avoided (or remedied) is to ensure that there is sufficient ventilation to remove moist air before the water vapour can condense on a cold surface. Until the 1960s, most houses were well ventilated. Houses were built with chimneys in most rooms; it was common to provide vents for small bedrooms without chimneys. When coal fires were burning, the room was ventilated by the action of the fire gases rising up the chimney and drawing more air into the room. Even without a fire burning, there was some natural convection in the chimney. Windows were single glazed and often quite draughty. Windows were more likely to be left open because of the lower risk of crime. Single-glazed windows acted as 'sacrificial' points. The condensation forming on the windows slowly reduced the humidity of the air. As long as windows and sills were wiped over regularly, the condensation was just a nuisance, not a risk to the building, or more importantly, to health. In addition it was quite common to keep coal fires burning for long periods in cold weather and this helped to keep the building fabric warmer for longer periods which reduced the risk of condensation forming on the walls. The mass of a solid wall also acted to some extent as quite an efficient thermal store which was slow to heat up but also relatively slow to cool down, thus evening out the temperature profile of the wall in houses where heating was relatively constant.

Houses today are very different. A modern double-glazed house is almost a sealed unit. Trickle vents in windows (where they exist) are often kept closed so ventilation is limited. The use of balanced flue boilers as opposed to open fires reduces natural ventilation. In addition, central heating systems tend to be used intermittently. This means that when the heating is not switched on, cold wall surfaces may sometimes coincide with high humidity levels. Even if the heating is fairly constant and walls have high levels of insulation, condensation can still occur if there is excess moisture in the air.

Where does condensation tend to occur?

There are a number of areas where condensation commonly occurs. Bathrooms and kitchens are the most obvious because this is where relatively large amounts of moisture are generated over a relatively short period of time because of the activities that occur in these rooms (but see point later about moisture transfer between rooms). Within any given room, cold surfaces – particularly those in poorly ventilated areas – are most at risk (if a cold wall is enclosed by a cupboard, for example, this lack of ventilation may be a deciding factor in creating conditions for condensation to occur).

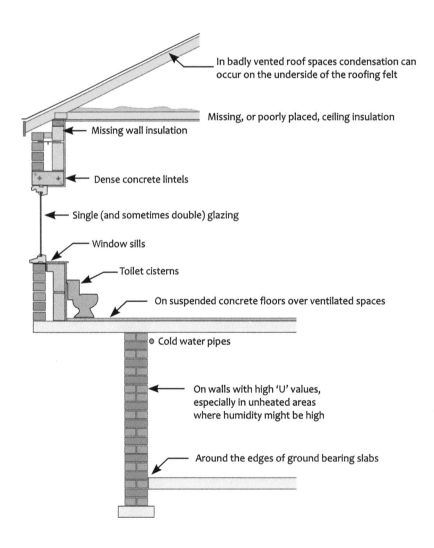

In badly vented roof spaces condensation can occur on the underside of the roofing felt

Missing, or poorly placed, ceiling insulation

Missing wall insulation

Dense concrete lintels

Single (and sometimes double) glazing

Window sills

Toilet cisterns

On suspended concrete floors over ventilated spaces

Cold water pipes

On walls with high 'U' values, especially in unheated areas where humidity might be high

Around the edges of ground bearing slabs

Examples of elements which may be susceptible to condensation include (uninsulated) solid walls, windows (particularly single glazed), cold water pipes and the underside of roofing in an insulated roof space (see Chapter 9). The diagram below shows these and other examples of danger spots for condensation.

Cold bridges

The poorly thought out or inadequate design and detailing of a house can create a situation where condensation is likely to occur through the creation of what is referred to as a cold bridge. Cold bridges are a problem, particularly in well-insulated buildings. They occur in localised spots where the nature of the construction allows heat to escape through the structure. Often these occur where there are breaks or gaps in thermal insulation creating a situation where there are quite low internal surface temperatures, thus encouraging patches of local condensation. These damp patches are often confused with penetrating damp, especially where they occur around window or door openings. A few typical situations are shown below.

Concrete lintels which cross the cavity will have cold lower and inner surfaces (early twentieth century).

Uninsulated box-section lintel (1960s to 1990s).

Lower steel member of insulated lintel creates cold bridge.

Improved detail ⟶

Cold bridges can be created at jambs and sills. Two examples are shown here. The left-hand example shows a cavity wall with the dense block inner leaf returned to the brickwork to support a tiled sill. The right-hand example shows a similar problem at the jamb. In modern construction insulated cavity closers avoid this problem.

Note that retro filling cavities can increase the risk of cold bridging – see text.

Condensation can also occur on floors.

The protruding concrete first-floor slab will form a cold bridge.

The concrete balcony slab and the exposed concrete floor slabs on this block of flats will cause a cold bridge – the effect of which may be made worse if the cavity wall is insulated.

Suspended timber floor

Ground bearing slab

Pre-cast ground floor

Potential cold bridge

Quilt supported on netting, or insulation boards on nails

The risk of cold bridging depends, to some extent, on the nature and position of any wall insulation and the insulating properties of the internal leaf.

Vapour control layer

Ventilated sub-floor void

Condensation on ground-bearing slabs usually occurs around the edges, if at all. In suspended pre-cast floors it is more likely, particularly if the sub-floor void is vented. A vapour control layer helps prevent it.

Condensation does not always occur where the moisture is generated. In general, moisture-laden air will tend to move from relatively warm to relatively cold areas and from relatively wet to relatively dry areas. As the amount of water vapour increases, so does the pressure of the air. The increased pressure encourages the water vapour to travel outwards through the external wall and to other parts of the property where the vapour pressure is lower. This often results in condensation in colder rooms well away from the source of the moisture vapour. A well-heated kitchen, even with high humidity levels, may have walls well above the dew point temperature (the temperature at which condensation occurs) but the 'escaping' air may condense elsewhere.

Where damp patches mysteriously appear in unheated or partially heated bedrooms, particularly where there is poor ventilation, the problem may be condensation. It is often confused with rising or penetrating dampness, particularly when it occurs, for example, on the bottom half of a wall, around the outer edges of upper floor ceilings (i.e. just below the eaves) or around openings. A later section in the chapter considers diagnosis in more detail.

Moisture generated in warm kitchen.....

...condenses in cold bedroom.

← In a heated room the top of the wall is usually warmer than the bottom.

No condensation

← – – – – – **Dew point temperature**

← *Condensation*

Condensation occurs on surface which is at or below the dew point temperature; in this example the bottom half of the wall.

If all the wall surface is below the dew point temperature condensation will occur over whole of the wall.

Condensation at high level is often confused with penetrating damp. →

Condensation on glass runs on to wall below.

Condensation near floor level is often confused with rising damp. →

Where does the moisture come from?

There will always necessarily be moisture in the air in a house. The amount will fluctuate almost continuously and this will relate to not only the 'natural' changes in moisture content of the air but also to normal day-to-day activities. Human beings will themselves contribute to the moisture content of the air through breathing and they will exhale different amounts of moisture when they are active compared to when they are resting. Also the normal routines of washing, bathing and cooking, etc. will introduce moisture – often quite intensely and for relatively short periods.

Typical moisture levels produced in a three-bedroomed house (per day) are shown in Table 16.1.

Table 16.1 Typical moisture levels produced in a three-bedroomed house (per day)

ACTIVITY	MOISTURE (KG)
A family asleep	1.5–2.0
A family during the day	2.5–3.5
Cooking	2.0–3.0
Washing and bathing	1.0–1.5
Washing clothes	0.4–0.6
Drying clothes indoors	3.0–5.0

In addition to moisture produced by the everyday activities of humans, family pets, such as large dogs, can produce significant amounts of water vapour.

In addition, problems of penetrating and rising damp can add significantly to the amount of water in a house. As the damp evaporates, it raises the amount of water vapour in the air.

Fuel-burning appliances

Nearly all fuels (coke is an exception) contain hydrogen. As part of the combustion process the hydrogen combines with oxygen to form water. In addition, some fuels contain water; coal or coke stored outside can be quite wet. As these burn, the water evaporates. Both of these factors combine to ensure that the exhaust gases from the fuels contain high levels of water vapour. As the water vapour rises up the flue, it may condense. In open fires substantial amounts of air are added from the room and, as long as the gases travel up the chimney quickly, condensation will be minimal.

However, in slow-burning appliances, the flue gases may rise quite slowly and the lack of additional air from the room will ensure that the exhaust gas retains high levels of moisture and can therefore condense readily on surfaces even if they are quite warm. If there is no flue, as in the case of portable gas or paraffin heaters, high amounts of water are introduced into the air, slowly increasing its humidity.

If moisture condenses in a brick, block or stone flue, there is the risk that the condensate can absorb chemicals which are a by-product of combustion. Sulfates and other soluble salts, deposited by the rising exhaust gas, go into solution and travel into the surrounding brickwork. As they slowly evaporate, the salts are deposited on the face of the wall (in the form of white crystals) where they can damage plaster and wall decorations. Some of the salts are hygroscopic; in other words, they absorb water from the air. These salts will not crystallise as they continually draw water from the air. Damp patches at the top of a chimney breast which seem to grow and shrink according to the weather may be caused by these hygroscopic salts. Damp patches like this are often confused with other problems such as flashing failure or rain water running down the chimney.

Slow-burning appliances are often fitted with stainless steel or terracotta linings to prevent migration of salts into surrounding brickwork. They can be insulated in order to reduce the risk of condensation occurring in the lining.

Mould growth

Mould growth may form where condensation occurs, particularly if humidity levels remain above 70 per cent for long periods. The presence of 'black spot' mould is a useful signifier that the damp problem is condensation rather than penetrating or rising damp. As observed earlier, condensation is common on cold walls particularly in corners or behind furniture where air flow is restricted which is why 'built-in' wardrobes positioned against external walls are often plagued by mould growth. The mould grows from spores which are present in the air all the time. In order to flourish, they need to feed on organic matter and water. The former is supplied by the dirt and grease which, even though it cannot often be seen, accumulates on porous or rough surfaces such as woodchip paper, emulsioned ceilings and tile grout. Leather goods, particularly when stored in cold, unventilated wardrobes are also affected by mould growth; the mould appears as mildew and occurs because leather goods are often quite dirty.

To the elderly, the very young and asthmatics, mould growth represents a health risk. To others it is just an unsightly inconvenience. It can be removed by dilute bleach though serious cases may require papered walls to be stripped first. Most DIY stores sell a spray which prevents the mould growth from occurring. However, this will not cure the underlying problem of condensation on the wall.

Black spot mould – on the ceiling, wall and single-glazed metal window of a bathroom that suffers from condensation.

Roofs

Condensation is quite common in roof voids. It usually manifests itself as water droplets on the underside of the underlay as shown in the picture below. In most roofs, the problem is caused by inadequate ventilation.

(Photo courtesy of Surevent.)

In a traditionally vented roof with insulation at ceiling level (i.e. a 'cold roof'), condensation is often a result of insulation blocking the flow of air at the eaves. A similar problem can occur in older roofs which are retro-insulated; it occurs because the original roof had no provision for ventilation.

Mould growth on the edge of the ceiling is often caused by condensation on the soffit of the plasterboard; usually a result of inadequate insulation cover.

Eaves-to-eaves or eaves-to-ridge ventilation   **Ventilation paths blocked**   **Gaps in insulation – cold surfaces**

In recent years, loft conversions have become very common. In a vented roof design, insulation can be provided in between the rafters ('warm roof') as long as a minimum ventilation gap of 50mm can be maintained. In addition, a vapour check (vapour control layer) just behind the new plasterboard will help to reduce vapour transfer. Condensation can occur on the soffit of the underlay if either is missing.

Warm roof – ventilated

To prevent condensation there must be a ventilated space of at least 50mm above the insulation. In addition, an effective vapour control layer should be positioned just above the plasterboard (i.e. on the warm side of the insulation).

In 'breathing roofs' a vapour-permeable underlay allows vapour to escape. These can be designed as 'cold roofs' or 'warm roofs' as shown below. In practice, breathing roofs are sometimes misunderstood, particularly the need for ventilation, vapour control layers and counter-battens. All three are not always necessary; but a vapour-permeable underlay on its own may not be enough to prevent condensation.

Cold roof – 'breathing'

Counter-battens to encourage ventilation

Vapour permeable underlay

Vents at eaves and ridge sometimes recommended

Vapour control layer

There has been much debate about 'breathing roofs' in recent years. If the wrong type of underlay is used or if the vapour control layer is not continuous (taped joints and seals around light fittings and loft hatches) condensation can occur. Counter-battens encourage ventilation; some organisations also recommend eaves and/or ridge vents to help ventilate the void above the underlay.

Warm roof – 'breathing'

Counter-battens

Quilt or board insulation

Vapour control layer

Vent (option)

To prevent condensation there should normally be a ventilated space above the vapour permeable underlay; hence the counter-battens. Some organisations recommend that this space is vented (eaves-to-ridge or just ridge). Any gaps or tears in the vapour control layer will increase the risk of condensation.

Two examples of condensation problems

These two examples show how condensation can occur, all in very different situations. Condensation and damp penetration are easily confused so careful analysis is important; money is often wasted because of incorrect diagnosis.

Gutter lined with bitumen

A pre-cast concrete gutter section closes the cavity at the eaves. On the inside face of the plaster there is evidence of dampness. The problem only became apparent after filling the cavities with insulation.

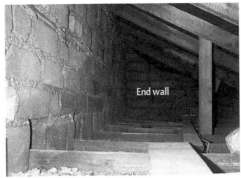

This problem was initially diagnosed as either penetrating dampness through the solid wall or rainwater running down the chimney flue. When the roof space was inspected, however, it was clear that the inner face of the end wall was quite wet and that water was running down the wall causing the damp patches shown above. We eventually discovered that a bathroom extractor fan which should have discharged through the roof had become disconnected and was actually discharging into it! The moist air was condensing on the cold end wall.

Analysis

Measuring the amount of water vapour in the air

As we have already established, the amount of water (in the form of water vapour) that air can hold or contain depends on its temperature. If, for a given temperature, the amount of water vapour the air can hold is exceeded, the air becomes saturated and condensation will occur (hence 'dew point' and 'saturation point' explained earlier). 'Relative humidity' (RH) is a term used to quantify the amount of vapour in the air. It can be defined (not a strict definition but one suitable for the purposes of this chapter) as the amount of water vapour that air holds relative to the amount it can hold at saturation point. Saturation occurs at 100 per cent RH. Relative humidity in itself cannot be used to quantify the amount of water vapour in the air unless the temperature is taken into account. For example, at different air temperatures, say, 10°C and 20°C, an RH of 50 per cent will represent different amounts of water vapour (Table 16.2). It should be clear that air at the higher temperature (20°C) is capable of holding higher quantities of moisture or water vapour.

Table 16.2 Water vapour amounts at an RH of 50%

AIR TEMPERATURE	RH (%)	AMOUNT OF MOISTURE IN THE AIR
10°C	50	4.0g/100g dry air
20°C	50	7.4g/1000g dry air

Relative humidity affects comfort. Very dry or very humid conditions can both provide an uncomfortable environment. An RH of 45–65 per cent is typical of a well-heated and ventilated room, with a temperature of 18–20°C. Levels below 45 per cent will feel very dry; levels over 70 per cent are likely to feel 'close' and will increase the risk of condensation and mould growth.

Establishing if condensation is taking place

A psychometric chart enables the incidence of condensation to be predicted (psychometric is from the Greek, meaning 'a measure of coldness'; in fact, it is the Greek name for a thermometer).

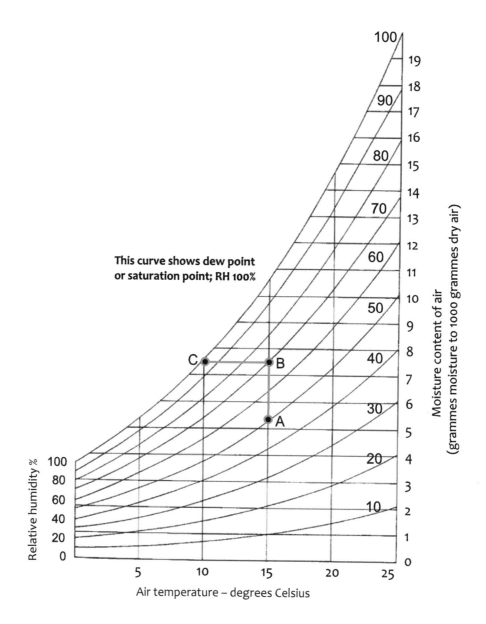

The horizontal lines show moisture content, the vertical lines show air temperature and the curved lines show relative humidity. In practice measuring the moisture content of the air during a building inspection is difficult but it is fairly easy to establish the RH and the air temperature. If these two variables can be measured, establishing the moisture content of the air, and therefore the dew point, is easy.

Example (see the chart above)

If the air inside a house is 15°C and it contains 5.3g of water per 1000g of dry air, its RH is 50 per cent (Point A). If additional moisture or water vapour is added, say, another 2.2g per 100g of dry air (by activities in the house such as washing or cooking), its RH will increase to 70 per cent (Point B). If this air comes into contact with a cold surface such as glazing or a solid uninsulated wall, the air will be cooled. As it cools, the RH increases. If the RH reaches 100 per cent (dew point), saturation will occur. From the chart it should be clear that this temperature (Point C) is 10°C. Therefore, any surface at 10°C or less will cause condensation. Condensation in the form of a fine mist, or droplets of water, will form on the wall's surface. Although warm air can hold more water than cold air, the dew point is directly related to the actual amount of water in the air. If the air contains 7.5g of water per 1000g of dry air, its dew point will always be 10°C, whatever the internal air temperature.

This last point is commonly misunderstood. It is often assumed that heating up the air, by possibly turning up the heating system, will prevent condensation because the warm air can hold more moisture. Turning up the heating will certainly reduce the RH, but however high the internal air temperature, condensation will still occur if the wet air comes into contact with a surface at 10°C or less. In fact, the value in turning up the heating system is to raise the temperature of the walls above dew point.

The psychometric chart identifies the point at which condensation will occur given a specific RH and internal surface temperature. But how do you measure these two items?

Inner surface temperatures

The temperature of the internal surface of glazing or walling will depend on three factors: the thermal transmittance of the material (the 'U' value – a measure of the rate at which heat is transmitted through a wall), the outside air temperature and the inside air temperature. Well-insulated walls (with low 'U' values) will have high internal surface temperatures. In a house the coldest surface will probably be the single glazing. In practice, surveyors nowadays use electronic instruments to measure internal surface temperature. A modern instrument with a digital readout will give an accurate surface temperature in a few seconds. Before these instruments were available, the internal surface temperature was established using the chart shown below. The chart assumes that 'U' values are known.

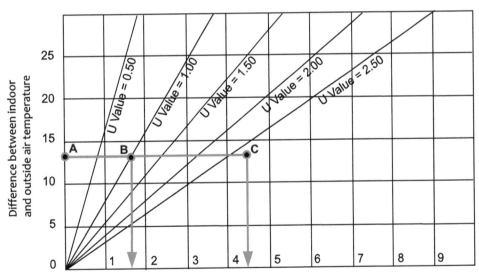

Difference between indoor air and internal surface temperature

Although this chart is virtually redundant these days, it does show how poor insulators result in lower internal surface temperatures. Consider a situation where the temperature externally is 7°C and the temperature internally is 20°C. The temperature difference is therefore 13°C. This is shown as point A on the chart. In addition, imagine we are considering two types of building: a 1920s house with a solid 1 brick wall ('U' value of 2.60) and a 1980s cavity wall house ('U' value about 1.00).

Point B is the 'U' value of the 1980s house. Point C is the 'U' value of the 1920s house. If the dotted lines are extended downwards, the difference between indoor air and the internal surface temperature can be established. These are 1.6 for the 1980s house and 4.3 for the 1920s house. Subtract these values from the internal air temperature to find the internal surface temperature.

Thus the internal surface temperature of the old wall is 15.70°C and the new wall is 18.40°C. As you would expect, the 1980s house, with its better insulation, has an internal surface temperature closer to the internal air temperature.

Now if we refer to the psychometric chart shown previously, we can assess the risk of condensation. If the RH is 80 per cent and the internal air temperature is 20°C, the dew point is about 16–17°C. In other words, any surface at 17°C or less will cause condensation. In our examples the cavity wall is above dew point (no condensation) and the solid wall is below dew point (condensation will occur).

Measuring relative humidity

Relative humidity can, nowadays, be measured quite simply. Thermohygrometers (sometimes electronic) measure air temperature and relative humidity. The dew point can then be read by reference to the psychometric chart or by reference to a simple table based on the chart.

Table 16.3 Dew point

AIR TEMP	R H 20%	R H 30%	R H 40%	R H 50%	R H 60%	R H 70%	R H 80%
10				0	2	5	7
11				1	3	6	8
12				2	4	7	9
13			0	3	5	8	10
14			1	4	6	9	11
15			2	4	7	10	12
16			2	6	8	11	13
17			3	6	9	11	13
18		0	4	7	10	12	14
19		1	5	8	11	13	15
20		2	6	9	12	14	16
21		3	7	10	13	15	17
22		4	8	11	14	16	18
23		4	9	12	15	17	19
24	0	5	10	13	16	18	20
25	0	6	10	14	17	19	21

If the internal air temperature is 20°C and the RH is 80 per cent, condensation will occur if the moist air comes into contact with surfaces which are 16°C or below. Thus 16°C is the dew point.

The blank spaces indicate dew points below freezing point. Internal surface temperatures below 0°C are almost impossible in a house.

Interstitial condensation

In most cases condensation will occur on the surface of a material. Most walls, however, are porous and will therefore allow the water vapour to slowly pass through the wall. If the internal surface temperature is above dew point but the interior temperature of the wall is below dew point, there is a risk that interstitial condensation will occur. The water vapour therefore condenses inside the wall rather than on its surface. Impervious wall finishes such as gloss paint will limit the passage of water vapour into the wall but emulsion paints and porous wallpapers will be less effective.

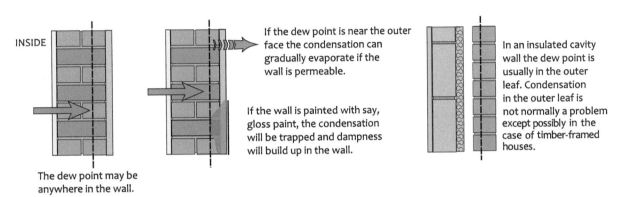

If the dew point is near the outer face the condensation can gradually evaporate if the wall is permeable.

If the wall is painted with say, gloss paint, the condensation will be trapped and dampness will build up in the wall.

In an insulated cavity wall the dew point is usually in the outer leaf. Condensation in the outer leaf is not normally a problem except possibly in the case of timber-framed houses.

The dew point may be anywhere in the wall.

Interstitial condensation is a common problem with one-brick thick solid walls. It is not normally a problem in brick and block cavity walls but it may lead to problems of rot in a timber-framed wall.

The temperature of the wall obviously falls towards the outside. This temperature gradient depends on the nature of the walling material and the presence, and position, of insulation. If a one-brick wall is insulated by dry lining, the temperature drops quickly and the wall itself will be quite cold. Where the wall is insulated externally, the wall will remain at a high temperature. These temperature gradients can be calculated accurately if the characteristics of the various materials making up the wall are known. However, this level of scientific analysis is beyond the scope of this book.

The low temperature on the face of the brick is likely to be below the dew point. If so, interstitial condensation will occur.

A vapour check on the warm side of the insulation prevents the moist air reaching the cold surface.

When external insulation is used a vapour check is unnecessary.

When external insulation is used the temperature of the main wall remains fairly close to the internal surface temperature.

As the temperature drops through the thickness of the wall, so does the dew point. This is because the wall's degree of porosity will provide some resistance to the passage of water vapour through the structure. The dew point will drop because air containing lower amounts of water vapour has a correspondingly lower dew point. Interstitial condensation will only occur if the dew point temperature, in other words, the temperature at which saturation occurs, is higher than the surface temperature.

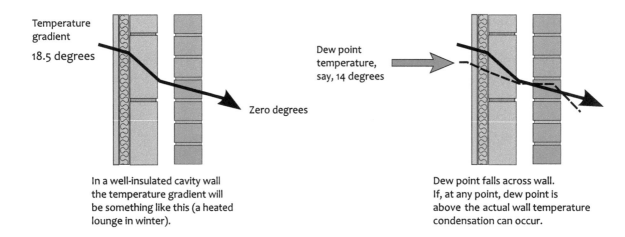

Temperature gradient

18.5 degrees

Zero degrees

In a well-insulated cavity wall the temperature gradient will be something like this (a heated lounge in winter).

Dew point temperature, say, 14 degrees

Dew point falls across wall. If, at any point, dew point is above the actual wall temperature condensation can occur.

Interstitial condensation may cause problems of dampness but it should not affect the durability of brick or block walling materials. However, as we observed previously, if interstitial condensation occurs in a material such as timber, action will be needed to prevent the risk of rot.

In timber frame housing the risk of interstitial condensation is minimised by placing a vapour control layer on the warm side of the insulation. The layer (usually polythene) should have taped joints and any holes for socket boxes, etc. need to be made with care.

Some occupiers of concrete pre-cast houses first noticed damp problems when their houses were insulated in the 1980s. Condensation occurred on the cold inner face of the external concrete panels and ran down to ground level. In extreme cases the build-up of water eventually soaked the base of the fibreboard/plasterboard linings; this is easily mistaken for rising damp.

Interstitial condensation can be seen on this steel joist which was encased in an external wall (photo courtesy of Tim Pearce).

'Reverse' or 'summer' condensation

This can occur when external warmer air with a high moisture level, and therefore higher pressure, moves 'backwards' through the wall (from the outside to the inside) in a situation where the interior of the house is relatively colder and drier. This phenomenon is known as 'reverse' or 'summer' condensation. Although it is relatively rare in houses in this country, it has been known to affect solid walls where vapour checks have been installed and where warm moist air reaches the back of the vapour check which may have a surface temperature below dew point (as illustrated in diagram below). It manifests itself as wet patches on the floor at the base of the wall and is caused by condensate running down the back of the vapour check.

INTERNAL   **EXTERNAL**

In some extreme conditions there is the risk of condensation occurring on the back of the vapour check.

In hot weather after rainfall the sun drives the moisture back into the wall and it can condense on the back of the vapour check. The cooler the internal room the more the phenomenon is likely to occur.

Hot, humid, air at high pressure.

Effect of wall structure and heating pattern

In a well-heated dwelling, with average levels of humidity, condensation will only occur locally after long hot baths, cooking or drying clothes. As you would expect, kitchens and bathrooms are most at risk. However, when houses are heated intermittently, condensation may also appear on living room and bedroom walls; even houses with central heating are at risk. In houses that are not occupied all day, a typical pattern is to have heating on for a few hours in the morning and a few hours in the evening. This pattern of heating can cause condensation at specific times of the day or night. For example, a house with one-brick walls has a fairly high heat capacity – in other words, heat is stored in the walls which cool slowly when the heat is turned off. If the heating has been turned off over night, condensation may occur in the morning as the house comes to life and moisture-generating activities cause condensation on the cold walls.

Modern lightweight construction, with, say, an insulated dry lining, acts in the opposite way. When the heating is turned on, the wall will rise in temperature very quickly so condensation in the morning should not be a problem. The problem with these walls occurs at night. When the heating is turned off, the wall cools rapidly. A family asleep produces large amounts of water vapour. Thus condensation occurs throughout the night. This problem of intermittent condensation makes its diagnosis difficult and is another reason why a broad understanding of condensation is required to help interpret instrument readings. It is possible to obtain a simple instrument which is designed to operate over a 24-hour cycle and show if condensation has occurred in any time over this period.

How can condensation be avoided?

Because condensation is a dynamic process which is affected by changes in the temperature and moisture content of air, it follows that it can be avoided and/or combated by consideration of heating, thermal insulation and ventilation – and by limiting the amount of moisture generated.

Moisture generation

As we have already seen, moisture generation in a typical family house can be considerable. In order to limit the amount of water generated, there are a number of simple precautions:

- *keep lids on pans, etc.*
- *dry clothes outside whenever possible*
- *ensure tumble dryers are ventilated externally*
- *do not use paraffin or bottled gas heaters*
- *put cold water in the bath before hot water*
- *keep the water vapour in the room where it is generated to prevent it from travelling to cooler areas (close bathroom doors, for example, after taking a shower)*
- *ensure that trickle ventilators are kept open, particularly at night*
- *open windows when cooking or taking a bath.*

Minor life style changes, as set out above, can significantly reduce the incidence of condensation. Because there are no cost implications, these are the areas to consider first.

Ventilation

Ventilation plays an important part in combating condensation by removing moisture-laden air from the house. However, ventilation alone will not prevent condensation from occurring and may, in fact, cause lower air temperatures, which could increase the risk of condensation. The BRE has established that ventilation rates of over one air change per hour have little effect on relative humidity but a marked effect on air temperature. Controlled ventilation will help to keep relative humidities below 70 per cent. This includes ensuring that trickle ventilators in windows are left open and that extractor fans are used in kitchens and bathrooms. Where fireplaces are removed, a hit and miss ventilator in the chimney breast will provide additional ventilation (though care should be taken to ensure that defective or capped flues cannot transfer water vapour into the roof space). Tumble driers should be vented to the external air and, wherever possible, clothes should not be left to dry in rooms without adequate ventilation.

Insulation and heating

These two areas cannot be considered in isolation. Wall insulation on its own will have little success in combating condensation unless heating is provided. However thick the wall insulation is, it can only be effective if heat is provided. If a house is poorly heated, well-insulated walls can still be below the dew point. If the internal air temperature is 10°C, a well-insulated wall will have an internal surface temperature of about 8°C. If the air contains 7g of moisture per 1000g of dry air (a fairly modest amount), the dew point is about 9°C; if the dew point is above the internal surface temperature, condensation will occur. Where insulation is retro-fitted, consideration must be given to problems of cold bridging and interstitial condensation. As already mentioned, external insulation is more effective in maintaining consistently high internal temperatures. It will also reduce the risk of cold bridging. However, external insulation is generally more expensive than internal insulation and there are a number of situations where it is not a suitable option. Where the external façade is quite decorative or where there are planning restrictions, internal insulation may be the only option.

External insulation being added to a house – the insulation will be finished with cladding which is fixed to the additional horizontal timbers battens which are shown in the photograph to the right.

Note wide window sill to accommodate cladding when it is fixed over the insulation.

Heating, on its own (assuming correct levels of ventilation and controlled moisture generation) can often prevent condensation from occurring. However, few people can afford to run their heating systems continuously. By providing some form of thermal insulation, occupiers can use their heating systems more effectively and with lower running costs. There are also obvious environmental savings.

Check list

A simple check list is shown below to help identify condensation or confirm a diagnosis.

Condensation

Stage 1 ⟹ **Pinpoint damp contours with moisture meter.**

Key features of condensation:

- Readings may occur at full height of wall; possibly higher readings at lower levels (see main text).

- Staining of skirting caused by water run-off from wall (particularly under windows).

- Moisture content of skirting normal

- Evidence of mould growth

- No sharp drop from wet to drier areas

In addition consider:

'U' value and exposure of wall
Levels of ventilation
Heating (particularly portable gas fires)
Levels of occupancy

Stage 2 ⟸ ⟸ ⟸

- Deep wall probes indicate low readings in centre of wall, dampness unlikely to be rising or penetrating damp.

- Salt analysis is negative (no chlorides or nitrates or sulfates) – suggests condensation.

- Measure internal wall surface temperature using surface thermometer and establish dew point temperature by measuring air temperature and RH. Surface temperature below dew point temperature = condensation.

- Consider use of Protimeter Damp Check (records incidence of condensation over 24 hours or so).

Conclusion

Rising damp, penetrating damp and condensation are easily confused. Correct diagnosis requires a systematic approach which recognises their individual characteristics and symptoms. It should also be recognised that there may be more than one source of dampness. In an old solid wall, penetrating dampness and rising damp may both occur at the same time. In addition, a wet wall conducts heat more readily than a dry wall; if the wall surface temperature drops below the dew point, condensation can occur.

Although carbide testing and moisture meters will help to determine the relative intensity and contours of the dampness, they will not, on their own, provide conclusive information as to its causes. A detailed building inspection to highlight any obvious building defects, together with chemical analysis and an assessment of humidity levels, wall surface temperatures and patterns of occupation are all of vital importance to a correct diagnosis. In practice, mistakes are common and can result in wasted effort, wasted resources and, in some cases, incorrect remedial work which exacerbates rather than resolves the original problem.

System building 17

Background

Introduction

Possibly the earliest example of a 'modern' system building in England, this concrete house on the Isle of Wight was erected in the 1850s (photo courtesy of John Lobb).

Traditionally, most of the construction process took place at the building site (though some elements were formed elsewhere, e.g. the hardwood members of timber-framed buildings were shaped and jointed in yards and taken to the site for erection). However, the first half of the twentieth century saw industrialised processes introduced into the UK construction industry. At their simplest, these involved the use of factory-produced components within the traditional building process, e.g. roof trusses. More wide-ranging methods of non-traditional or system building were also introduced. These ranged from the use of in-situ factory techniques, such as shuttered and poured concrete walling, to the site assembly of pre-fabricated components, e.g. timber, steel or concrete frames, and cladding. It is worth noting that most system building involves elements of both traditional and industrialised construction methods. In particular, the substructure is usually formed in situ rather than assembled from delivered components.

There were a number of apparent beneficial factors that led to the introduction of system building:

- **Efficiency.** *Factory processes would make the best use of available labour and materials to produce as much of a building as possible.*

- **Speed.** *Factory prefabrication combined with simple site assembly would result in considerable savings in time.*
- **Specialisation.** *Standardisation of components together with mass manufacture would give economies of scale.*
- **Better quality.** *Use of the workshop production line rather than in-situ construction techniques would allow higher levels of quality to be achieved.*
- **Economy in cost.** *The combination of the above factors would lead to cheaper housing.*

Unfortunately, there were also a number of problems that needed to be avoided if the system-built housing stock was to perform adequately over its intended life span. This chapter examines some of the more important of these problems and their effects upon the buildings involved.

Brief history

The aftermath of the First World War saw an urgent need for thousands of houses. The immediate reasons for this were a lack of any new construction and maintenance during the war years combined with the need to house large numbers of demobilised soldiers and their families. Post-war shortages of materials, such as bricks, timber and labour, in particular skilled labour (as the result of war casualties and the lack of training), meant that new methods of construction were encouraged by the government. System-built housing was undertaken by a number of local authorities throughout the UK, though the government did not offer any special financial assistance towards its construction.

Ultimately, some 50,000 system-built domestic units were constructed during the inter-war period, about 1 per cent of the total number of houses built between 1919 and 1939. Large numbers of different systems were approved, including timber frame, steel frame, concrete frame/slab and in-situ wall slab. Most have subsequently been demolished.

The Second World War resulted in similar shortages of building materials and skilled labour. The situation was exacerbated by the loss of some 200,000 houses due to enemy bombing, as well as damage to about 25 per cent of the entire building stock. An increasing population (the post-war 'baby boom') caused further pressures.

Government response was much more positive, there being generous funding of both traditional and non-traditional (system) building by successive Labour and Conservative Governments between 1945 and 1955. During this period about 20 per cent of the new houses in England and Wales and 50 per cent in Scotland were system-built – approximately 500,000 units in total. Initially, there were two approaches – a permanent housing programme and a temporary housing programme (below). The former produced houses of two or more storeys, many thousands of which are still in use. However, relatively few remain of the 150,000-plus

Left: Typical pre-fabs in the 1950s. Note the corrugated asbestos roofs (photo courtesy of Bristol Record Office). Right: A mixture of modernised and un-modernised pre-fabs.

temporary houses (or 'pre-fabs'). A large number of different systems were produced including many with frames and/or claddings of steel or pre-cast reinforced concrete (PRC) and others were formed with cast in-situ concrete walls. The numbers of each system ranged from a mere handful to many thousands. Some local authorities invested very heavily in system building to meet their housing needs, e.g. Birmingham and Leeds Metropolitan Councils, Bristol City Council.

Between 1955 and the late 1970s, the techniques learned in low-rise system-built housing were developed and used in the construction of medium-rise and high-rise flats. During this period, it is believed that between 750,000 and one million 'industrialised' (system) dwellings were constructed, of which some 500,000 were low/medium rise and 140,000 high rise (available statistics are, unfortunately, incomplete). However, the 1970s saw a reaction against system building in general. This was mainly as a result of problems of maintenance and repair due to one or more of the following causes:

- *poor design*
- *inadequately controlled prefabrication processes*
- *poor construction as a result of the use of unskilled labour and/or poor site management.*

These problems led to considerable failure of structural and cladding components. Many houses and flats also suffered from condensation as well as poor thermal and noise insulation.

The early 1980s produced another problem. The Conservative Government of that period had introduced 'Right to Buy' legislation which led to the purchase of many thousands of local authority houses by sitting tenants. This included a great number of system-built houses. At the same time, but too late to prevent such properties being purchased by the tenants, it became apparent that those houses constructed with pre-cast concrete components were often affected by carbonation or chloride attack (see following section for details). Building societies refused to mortgage or remortgage these houses and they were, effectively, blighted. The government responded to the resultant outcry by introducing the Housing Defects Act of 1984, which provided repair grants for those houses that were designed before 1960 (i.e. not high-rise homes, which, mostly, tend to be of later design), were of certain designated classes (all of pre-cast concrete construction), and had recognised qualifying defects. These grants were controversial as they were only available to private owners of such properties and not to local authorities. In addition, the grants were available even if the houses were not themselves defective, but were of a potentially defective type.

Left: A pair of semi-detached Cornish Unit system-built houses. The right-hand house is as originally constructed; the left-hand is privately owned and has been repaired under a Housing Defects Act grant. Right: A detail of the party wall junction between two similar houses.

Limited numbers of steel or concrete-framed/clad houses have been constructed since 1970, but the following decade saw a rise in the popularity of timber-framed houses. Unfortunately, many of those built in the 10–15 years leading up to 1984 were found to have inherent defects as a result of poor design/construction. This led to a short period of decline. However, since the mid-1980s, improved design, construction techniques and site control have resulted in timber-framed housing regaining its popularity.

SECTION 2

Defects in concrete houses and steel-framed houses

Introduction

It would be impossible in a textbook of this nature to detail the problems experienced by the complete range of concrete and steel houses currently in use. The following section, therefore, initially describes some important defects that are commonly found across a number of different types of system-built houses and then briefly examines, in more detail, a representative sample of systems. It should be noted that the individual housing systems frequently had variations in design and/or construction that could result in minor changes of detail and the use of different materials.

Defects in system houses constructed between 1945 and 1955

This sub-section concentrates on those houses constructed with components of pre-cast reinforced concrete (PRC), cast in situ in concrete or steel. Many thousands of these were erected in this period. A great number have subsequently been demolished for the reasons briefly referred to above. The remaining houses, which number in their thousands, are still in use. Some are in good condition, though many suffer from problems similar to those already demolished. Many types of PRC houses designed before 1960 were designated by the government in the 1980s and 1990s as 'Defective by reason of their design and construction'. Their continuing use is normally due to the owner (usually a local authority, sometimes a private person) being unable to replace them because of a lack of finance.

Carbonation of concrete

This is a problem that affects reinforced concrete and has been discovered across a broad range of PRC houses. Carbonation is a natural process that takes place in all concrete, but, where insufficient allowance has been made for its effect, it can have disastrous consequences on reinforced concrete, as it can lead to corrosion of the steel and cracking/spalling of the concrete.

Fresh concrete contains calcium hydroxide which is highly alkaline and protects the steel reinforcement, preventing it oxidising. The concrete will, over time, react with carbon dioxide in the air to slowly form calcium carbonate, which is insufficiently alkaline to protect the steel. This process will slowly penetrate the concrete to a depth of 50mm or more. Where the outer layer is also affected by moisture, this can accelerate the process. The effect is to lead to the rusting of any steel within it.

The rate of carbonation will depend on a number of factors:

- **Quality and density of the concrete.** *Improperly compacted lightweight concretes are particularly prone to the problem.*
- **Exposure of the building.** *The process requires moisture.*
- **Relative humidity of the atmosphere.** *Carbonation is encouraged where relative humidity is between 25 per cent and 75 per cent, in particular, the higher range of 50–75 per cent (this often results in internal components being affected more quickly than wetter external members).*

The design of any reinforced concrete component should ensure that the steel is placed at a sufficient depth to prevent carbonation reaching it during the anticipated life of the building. Similarly, the manufacturing process must strictly follow this criterion.

Unfortunately, many prefabricated and cast in-situ reinforced concrete components of system houses (and medium- and high-rise blocks of flats) were poorly manufactured, with insufficient depth of cover to the steel as well as poor-quality concrete. These defects inevitably led to carbonation problems over a period of time. Inadequate thermal insulation was also a common problem with these buildings and, when combined with the former defects, tended to accelerate the carbonation process. Furthermore, the presence of chlorides in the concrete (see the next sub-section) increased the depth and rate of carbonation.

In appearance, the affected component will tend to show longitudinal cracking along the line of any steel reinforcement. Its initial appearance will be in the form of hairline cracking, which can occur as early as a few months after construction. Over time, expansion of the rusting steel results in the column cracking along its length, as well as spalling of the surface concrete.

The presence of carbonation can be determined by in-situ testing of the concrete. A chemical indicator (manganese hydroxide or phenolphthalein solution) is applied to the surface of the suspected area and will show both the extent and depth of any attack. Determination of the effect upon the steel is then necessary. It is normal to assume that the current rate of carbonation will continue and that, once it reaches the steel, corrosion and longitudinal cracking will commence immediately.

Two examples of carbonation: one in a concrete lintel and one in the load-bearing column of a Cornish Unit. Many Cornish Units erected in south-west England had high levels of chloride accelerators added to the concrete members during manufacture.

System buildings that have been shown to have instances of carbonation problems include:
- *Airey Houses*
- *Boot Pier and Panel Cavity Houses (in England, where breeze aggregate was used)*
- *Cornish Units (in south-west England, where high levels of chloride additive were used)*
- *Easiform Cavity Wall*

- *Orlit Houses (for similar reasons to Cornish Units)*
- *Parkinson Framed Houses*
- *Reema Hollow Panel Houses*
- *Smith System Houses*
- *Unity Houses*
- *Woolaway Houses.*

Cornish units are easily recognised. They have mansard roofs and were made in 2-, 3- and 4-storey versions.

Chloride attack of concrete

Calcium chloride was commonly added to concrete up to the late 1970s to accelerate its curing time, especially during cold weather, in order to speed the construction process. Unfortunately, it can break down the alkaline content of the concrete, especially where it has been introduced as an on-site additive (rather than during cement manufacture). Quality control of on-site additives is always difficult and this practice often resulted in uneven distribution of the chemical throughout the concrete, which tended to exacerbate the problem.

The loss of alkalinity within the surface of the concrete removes its protective capability to stop the encased steel from oxidising. Where carbonation is present, chloride attack can increase the rate of oxidisation of any steel reinforcement. However, chloride attack can lead to the steel suffering corrosion even if carbonation is not present.

The appearance of a chloride attack differs from carbonation in that it tends to induce large cracking or bulging within the concrete of a more localised nature. The steel can also suffer sudden failure, especially in the presence of both chloride attack and carbonation, as it can become relatively brittle due to extreme corrosion. Low levels of chloride ions (below 0.4 per cent by weight of cement) are not considered to be of concern unless carbonation is present; with between 0.4 per cent and 1.0 per cent chloride ions by weight, cracking is assumed likely to occur within 5 years, even quicker if carbonation is also present. Where high levels of the chloride are present (above 1 per cent by weight), corrosion of steel can occur even if the concrete is highly alkaline.

System-built houses experiencing such problems include:

- *Airey Houses*
- *Cornish Units (in south-west England)*
- *Orlit Houses*
- *Reema Hollow Panel Houses*
- *Unity Houses.*

Thermal problems

Very little consideration was given before the early 1970s to the level of thermal comfort enjoyed by the occupants of dwellings. New houses, including system-built housing, had extremely low levels of insulation incorporated into their construction. Generally, system-built houses constructed with a steel or concrete frame suffered more severe problems than houses of solid or cavity wall construction because their structural form and the lightweight materials used were extremely thermally inefficient. This was a situation that was normally exacerbated by a lack of any central heating system.

Specific systems and their defects

Airey Houses

Approximately 26,000 of these houses were constructed between 1945 and 1955. They were built with storey-height pre-cast concrete columns at 450 centres. These had tubular steel centres and were erected over a cast in-situ ground-floor slab. They were connected at roof and first-floor levels by timber or steel lattice joists.

This Airey House has undergone some modernisation, e.g. new roof tiles and windows, but the original shiplap cladding is still in place and is generally in good external condition.

The external elevations were finished with horizontal pre-cast concrete slabs which were either smooth-finished or pebble-dashed and secured to the columns with copper wire in shiplap fashion. Internally, the construction was usually of plasterboard at both floor levels, though later houses incorporated clinker blockwork for lining kitchens and bathrooms.

Common problems include:

- *Cracked columns and first-floor beams as a result of poor design and site processes.*
- *Corrosion of the tubular steel reinforcement due to water penetration at the joints between the horizontal planks or insufficient concrete cover.*
- *Similar problems as a result of condensation due to a lack of any vapour check.*
- *Poor thermal insulation made worse by draughts, as the shiplap construction was not windproof.*
- *The original windows were of poor-quality steel and have often been poorly maintained.*
- *Structural problems caused by the removal of the vertical posts when replacing windows (the original posts acted as mullions which were too closely spaced to suit the new windows).*

Left: An Airey House during repair. The external cladding is tied to the columns with twisted copper wire. A timber batten on the inner face of the column will provide a fixing for the plasterboard and insulation. Top right: This photograph (1950s) shows a newly constructed Airey House. Bottom right: The left-hand 'semi' has had the external walls replaced with load-bearing cavity work. Much of this work was funded through the Housing Defects Act 1984; funds were only available to those who had bought their houses.

British Iron and Steel Federation (BISF) Houses

Some 30,000 of these houses were erected in England and Wales together with over 4,000 in Scotland. This is out of a total of about 140,000 metal-framed houses, in all, of some 30 different types, the vast majority being built after 1940. There were three different types of BISF house, though the essential feature of each was a frame of either hot-rolled or cold-formed steel. A number of cladding systems were used, of which the most common consists of an outer cladding of profiled steel sheeting at first-floor level, cement rendering applied to metal lathing at ground-floor level, and internal facings of plasterboard on timber battens. Steel trusses supported roof coverings of asbestos or metal sheeting.

These houses have suffered from a number of problems, but their overall performance is better than many other types of system-built housing. Common problems include:

- *Corrosion of the profiled sheeting at first-floor level and, in some cases, the horizontal (or sheeting) rails that support the cladding. This can be as a result of condensation (the steel sheet is impervious to moisture, thus trapping vapour within the structure) or the failure of the weather stripping on the gable ends.*

- *Corrosion of the vertical steel stanchions, especially corner members where the cladding of steel or render may be less effective.*
- *Impact damage to the ground-floor render can lead to rusting of the metal lathing and progressive failure of the render.*
- *The asbestos roofing tends to become brittle and crack with age. There is probably no health hazard unless it starts to shed fibres, particularly into the building. Its removal is strictly regulated. (See Chapter 18 for a more detailed discussion of asbestos products and their safe removal.)*
- *The large void between the internal wall lining and the external cladding is now recognised to be a potential fire risk, allowing an easy spread of fire throughout the house. This can be exacerbated where internal linings are of fibreboard or similar combustible material.*

Left: A pair of BISF houses with original asbestos roof and steel cladding, but having replacement windows installed. Right: BISF houses with a variety of remedial works in evidence, e.g. replacement of roofs, external cladding and windows.

Wates Houses

These were based on pre-fabricated, load-bearing, reinforced, tray-shaped concrete panels erected around a special jig. The horizontal joints between the panels interlocked into a mortar bed, while the vertical joints were hollow and required filling with fine concrete after erection. Steel reinforcement was sometimes inserted into the horizontal and vertical joints.

Wates House

Detail over porch

In-situ column — Pre-cast ring beam

Storey height precast panels —

Pre-cast ring beam

Continuity reinforcement

Foot of panel bedded on foundation

Some 30,000 Wates Houses and low-rise flats (two- and three-storeys) were constructed in the UK in the 10-year period up to 1955. This panel system was subsequently developed for use in high-rise blocks of flats of the 1960s and 1970s, though the panels were not load-bearing for buildings above six storeys.

Initially, the panels were lined internally with wood-wool slabs, though these were superseded by lightweight concrete blocks and plaster. Similarly, the original prefabricated sectional pitched roof was replaced by 'cut' rafters in later roofs.

Common problems include:

- *Corrosion of the in-situ steelwork in the joints and above windows – a problem that led to cracking and spalling of the concrete and is often due to lack of concrete cover.*
- *Corrosion of the reinforcement in the flange of the panel. This may affect either the front face of the concrete or the internal lining.*
- *Cracking along the vertical joints between the panels as well as the panel corners. Both these defects may lead to damp penetration.*
- *The use of clinker blocks for party walls and the inner lining of the external walls. The use of this material was common in the North, where Wates load-bearing panel system housing estates were developed by the National Coal Board. In the long term, these blocks may become a problem because of chemical expansion of the clinker.*

Unity Houses

There were approximately 15,000 Unity Houses built, most of them between 1953 and 1955. The structure comprises a series of pre-cast columns, usually at 3ft centres with pre-cast (but not reinforced) panels forming internal and external skins. In early forms, the inner and outer panels were tied together with metal ties, but later models had the external panels tied independently to the columns. The inner panels were of lightweight concrete to provide some measure of thermal insulation. The outer panels were profiled along their top and bottom edges to provide weather protection. In addition, they were separated from the columns by a vertical DPC. The panels helped stiffen the frame, though additional bracing was provided by steel straps running diagonally between the columns. On completion, the outer panel joints were pointed. At first-floor and roof level, steel beams spanned across the building and also joined the ground- and first-floor columns.

Most Unity Houses were built as two-storey houses. However, there were also a number of bungalows and some medium-rise flats. The party walls were formed in cavity construction and comprised two 75mm leaves with a 50mm cavity. Ground floors were usually of solid construction with splayed timber bearers bedded into the screed or slab and supporting softwood boards. Internal partitions were formed from storey-height 50mm reinforced plaster panels or from 50mm breeze blocks. Windows were mostly metal. During the late 1940s and early 1950s, there were a number of changes aimed mostly at speeding up construction times.

Common problems include:

- *Corrosion through carbonation and impact damage. This problem can result in steel expanding and columns spalling.*
- *Chloride attack of concrete due to over-use of chloride accelerators during manufacture. A problem made worse by the presence of carbonation.*
- *Corrosion of ties that connect cladding panels to the columns or internal leaf. A potential problem where claddings contain high levels of chlorides.*
- *Poor sound insulation. Tenants have complained about poor sound insulation between units.*
- *Damage to the external cladding panels. This problem has occurred in some areas of the country, probably through frost action.*

Unity Houses are similar in principle to Airey Houses. They both comprise pre-cast concrete columns with a pre-cast concrete cladding. Unity Houses also have steel braces to prevent 'racking'.

'Pre-fabs'

At the end of the Second World War, the government introduced a temporary housing programme in an attempt to meet immediate emergency housing needs. This involved the erection between 1945 and 1948 of temporary dwellings, nicknamed 'pre-fabs', with only a very short designed life span of 10 years. These buildings, of which some 157,000 were constructed, including 32,000 in Scotland, performed better than expected and, though most have subsequently been demolished, there are several thousand still in use.

The temporary houses were normally single-storey buildings which were site-assembled in 2–3 days from a number of prefabricated components, including those manufactured by former aeroplane factories using war-surplus materials, such as aluminium and steel. A number of other readily available materials, such as asbestos-cement sheeting and plasterboard were also used.

The two outer photographs show the simplicity of the 'pre-fab' design. The left-hand pre-fab was of aluminium construction; that on the right was of steel. The middle photo shows the fitted bathroom of the aluminium pre-fab.

Eleven different types of temporary house were produced, of which six involved numbers of fewer than 2,500 units each. Of the remaining five, the following three were produced in the highest numbers:

- *Arcon Houses – 39,000 houses of steel frame, asbestos-cement cladding and plasterboard/building board linings. The Arcon Mark V was the most common of all 'pre-fab models'.*
- *Aluminium Bungalows – 54,500 houses of aluminium-alloy frame, cladding and roof lined with plasterboard.*
- *Uni-Seco Structures – 29,000 houses with timber frame, asbestos-cement cladding and wood wool linings. They had a flat roof.*

Problems include:

- *Poor insulation leading to low comfort levels.*

- *Condensation affecting the frame, claddings and linings due to the low thermal insulation of the original construction.*
- *The aluminium sheeting was often formed from re-salvaged scrap from shot down and obsolete planes. There was often no attempt to separate different metals and this resulted in corrosion of the metal sheeting and frame.*
- *Poor detailing, leading to water penetration of joints and gradual deterioration of the materials used, resulted in damp problems of the frames, claddings and linings.*
- *Deterioration of the asbestos-cement sheeting can lead to a dangerous health hazard if individual fibres are released into the environment. (See Chapter 18 for a more detailed discussion of asbestos products and their safe removal.)*

Defects in system houses constructed after 1955

Many of the defects associated with earlier system buildings are found in dwellings of this later period. Problems caused by carbonation, chloride attack, lack of thermal insulation, damp penetration and inadequate sound insulation are all too common across the broad range of industrialised (or system) buildings constructed – low, medium and high rise. Over 150 different systems were authorised in England and Wales alone, with the individual numbers erected ranging from single figures to over 128,000 for Wimpey No-Fines.

Specific systems and their defects

Cross-wall construction

This form of construction became popular in the 1960s and 1970s as it enables standardisation and repetition of the construction processes for both the structural and non-structural elements of a building. It is based upon the premise that all structural loads are carried on regularly spaced parallel walls; all other walls, either in-between or at right angles, merely being non load-bearing partitions or cladding panels. The cross-walls are constructed at approximately 3.5–4.5m intervals in brickwork, blockwork or in-situ concrete; they form the gable and party walls of semi-detached or terraced housing. In medium-height blocks of flats, where reinforced concrete cross-walls form the gable and intermediate walls, the process is known as 'box-frame construction'.

Cross-wall contruction

Weather boarding or tile hanging.

Tiling on trussed rafters

Gable walls – concrete with brick facing or brick and block cavity construction.

Non load-bearing studwork infill panels with render or tile hanging.

Return on gable wall to provide resistance to longitudinal deformation.

Trussed purlin carries load of first floor joists.

The cross-walls not only provide the structural element of the building, but also offer fire resistance and sound insulation between separate dwellings. The loads from upper floors and roofs are transferred onto the cross-walls either directly, where the floor/roof is built into the wall, or indirectly, where the floor/roof is supported on transverse beams or lintels that are built into the cross-walls. The cross-wall requires lateral bracing against wind pressure at right angles to its length and this is achieved by a number of means; in-situ reinforced concrete floors; longitudinal beams or walls; returning the ends of the cross-wall in the shape of an 'L' or 'T'. (A typical system is shown on the previous page.)

The flank and party walls of this house are the cross-walls. The brick panel on the front elevation is non-structural.

Problems commonly experienced include:

- **Water penetration.** *This is a particular problem at the vertical junction between the end of a cross-wall and the adjoining cladding/infill panel. This detail is often poorly formed – the junction is often incorrectly finished with just mastic/sealant plus, possibly, a coat of paint.*
- **Movement as a result of a lack of lateral restraint.** *This may occur where the floor or roof joists run parallel to the cross-walls. In this case, no intermediate restraint is given to the walls, which are dependent for stability only upon the end beams receiving and transferring the floor/roof loads to the walls.*
- **Expansion of joists that are at right-angles to the cross-walls.** *This can cause lateral movement which may be as much as 40mm and be due to either thermal or moisture movement.*
- **Condensation on poorly insulated and exposed dense concrete walls.** *This may occur on flank/gable walls and at the outer ends of party walls, i.e. a form of cold bridging.*

Typical 1960s/1970s concrete cross-wall construction

Concrete roof panels covered with tiles or felt.

Pre-cast upper floor panels.

Pre-cast concrete stairs.

Pre-cast or in-situ cross walls.

Timber frame cladding to front and rear elevations. Tile hanging between floors.

Wimpey No-Fines Houses

This form of construction involves the use of no-fines concrete, i.e. concrete containing cement and graded aggregate (often 12 or 19mm in size), but no sand or fine aggregate. The lack of fines results in a wall that can be cast more economically than a normal concrete mix and which contains many air gaps, giving some improved insulating capability. It is also more resistant to damp penetration as its wide pore structure reduces capillary action detailed (see Chapter 16 for a more discussion of damp problems).

The first recorded no-fines house was constructed on the Isle of Wight in 1852 and several hundred were built in England and Scotland during the inter-war period. Over 72,000 of these low- and medium-rise houses were constructed by Wimpey in England and Wales from 1945 to the early 1960s and a further 128,000 between 1964 and 1979. Several thousand were erected by a number of other contractors, while in Scotland, the Scottish Special Housing Association, Wimpey and other contractors commissioned or constructed several thousand more. The majority of the dwellings were low/medium rise, though a certain number of high-rise buildings were constructed.

The Wimpey version had solid external walls formed with 200–300mm thick no-fines concrete poured on site into prepared shuttering containing continuous reinforcement. External elevations were finished with a wide range of materials, though cement render was very commonly applied. Internal walls were of breeze blocks or timber studwork. Very little prefabrication was involved except for the shuttering panels which were re-usable. The result was a functional, though aesthetically uninspiring, building.

Left: An early 1970s no-fines house. These particular units had long-term problems of rain penetration at roof level and defective render. They were demolished in 2001. Right: Two residential tower blocks of no-fines construction are being erected (1950s photo).

Common problems include:
- **Rotting windows and external doors.** *Caused by poor joinery design and/or inadequately sealed gaps between joinery and structure.*
- **Excessive heating loss due to poor thermal insulation.** *Often, this problem is combined with an inadequate or expensive-to-run heating system.*
- **Surface condensation and mould.** *This problem is as a result of poor ventilation, cold bridging, and inadequate thermal insulation/heating.*
- **Rainwater penetration problems.** *The wall construction should reduce capillary action, but there have been rainwater penetration problems.*
- **Cracking and spalling of render on external wall faces.** *Often caused by lack of suction in the no-fines concrete, or unsuitable mortar mixes, e.g. strong over weak.*

Pre-cast concrete houses and flats

A number of pre-cast concrete houses were manufactured in the 1970s. These were often adaptations of high-rise buildings. The problems experienced (i.e. damp penetration, condensation and high heating costs) have led to many units being demolished.

Bison Wall frame

External wall panels: 75mm facing concrete attached to 150mm structural wall with stainless steel ties.

25mm polystyrene insulation is sandwiched between concrete panels.

Floor units are pre-cast concrete with 12mm polystyrene insulation and 50mm floor screed.

The internal surface of the panels was designed for direct decoration.

Many panel systems were developed during the 1960s.

The Bison system was originally intended for high-rise construction. Later developments included medium-rise flats and two-storey houses.

Some manufacturers concentrated their efforts producing pre-cast concrete low-rise houses. A number of systems were available in the 1960s and 1970s. The system shown below is based on panel construction.

Apart from the problems noted above, many of the buildings were incorrectly assembled; panels were sometimes insecurely fixed, joint gaskets were omitted and the properties suffered raised levels of dampness. Problems of penetrating dampness were, therefore, common.

Proprietary system using pre-cast concrete panels – houses or flats up to four storeys

Party wall or gable panels

spine wall

Party wall or gable eaves filler

Trussed rafter roofs prefabricated and hoisted in position – supported on pre-cast concrete wall plates which also act as parapet gutter. Gutter factory-finished in asphalt

Gable walls clad in brickwork

Spandril panel with weather boarding or tile hanging finish

In situ or pre-cast ground floors Pre-cast concrete or timber upper floors

Ground beam

Load-bearing concrete panels with render or other cladding

Conclusion

Historically, prefabricated reinforced concrete and steel systems have promised rather more than they have delivered. The present stock of such houses contains many that have under-performed. However, a number of building companies are currently experimenting with system buildings. Some are investigating steel-framed housing, while other organisations are looking into buildings formed with in-situ aerated concrete or timber framing. The underlying factor is no longer simply about increasing the speed of erection, but is also about seeking prefabrication methods that reduce dependency on skilled site trades.

SECTION 3

Timber frame housing

Introduction

In this country, a modern house is still usually thought of as having a cavity wall with an outer leaf of brick and an inner leaf of blockwork. However, an alternative approach is to construct the house from a framework of timber protected by weather boarding, tile hanging or brickwork. This section briefly describes the development of timber-framed housing and explains those defects that are associated with this approach.

Before the ascendancy of brick and stone, most houses of substance would have been constructed from timber. These buildings were generally constructed in one of two structural forms: cruck or box frame. By the end of the sixteenth century, timber buildings, particularly box frames, had evolved into a very sophisticated form of construction.

A cruck frame and a later box frame (seventeenth century).

In these traditional timber-framed buildings the structural members were infilled with panels made of wattle and daub. The wattle consisted of oak staves, hazel wattle, and a daub of lime, mud, straw and cow dung. Later, the infill panels were often replaced with brickwork. In later timber-framed buildings bricks were sometimes used as the original infill for the panel.

The twentieth century saw a revival of timber frame construction, albeit using different methods to those employed traditionally. The timber frame houses of today have evolved from the approach pioneered in Scandinavia and North America where there has been a long tradition of timber construction. Both Scandinavia and North America had plentiful supplies of renewable timber, and, in the latter case, the carpentry skills that the early English settlers took with them. Mechanised sawmills, which enabled speedy production of uniform cross-sections of timber, together with industrial techniques which produced low-cost nails, both played their part in developing this form of construction.

Modern timber frame housing

The system-built timber frame houses constructed today have evolved from a number of earlier forms including:
* *platform – where walls are constructed as storey height elements*
* *balloon frame – where walls are constructed as complete entities that are continuous from sole plate to eaves*
* *post and beam – where, like traditional timber-framed buildings, the loads are carried by columns and beams*
* *volumetric box – where the units are delivered and placed on site as a series of pre-constructed boxes.*

Balloon frame

Full height studs framed in factory or on site. Floor is attached to studs.

Early balloon frame

From Specification 1945

An example of a balloon frame used on a 1940s house. Note the flat roof.

Balloon-framed construction was common until it was replaced by the platform method. This is usually more economic and better suited to the demands of factory-based prefabrication. An example of a platform frame is shown as follows.

Typical platform frame
(There are a number of minor variations)

Head binder (if required)

Timber lintel

Cripple studs (support lintel)

Upper frame

Header joist

Joists

Head binder
(if required)

Head plate of frame

Studs

Outer face covered in
ply or OSB sheathing
(oriented strand board)

Bottom plate
of frame

Lower frame

Sole plate (if required)

History of modern timber frame housing in the UK

Although not significant as a percentage of total houses built, there were a number of timber-framed houses constructed in the period between the two World Wars. Many of these were constructed from timber studwork though some were constructed from what were effectively solid timber walls, usually clad on the outside with timber boarding.

Solid core timber frame

Plasterboard
or fibreboard
lining

Weather boarding
on breather membrane

Solid core

In these solid wall structures there was usually a protective breather membrane sandwiched between the cladding and the frame. Its purpose was to prevent rain penetration while, at the same time, allowing moisture in the frame to evaporate. Vapour control layers (vapour checks), which limit interstitial condensation, were not usual at this time.

Following the Second World War the increasing interest in industrialised systems revitalised timber frame systems. Apart from the reasons for embracing system building that have already been mentioned earlier in the chapter, timber frame became a focus of attention because the experience of several severe winters raised concerns about levels of heating and insulation in both traditional and PRC domestic buildings. An export drive by the Canadian Government in the early 1960s helped to sell the idea that timber-framed construction produced warm buildings; this coincided with the UK Government's policy to further develop industrialised systems of building. The picture below shows a Canadian Trend House first exhibited at the Ideal Home Exhibition in 1957.

From The Builder June 21st, 1957

Skirting
Foil-backed plasterboard
Window over
Cedar
Cedar siding on building paper and ply

Timber frame housing is by no means new. However, systems used for the last 50 years or so have nothing in common with the traditional timber framing of Medieval England.

In 1957 a Canadian house was exhibited at the Ideal Home Exhibition. For the first time there was a quick and cost effective alternative to traditional housing and existing system building (mostly pre-cast concrete). However, not until the 1960s did it become popular and then in cross-wall form rather than true timber frame.

Nowadays most timber frame houses are not clad in Cedar (as this example is) but brickwork. So there is a load-bearing internal timber frame and an external single leaf of brickwork. Many of the defects in modern timber frame housing are, in fact, caused by the problems of building in these two very different materials.

In the immediate post-war period most timber frame housing was largely based on timber stud walls which were covered with a variety of claddings including; timber boarding, tiles/slates, render on a plywood backing, brick, and asbestos cement sheets. These lightweight claddings would normally be fixed directly to the timber frame, or indirectly using timber battens. In these designs there was no cavity to protect the inner load-bearing frame against water penetration. There were also hybrid versions where, for example, there would be a brick cladding at ground-floor level and one of the directly fixed claddings on the upper floor. Some of the systems were thermally insulated but not to a very high level. A breather member was normally incorporated between the cladding and the studs though vapour checks were not normally included (foil-backed plasterboard was used from the mid-1960s). Some houses had an external cladding of 100mm brickwork; a cavity separated the brick and timber studding. This approach later became the norm.

Some cross-wall houses (see previous section) were constructed with brick party walls but with timber frame panels to the front and rear. These were usually clad with timber boarding and /or tiles.

Left: In the 1960s timber frame houses were often clad in tiling, slating or weatherboarding. Right: Modern houses are usually clad in brick.

From the mid-1960s to the early 1980s there was a steady increase in the number of timber-framed houses. Most of these were in the public sector though many volume house builders were developing their own timber-framing systems to cope with the increased demand for private housing.

Advantages of timber frame housing

The main advantages of timber frame houses, when measured against brick and block construction, are generally argued to be related to cost savings that can accrue from industrialised processes, including standardisation and speed of on-site construction. Speed of construction is related to the possibility of quick erection times and the ease of making connections between uniform materials. Additionally, there can be a significant reduction in time spent waiting for the finished structure to dry out, because fewer 'wet trades' are involved.

Other advantages of timber frames are generally perceived to be:

- *the ability to, relatively easily, increase the amount of thermal insulation incorporated in the walls*
- *a relatively lightweight structure that allows some savings in foundation size*
- *the ability to accommodate settlement or other foundation movement more readily than brick/ block construction.*

The usual method of constructing timber frame in the UK is, as mentioned, the platform frame, i.e. the assembly of the ground-floor wall panels on the substructure, followed by the first-floor structure, which, in turn, provides a platform to erect the next storey, and so on. The timber frame is the main load-bearing element in this type of construction; the brick cladding's primary purpose is weather protection though it may also have been chosen for its appearance. The brickwork is separated from the timber frame by a cavity and carries none of the building load apart from its own weight. In order to provide stability, the brickwork is tied to the frame at regular intervals with metal ties. The other key elements of the timber frame walls include:

- **Breather membrane.** *This is intended to prevent wind-driven rain reaching the timber frame. It should also allow any water trapped in the frame to escape, i.e. it should be 'breathable'.*
- **Sheathing.** *The purpose of this is to add structural stability to the frame by helping to prevent 'racking'.*
- **Thermal insulation.** *This is usually placed within the timber frame (rather than in the cavity).*

Sheathing — Thermal insulation

Breather membrane

Timber frame

Plasterboard

Vapour check or vapour control layer

The left-hand graphic shows the construction of a modern timber-framed wall. The photo above shows a party wall – note there are two timber frames separated by a small cavity. The strand boarding provides stiffness against racking.

This photograph shows the brick skin being built as a cladding to a modern timber frame. The frame itself is hidden by the protective breather membrane. Vertical fire barriers can be identified here by their red polythene sheathing.

From the mid-1970s the Building Regulations required external walls to meet certain levels of thermal insulation. In timber-framed housing this could be achieved by insulating the spaces between the vertical studs. A vapour barrier, now usually referred to as a vapour check or vapour control layer, is placed on the inner surface of the insulation (its warm side) in order to limit the amount of vapour moving towards the cavity where it might condense on the inner face of the sheathing. The vapour check usually takes the form of a polythene sheet though foil-backed plasterboard has been used.

Polythene vapour check.

The breather membrane, which is fixed to the external face of the sheathing, has the function of protecting against any wind-driven rain which may pass through the outer cladding. The membrane also protects the timber frame during the construction period. Therefore to be effective the membrane has to prevent water penetration but, at the same time, allow the passage of water vapour.

The breather paper (membrane) can be seen in this photograph prior to the brickwork skin being built. The membrane is usually applied in the factory. If this is not the case, it should be fixed in position as soon as the frame is erected in order to prevent the sheathing getting wet. Breather membranes vary in colour depending on the manufacturer.

Identifying brick-clad timber frame houses

It can be difficult to differentiate brick-clad timber frame houses from brick/block cavity walled houses. However, there are a number of features which, while they may not confirm timber frame (because the same features could be on brick/block construction), will provide cumulative indicators. Examples include:
- *Windows fixed to the inner rather than the other leaf.*
- *Weepholes incorporated just above or just below damp proof course level. These were provided to allow drainage of moisture away from the vulnerable timber frame.*
- *Movement joints formed around windows, verge and eaves (or signs of distress if not incorporated). The movement joints accommodate the different rates of movement of brick and timber.*
- *Weepholes at first-floor level to indicate the presence of a cavity tray situated over horizontal fire stops.*
- *Cavity vents – by no means universal but cavity vents in the form of airbricks were sometimes incorporated to help disperse moisture vapour.*

An internal inspection might confirm whether or not timber frame has been used by reference to:
- Separating walls in the roof space. *If the party walls are finished in plasterboard, this will be indicative of timber frame. (The plasterboard provides the necessary fire/sound insulation to the timber studs. It should be noted, however, that some early timber frame houses were of hybrid construction and used brick or block for the party walls.)*
- Gable walls. *The inner gable leaf will probably be built in timber.*
- Eaves. *The cavity closer may be timber or mineral wool and the wall plates supporting trussed rafters may be planed (in the factory) rather than 'sawn'.*
- *The* timber studding *might be revealed by removing electrical sockets or by lifting floor boards or skirting boards.*
- Dry lining. *Although dry lining is often used on blockwork, with timber frame the position of the fixing nails may indicate the standard centres of the vertical timber members.*

Defects in timber frame housing

Timber frame became so popular that by the early 1980s nearly a quarter of house building starts used this method of construction. However, around this time reports began to emerge of a number of defects in relatively new timber frame houses. These reports, which included a damaging television broadcast, led to a decline (since reversed) in the construction of timber frame houses. Volume house builders withdrew from timber frame construction on the assumption that adverse publicity would affect sales. It appears, however, that public reaction was muted: either because people were not aware of the adverse publicity, or perhaps because occupiers of timber frame did not realise how their houses were constructed.

Many of the reports highlighted defects that arose due to either mistakes and/or ignorance, at both the design and construction stage. Other reports raised more fundamental questions about inherent problems in the use of timber frame. It is clear, however, that many of these reports, at least in the press, were related to faults that could have been found in any house type. Because they were also found in timber frame houses, the faults were erroneously linked to this construction method. For example, reference to rusting wall ties, missing thermal insulation, poorly supported water tanks and defective brick garden walls could all be found in contemporary articles about timber frame, written as though such defects were unique to the method of building. Nevertheless, some criticisms were valid and though the reports made reference to defects that were not specifically related to the construction type (such as mortar droppings on wall ties), they were pertinent because, for example, they could result in the timber frame becoming wet. More specific criticisms were directly related to the design principles of timber frame housing.

With hindsight it is now clear that the incidence of failures from the timber frame 'boom' of the 1970s and 1980s has been less significant than predicted. However, theoretically there are some potential problems inherent in timber frame houses from this period – and indeed in those timber frame houses constructed where these issues have not been addressed.

The problems identified in the reports were caused by a number of factors acting both singly and jointly. These include:
- *poor design and specification*
- *poor site practices*
- *a lack of familiarity with the function of key elements in timber-frame design*
- *a lack of experience about how theories of timber frame design would apply in practice.*

Problems also sometimes arose because of the difficulties inherent in constructing factory-made timber frames on top of a substructure formed on site. Specific concerns centred on a number of defects which may also be found in current timber frame construction of poor quality. Potential problems include:

- **Excess dampness in the timber frame.** *The normal risks associated with rising damp, penetrating damp and condensation are increased because of the vulnerability of the timber frame and sheathing to fungal attack. Excess dampness allows fungal attack to occur, and the risk is increased if the timber is not treated and/or the timber is of low natural durability (for example, where it contains a high proportion of sapwood). In some cases timber treatment is targeted on those areas, such as sole plates, that are considered most at risk. Where timbers are cut on site (perhaps in the fixing of sole plates, or to overcome assembly problems), any treatment becomes less effective.*

- **Fire.** *Potential defects in design or construction can lead to rapid spread of fire.*

- **Movement.** *Differential movement of the timber studding in relation to the brick cladding can produce stresses which may lead to damage of the wall and/or associated elements.*

Condensation

A key element in the design of most timber frame houses is the vapour barrier (vapour check or vapour control layer). Its purpose is to prevent the passage of moisture vapour from the warm side of the insulation to the cold side where there is a danger that it might condense within the wall structure. This phenomenon is known as interstitial condensation (see Chapter 16). Where timber frame houses are not insulated, the risks of interstitial condensation are much lower as heat losses through the frame keep the timber structure above dew point. If a frame is insulated, the temperature of the studding remains high but there is a rapid drop towards the outer face of the insulation.

Where there is a risk of interstitial condensation, the performance of the breather membrane is crucial if it is to successfully allow trapped moisture to escape. Unfortunately there have been reports of impermeable materials such as polythene being used for breather membranes.

Plasterboard

Vapour check

The vapour check should not be punctured in the course of fitting electrical sockets or central heating fittings.

WARM
(inside face)

Breather membrane
(vapour permeable)

COLD (outside face)

Sheathing

The breather membrane should be fixed as soon as the frame is erected. It prevents the ply sheathing from getting wet. In most modern systems, it is applied in the factory.

There is a danger of condensation occurring if the vapour control layer fails to perform adequately. This may occur if it is:

*   *missing – either not included as part of the design or omitted on site*
*   *placed on the cold, rather than the warm, side of the insulation*
*   *incorrectly lapped to adjoining sheets or with other elements, such as damp-proof courses*
*   *damaged during (or before) the construction process*
*   *placed in position while the timbers' moisture content is high.*

Additionally, with aluminium foil-backed plasterboard vapour control layers there are two potential problems:

1   *the gap at the junction between the boards allows water vapour penetration*
2   *the aluminium foil can corrode.*

Condensation can also affect the timber frame where it occurs on, say, single-glazed windows, and the water run-off penetrates through poorly sealed joints.

Rising damp

Although incorrectly specified, incorrectly located or damaged damp-proof courses can result in damp problems in any type of house, it is the vulnerability of timber which is of particular concern with timber frame. The bottom section of the timber frame is most at risk, particularly the sole plate or bottom rail. As the sole plate is often fixed in position by nailing through the DPC, this is a potential area of damp penetration.

The panel is nailed to the sole plate

Sole plate clip

Sole plate shot-fired to blockwork - only effective where polythene DPCs are used and where blocks are strong enough to take shot-fired nails.

DPM

Many prefer galvanised or stainless steel sole plate clips because they do not puncture the DPC.

Rain penetration

In addition to the problems described in Chapter 16, the likelihood of rain penetration occurring in timber frame housing is exacerbated by the following factors:

*   **Reduced cavity width.** *Site adjustments may be required while trying to marry a factory made wall to an in-situ base or foundation.*
*   **Breather membrane that is defective or missing.** *Note: in early designs bitumen impregnated fibre board may have been used instead of plywood sheathing with a separate breather membrane.*
*   **Distortion of wall ties caused by differential moisture movement of the timber studding and brick cladding.** *Shrinkage of the frame may result in ties sloping towards the inner leaf, which will allow water to be transported towards the timber frame.*
*   **Misplaced insulation.** *When thermally insulating a timber frame the intention is to leave a clear cavity, and to place the insulation only between the vertical timber studwork. There have been reports of the cavity being filled. This poses the danger that the insulation may provide a path across the cavity to the timber frame.*
*   **Lack of weepholes.** *These are required above doors and windows and immediately below DPC level; their purpose is to allow moisture, from whatever source, to drain away. In properties constructed before the early 1990s, weepholes were also required at upper floor level just above the cavity fire barriers.*

The breather membrane is normally stapled to the sheathing in the factory. Laps or hems are left to provide an overlap at adjacent panels and over the substructure. These hems should be re-stapled as soon as the panels are erected. Membranes are also easily torn on site. Repairs should be carried out immediately.

External timber claddings may, of course, also be vulnerable to fungal attack. In those designs where the cladding is fixed directly to the frame, the danger may be increased because there is no ventilated cavity. For those systems with no cavity, the performance of the cladding and the breather membrane is, of course, critical.

If breather paper is not lapped over cavity tray frame may become wet.

Cavity tray

Weepholes for drainage

Heavyweight lintel

Gap for movement

Lintel fixed back to frame with clips. A small gap allows for vertical frame shrinkage.

Lightweight lintel

Some lintel shapes require cavity barriers (to limit the spread of fire).

This lightweight lintel does not have a cavity barrier below it (i.e. at the head of the opening). In addition, the breather membrane should lap the lintel rather than be dressed behind it as shown here (otherwise water can run down between lintel and membrane).

Movement

Timber frames are likely to shrink, particularly upon initial occupation when heating systems are first used. Failure to provide movement joints at positions where this shrinkage occurs may cause damage to the structure. Timber shrinks across the grain, rather than along it. It is therefore the horizontal members that mainly cause the problem.

Storing panels in the open will allow their moisture content to increase. The subsequent expansion may cause cracking and problems of condensation as the panels slowly dry out in the completed building. Careless stacking can also tear the breather paper.

A two storey frame will shrink about 15mm or so in overall height. A gap between rafter and brickwork is essential to accommodate differential movement.

Movement gaps are also required around window frames.

Adequate gap and flexible sealant

Vertical DPC

No flexible sealant, inadequate gap.

Horizontal DPC

The most vulnerable areas are as follows:

- *where there is a direct connection between the brick and the timber elements, e.g. wall ties, which, if not flexible, may snap*
- *those elements attached to the timber frame which pass into or across the brickwork, e.g. windows, roof structure, balanced flue boilers*
- *those elements such as flashings and sealants, as they may be disturbed or damaged. This in turn may lead to rainwater penetration.*

As the panels dry they will shrink slightly. The brickwork, on the other hand, may expand slightly. To allow for this differential movement, the ties should be fixed with a slight outward fall. If the timber panel in the left-hand photo shrinks, the horizontal part of the tie may slope backwards towards the inner leaf. The right-hand photograph shows better construction.

Poor structural connection

One of the advantages of timber framing is the ease of connecting the various timber elements. As long as the timbers are nailed in accordance with the appropriate specification, structural integrity will be assured. In some instances, however, connections such as simple nailing are ineffective because they have been carried out carelessly. Mistakes on site such as wall tie fixing nails passing through the sheathing layer but missing the stud framing have been observed. Reports have also revealed that similar poor connections have occurred in the factory.

Factory nailing should ensure this problem does not occur. The nails are not located in the studs.

There have also been instances where the omission of head binders has resulted in distortion of the panels.

Head binder keeps the panels in alignment. If it is omitted the panels can twist and crack the internal finish.

Head binders can often be omitted, on the top storey, but only if trusses sit directly over studs.

Panels nailed together

Head binder

Fire

In timber frame buildings with an external cladding of brickwork, there is a danger of fire spreading in the cavity. The fire can start within the building but spread into the cavity through, for example, an incorrectly fixed internal plasterboard lining. Fires can also result from careless soldering of pipes adjacent to the cavity. The breather paper in older timber frame houses was flammable. In more modern timber frame houses, where flame-retardant breather membranes are specified and used, this particular fire risk should not be a problem.

Cavity barriers are required to limit the spread of fire within the cavity itself. In modern construction they are required at the positions shown in the drawing below.

Cavity barriers in houses

Semi detached

Detached

Mineral wool fire stops are also required around the open perimeter of the party wall (see next graphic).

Flexible cavity barrier

Provision of cavity barriers (England only – In Scotland and Northern Ireland the requirements are more onerous)

Before the early 1990s cavity barriers were also required vertically at 8 metres intervals and horizontally at first floor level (look for weep holes over the cavity tray).

In addition, fire stopping is required at party walls. This usually takes the form of a mineral quilt.

Plan at party wall

Cavity barriers either side of fire stopping

31mm plasterboard (1x19mm & 1x12mm)

Sound insulation – in cavity or in panel.

Fire stopping in cavity perimeter and over party wall.

Photo taken from first floor looking down at the party wall.

Fire stopping at party wall

There is a danger of fire spreading within a cavity if:

- *fire barriers are missing or placed in the wrong position;*
- *fire barriers do not completely span the cavity, thereby allowing the passage of fire and smoke;*
- *fire barriers have deteriorated.*

Some early types of barrier were constructed from sheet metal or timber battens. With both of these materials it is relatively easy to fix them to the plywood sheathing but it is likely that some gaps will occur at the junction with the relatively irregular inner face of the brick cladding.

To overcome this problem modern fire barriers are made from a resilient material as shown below.

Early cavity barriers were often formed using 38 x 38mm timber battens. A DPC protected them (and the frame behind) from penetrating dampness. If the brickwork was slightly out of plumb the barrier would not fill the cavity - mortar packing was rarely successful. Modern barriers are more flexible and are usually made from polythene sheathed mineral wool.

Timber batten with vertical DPC

This small block of four flats has vertical and horizontal cavity barriers to separate dwellings.

Sound insulation

In timber frame houses incorrectly constructed party walls can lead to problems of sound transmission. The party wall normally comprises two leaves of timber studding (some early houses have a hybrid form of construction with dense concrete block party walls). A sound-deadening quilt and extra layers of plasterboard help reduce the transfer of airborne sound. If this quilt or additional plasterboard is missing, or if heavy metal ties bind the two leaves together, the sound insulation will be reduced.

Table 17.1 Summary of potential defects during construction process

POTENTIAL PROBLEM AREAS	EXAMPLES	IMPLICATIONS/MANIFESTATION
Fabrication errors	Careless nailing of studs and sheathing	Loss of strength
Incorrect storage or handling on site	Storage in the open will increase moisture content of panels Incorrect stacking can damage breather paper	Panels will shrink, twist and may affect plasterboard finish Torn paper will allow moisture penetration
Base out of square or out of level	If the base is out square panels will not fit properly without additional work on site If the base is not level the top of panels will not line up.	Panels which are a bad fit are more likely to deform Window openings may distort, floors may not be level
Construction errors	Incorrect fixing of sole plate and DPC Inadequate nailing Failure to provide bracing on internal partitions If panels are not vertical it may affect the brick cladding Omission of head binders	Rising dampness Racking of finished building Maintaining correct cavity may be difficult. Brickwork may not be plumb Panels may be out of alignment
Finishing the panels	Breather paper is non-continuous Insulation must be a good fit to avoid risks of cold bridging Vapour check must be on warm side of insulation and must not be torn Brick cladding not tied back to studs with flexible ties, or ties incorrectly positioned (i.e. no back fall on ties) Lintels not clipped back to frames in a way which permits differential movement Cladding fixed without allowance for differential movement at eaves and around windows	Damp penetration Condensation Interstitial condensation Damp penetration, lack of restraint for cladding Cracking and distortion of window heads. Cracking and distortion
Fire protection	Header joists omitted between joist ends Cavity barriers non-continuous and/or in incorrect position Cavity fire stopping in party walls non-continuous – does not run up to roof apex	These three problems can all lead to a reduction in fire protection and a subsequent risk to life

Conclusion

The timber frame houses built in the 1960s, 1970s and 1980s were subject to a significant amount of adverse publicity from the media. In retrospect, the number of houses affected by serious defects is probably less than was implied by much of the rhetoric of the time. Nevertheless, when dealing with timber frame housing, one should be aware of the potential problems explained in this text. Because of the risks of rot and fire, the need for good site practice and thorough site supervision cannot be over-emphasised.

Environmental and health-related building problems

18

Introduction

There are an increasing number of environmental and other problems affecting property in the UK. These are in part caused by factors that are beyond the control of humans (e.g. radon gas emitted by radioactive rock), but a great many are linked to human behaviour (e.g. contaminated land). Some are the result of a mixture of both natural and human causes (e.g. flooding). This chapter discusses some of the more common problems that affect residential and other buildings.

Historically, buildings have always been affected by environmental problems, for example, the Roman city of Pompeii was overwhelmed by the eruption of Vesuvius in 79 A.D. and the Japanese tsunami of 2011 caused havoc. Similarly, in the UK, strong winds, severe floods, massive coastal erosion, landslips, earthquakes (generally of a minor nature – the most recent in February, 2013), tsunamis (e.g. in the Bristol Channel in 1607) and other natural phenomena have all occurred in recorded history, i.e. over the last 900 or so years. There is also evidence that weather patterns are changing due to climate change, which is (arguably) entirely due to man-made industrial pollution affecting the Earth's climate, but may be due to natural factors such as the Earth's climatic cycle, or a mixture of both.

The body responsible for the environment in England and Wales is the Environment Agency (EA). This is a non-departmental public body of the Department for Environment, Food and Rural Affairs in England and an Assembly Sponsored Public Body of the National Assembly of Wales. It is responsible for: (1) the protection of the environment in England and Wales including flooding and pollution; and (2) enhancing the environment as a whole. Its remit is extremely wide-ranging and includes monitoring and regulating the environment and how we use it as well as prosecuting those businesses and individuals who cause damage to the environment. The EA states that its 'principal aims are to protect and improve the environment, and to promote sustainable development'.

On its web-site, the EA offers advice and guidance on a wide range of environmental problems and issues including air and water pollution, climate change, coastal change, energy efficiency, flooding, radiation, septic tanks, weather, waste and recycling (including hazardous waste), water environment.

This chapter discusses some of the more important environmental and health-related building problems that may be experienced by residential properties.

Environmental and health-related building problems

Asbestos

Background

Asbestos is a naturally occurring material that has fire-resisting properties as well as resistance to alkalis and acids. As a result, it was widely used within the construction and other industries from the 1930s until the1970s. Its use in textiles and for pottery goes back much further – some 4,500 years.

During the twentieth century, it became increasingly apparent that asbestos presented a health hazard to humans. The first diagnosis of asbestosis – an asbestos-related lung disease – was made in 1924 and the first asbestos-related laws were enacted in 1931. Unfortunately, although there was increasing evidence of its danger to health, asbestos component manufacturers chose to ignore this. In the 1940s, some 3,000 products made use of asbestos ranging from brake linings to cosmetics. Asbestos was used in many materials and components that were commonly used in both residential and commercial buildings throughout the twentieth century. Some 5.3 million tonnes of asbestos were imported into the UK between 1940 and 1998 and its use was not totally banned until 1999 and only stopped completely in 2005.

Asbestos can be found in many materials and components that were installed in residential property before 1999. They include:

- *linings for walls, ceilings and doors – often called asbestos insulating board (AIB) – a high hazard asbestos product – and includes brands such as 'Asbestolux' – used until about 1982*
- *window boards – frequently made up from AIB*
- *floor tiles – vinyl (PVC) and thermoplastic floor tiles as well as cushion floor frequently contained asbestos*
- *loose packing between floor joists and in partition walls*
- *finishes for ceilings, e.g. textured coatings such as 'Artex'*
- *bath panels*
- *fascia and soffit boards*
- *verge boards and undercloaks*
- *gutters and pipes, e.g. for rainwater and soil vapour pipes and for heating flues*

Recent examples of asbestos in current use: Left: The house in the centre has a roof of corrugated asbestos sheets. Right: A very aged soil vapour pipe (with a modern plastic connection) and a gas flue can be seen.

- *night storage heaters and warm air heating systems*
- *boilers and associated laggings*
- *asbestos cement corrugated and flat panel roofing and linings – banned in 1999*
- *pipe laggings – occasionally found in flats, but more normally used in public and commercial buildings – banned in 1986*
- *some paints*
- *roofing slates and profiled roofing sheets.*

The important point to note is that any building constructed before the year 2000 is likely to have some components that contain asbestos. Some experts feel that there are very few buildings constructed before as recently as 2005 that do not contain asbestos.

The hazard to health

It is now known that there are a number of diseases caused by asbestos, currently resulting in approximately 4500 deaths per year, and it is now the single greatest cause of work-related deaths in the UK. The diseases that it causes include:

- **Asbestosis** – *a scaring of the lung that progressively reduces lung function.*
- **Lung cancer** (*especially if a person smokes as well*).
- **Mesothelioma** – *a cancer of the chest lining or Pleura, the tumour cannot be treated.*
- **Pleural thickening** – *a thickening of the Pleura.*
- **Pleural Plaques** – *a thickening in spots of the Pleura leading to ossification, i.e. hardening.*

These diseases are caused by breathing in microscopic asbestos fibres which can be released during manufacture or by cutting the material or by heavy wear once it is installed. In some cases, only one fibre needs to be ingested to lead to ill-health. Normally, the ensuing disease will take many years to manifest itself, often 30–40 years or more. Those most at risk are construction workers who have been involved in the installation or remediation of asbestos in the past. Residential and commercial occupiers are not normally at risk unless fibres are released into the atmosphere. This can occur when asbestos becomes wet or when it is drilled or cut or removed. There is normally no danger to health when the asbestos components are stable, i.e. asbestos fibres are not released into the atmosphere.

The three types of asbestos that have been most commonly used in the UK for manufacturing products are:

- **White asbestos (Chrysotile)** – *used in UK manufactured goods until 1998.*
- **Blue asbestos (Crocidolite)** – *last used in UK manufactured goods in the 1960s.*
- **Brown asbestos (Amosite)** – *last used in UK manufactured goods in the late 1970s.*

White asbestos is the most commonly found of the three variants, all of which present a health hazard. If ingested, white asbestos can sometimes be eliminated by the body. Brown and, particularly, blue asbestos are the more dangerous and more likely to lead to asbestos-related diseases. Unfortunately, it is impossible to determine the type of asbestos being inspected by colour alone as the colour changes with the original manufacturing or treatment process.

Regulation, investigation and removal

For some years, the investigation of any suspected asbestos product and, if discovered, its removal have been required to be carried out under a very strictly regulated procedure. The regulations are frequently updated and are currently the Control of Asbestos Regulations 2012. If asbestos is discovered in a building, it should be left alone and, if required, advice sought from the Health and Safety Executive (HSE). This is a government body that has sole responsibility for health and safety in the workplace in England, Scotland and Wales – it promotes compliance with health and safety legislation by means of advice, inspection, enforcement and prosecution. Apart from the HSE, advice can also be sought from the relevant local authority or from a licensed asbestos removal company. It is also mandatory that, initially, a survey of any property suspected of containing asbestos should be carried out by an accredited or certified surveyor approved to undertake such an inspection.

Left: The bottom corner of the asbestos sheet is damaged. Although not currently presenting a problem, it will do if it sheds fibres or is replaced. Right: The asbestos sheet has broken at a fixing nail; removal of the sheet because of a possible lack of structural integrity will come under the Control of Asbestos Regulations 2012.

If the asbestos that is discovered is undamaged and it is intended to be left in place, no further action is required other than to record its presence. Minor damage can be repaired by sealing or painting over, but the asbestos must not be sanded or rubbed down before painting as this would release the fibres. Similarly, it should never be cut or drilled. Where asbestos is left in place, a building owner needs to develop an asbestos management plan. This should identify how they will manage any risks and communicate relevant information to building occupiers and users.

However, when the asbestos is badly damaged or deteriorating, it must be removed and only a licensed asbestos removal company is permitted to undertake asbestos removal. The same applies if the intention is to remove undamaged asbestos at any time including as part of the remediation or the redevelopment of a building. The removal process will involve sealing the building or that part where there is asbestos so that any fibres released by the work can be removed entirely by mechanical means.

Climate change

It is now a recognised fact that we are experiencing climate change, even if its underlying causes are much argued about. Its effects can be seen in changing and more extreme weather conditions, overall temperature increase, more intense sunshine and associated ultra-violet levels, glacial thaw, rising sea levels and storm-related flooding as well as stronger winds.

The full effects of climate change are not yet clear and this presents indeterminate challenges to the building designer with regard to the future performance of both existing and new buildings. All buildings are required to be weather-tight; they also need to withstand anticipated extremes of weather. The building designer (and user) need to recognise that the potential impact of climate change will be hotter and drier summers, milder and wetter winters and an increased intensity of rainfall, i.e. leading to larger volumes of water in shorter periods. This will impact directly on river and sea defences, the latter creating further problems because decontamination of sea-water-affected buildings is more difficult that those affected by river water because of its salt content.

Left: The experience of Venice where its inhabitants regularly need their waterproofs and boots may seem extreme to us, but flooding due to climatic change is now just as much a problem in England and Wales. Right: A gathering storm over Bristol in the winter of 2012–2013 eventually resulted in lightning, torrential rain and some very large hailstones, but fortunately no flooding in this instance.

In recent years, there have been greater numbers and also increased frequency of storms in the UK. There has also been more frequent increased intensity of rainfall. Such storms have often resulted in high winds and flooding, both of which have caused damage to buildings. Predicting future rainfall patterns is difficult. Whilst the total annual rainfall is projected not to change significantly, changes in the pattern of seasonal rainfall are expected to occur in the future.

Wind damage to houses in Bristol.

The River Avon in Malmesbury, Wiltshire, still running high after flooding the lower town in January 2013.

Heavy snowfall can lead to flooding problems when it thaws.

It also needs to be recognised that the UK is one of the windiest countries in Europe. This is because of its position in the North Atlantic. The prevailing wind in the UK is from the south-west and the western regions of England and Wales experience a greater level of windstorms than eastern regions. These can lead to building damage or collapse, including falling chimneys, roof coverings being lifted off and occasionally complete roofs being removed. There is evidence from the National House Building Council (NHBC) that, where more recently constructed roofs and their coverings have failed due to wind forces (including in the Great Storm of 1987 (a once-in-200-year event), this has been due to the affected roofs not being constructed in accordance with manufacturers' recommendations and the relevant British Standards.

Increased levels of wind will also impact on the level of exposure that a building will have to deal with. The level of exposure affects the choice of materials used in its construction as well as the construction technique selected. The result of climate change may be to increase the effect of driving winter rain especially in the more exposed western half of the UK. Existing buildings in western England and in Wales are, therefore, at increased risk of experiencing rain penetration of external elements. Older buildings with solid wall construction, poorly maintained roofs and parapet and valley gutters are more likely to suffer as are buildings with insulation-filled cavity walls that are situated on south-western-facing slopes, especially when built on higher ground.

Another potential problem of climate change is the suspicion that increasing temperatures may result in a greater number of longer drier periods and these will have an adverse effect on the shrinkable clay sub-soils of south-east England, leading to greater likelihood of subsidence. Examples of recent dry periods when increased levels of subsidence took place are 1975–76, 1989–90 and 1995–96. This means that, where a new building is planned or where an existing building is experiencing sub-soil-related problems, greater consideration needs to be given to one or more of the following: deeper foundations or piles; the use of ground remediation and improvement techniques; the use of compressible materials to cope with greater ground movement; management of trees to avoid desiccation of soils.

Higher temperatures linked to increased levels of ultra-violet radiation may affect the colour and performance of external materials, in particular paints, plastics and other synthetic materials. Higher temperatures will also provide a physical challenge to materials with high coefficients of expansion, resulting in increased allowances for movement. There will also be the need for greater attention to construction detailing, including the width and protection of any expansion joints as well as the use of adjoining materials with different coefficients of expansion.

It needs to be recognised that existing buildings will have been designed and constructed in accordance with relevant knowledge and guidance including regulatory control that was applicable at the time of construction. This means that many existing buildings do not meet recent building standards and codes of practice and/or manufacturer's recommendations that were brought in to meet the demands of the effects of climate change. For example, some of the design wind loads used in codes and standards have increased by a factor of more than 3 to reflect recently observed wind speeds. Ageing and poorly maintained buildings are particularly at risk.

Contaminated land

Introduction

Land contamination in the UK is mainly due to its long industrial history, with the mining of lead and other materials, for example, pre-dating the Romans. Mining, initially for ores and subsequently for coal as well as ores, together with various industrial process has been taking place over the last 2,500 years at least. The country was the birthplace of the so-called Industrial Revolution in the eighteenth and nineteenth centuries, during which many heavy and other industries including iron and steel making and chemical processing developed. Other more traditional industries such as leather tanning and lead smelting also created highly polluted sites. However, mines become worked-out and industrial processes become redundant for technical or financial reasons; the usual consequence of either is that the worksite of the industrial activity is closed down and often gets abandoned.

Redundant industrial sites can be found in urban and rural areas right across the UK and it needs to be recognised that contaminated land from a great number of industrially polluted sites exist. Some of these contaminated sites were used for residential development in the nineteenth and twentieth centuries before the potential health hazards of such land were properly recognised. Many more such contaminated sites, now known as brownfield land, are likely to be developed in the future for residential and other uses because of the limited amount of land available in the UK for development.

Unfortunately, until relatively recently, once an industrial site became redundant, it was the norm to carry out a minimal clean-up, sometimes merely knocking down the buildings, often leaving them to collapse over time. Removing residual industrial waste was hardly ever done. In many cases and certainly before the twentieth century, very few records of industrial use were made and even fewer retained after redundancy.

These two photographs are of a disused Victorian gas works that was within a mixed use redevelopment zone. The level of ground contamination was extremely high and very expensive to clean up.

The result was land that was contaminated with organic and inorganic pollutants including:

- *heavy metals (e.g. arsenic, cadmium, copper, lead, mercury, zinc);*
- *smelting slag including tin and other slags (some of which can be mildly radioactive);*
- *hydrocarbons from oil-based products;*
- *coal tar and other obnoxious substances;*
- *landfill gas (e.g. methane and carbon dioxide);*
- *tannery waste.*

Such contaminants may lead to death, injury or illness as a result of: asphyxiation; cancer; combustion; corrosion; explosion; radioactivity; organic and inorganic toxins.

The legacy of our industrial history is, therefore, a vast number of contaminated sites in both urban and rural areas. To make matters worse, where quarries were created in order to work minerals for industry and construction, they were often subsequently used for landfill purposes with no control over the type of waste used. Currently, the estimated area of land in the UK affected to some degree by contamination is 300,000 hectares.

Investigation of a potentially contaminated site

Until the end of the Second World War, there was little concern about contaminated land, most new housing being constructed on green-field sites on the edge of existing urban areas, but some re-development of previously used land did take place. However, some legislation did start to recognise the problems associated with contaminated land, e.g. the Public Health

Act 1936 included provision to mitigate the impact of old landfills. The increased demand for development land from the late 1940s onwards led to official recognition that there was a need to develop old industrial sites as well as bomb-damaged sites for housing and other purposes. Many houses were constructed on former industrial sites without proper account being taken of the implications of contamination and only slowly did it become evident that there needed to be geo-technical and geo-environmental assessment and, where appropriate, remediation of the land before any construction was undertaken.

Since the introduction of the Environmental Protection Act 1990 (EPA), all local authorities are required to assess the land in their areas in terms of identifying sites which are or could cause harm or pollution in terms of a risk to health or the environment as well as pollution of watercourses and groundwater. The dangers to humans from contaminated land include inhalation of obnoxious dust or gases, contact with toxic substances within the soil itself or through food grown in the soil and the pollution of water courses. When contaminated land is identified, the local authority is then required to take action to ensure that the land is made safe, i.e. ensure that the original polluter, the landowner or other responsible person (including residential property owners and, where appropriate, developers) undertakes the necessary action to clean up a site and restore it to a state in which it is 'suitable for use'.

Each local authority is also required under the EPA to review all planning applications in terms of any need for investigation of potential pollutants as well as any necessary remediation. It should be noted that a site previously having industrial use may not be identified as contaminated in terms of the legislation until an application for a change of use to residential is made.

Investigation of a potentially contaminated site involves two stages:

Stage 1 *An initial desk top study is undertaken. This is based on old maps and other available sources with a focus on seeking historical evidence of potential contamination and includes information on the geology, hydrogeology and potential contamination of the site. A walk-over survey is also undertaken in order to identify any continuing contamination processes taking place.*

Stage 2 *The undertaking of intrusive procedures on site. These include drilling and sampling as well as the chemical analysis of soil, water and other liquids. This will enable an evaluation of the range and type(s) of pollutants and contaminants affecting the site as well as their concentration level(s). Finally, decisions based on the evidence accumulated will determine the level of clean-up and remediation required to achieve a state in which the site is 'suitable for use'.*

Remedial treatment

If land is identified as contaminated, it will require remedial treatment before any further development can take place. A Remediation Notice may be issued under the EPA to ensure that, where the land is in existing use, e.g. a previously developed residential site, it is properly treated. Where the land is in current use, it may be necessary to prevent access to any part that is contaminated, e.g. the garden of a house, until such time as it has been remediated. At its worst, the contamination of the land may require the evacuation of the building(s) on it as well until remediation has been completed.

Remedial treatment is very site-specific and depends on the type and extent of any contamination identified as well as the physical profile and the location of the site. It can range from full removal of the contaminated soil (i.e. 'dig and dump' where the soil is removed to a properly licensed landfill site) to treatment of the contaminated soil and water in situ. In the latter case, such treatment may remove or destroy the contaminant(s) or, alternatively, reduce their amount or their potency to a safe level for current use (where required) or for future development.

Remediation treatments for the clean-up of contaminated soils include:

- **Bio-pile remediation.** *Where naturally occurring micro-organisms are used to clean up hydrocarbons, i.e. oil-based contaminants, by means of aerobic microbial activity. This process may take up to 6 months to achieve.*
- **Chemical oxidisation.** *This uses hydrogen peroxide to oxidise organic contaminants eventually leading to their complete breakdown/removal. There can be resultant gases given off and these are captured mechanically at ground level and removed.*
- **Phyto-extraction.** *Specially selected plants are used to remove dangerous elements such as heavy metals from soil or water. In some cases, this process can be effective in a matter of hours.*
- **Soil vapour extraction.** *Volatile light hydrocarbons such as petrol and solvents can be removed using a technique that involves drilling vertical wells or horizontal lines. A vacuum is then induced in order to extract polluted air and vapour for surface treatment.*

Where only part of a site is contaminated, remediation may not be considered appropriate, especially if it is of considerable area or depth. In this circumstance, it may be decided that a permanent barrier can be installed to isolate and contain the contaminated soil. The barrier can be constructed with steel piles, modified clay, slurry or non-permeable plastic sheeting. Treatment gates may also be installed in order to allow controlled groundwater movement.

Flooding

The extent of the problem

Flood risk has become an increasingly serious problem in the UK as a result of climate change over the last 20 years or so. The scale of damage to property and contents is considerable. For example, in 2007 alone, when summer flooding affected large parts of England, there were some 165,000 separate insurance claims due to flooding with a cost to insurers of over £3 billion. The average claim was £40,000 and total losses including uninsured risks amounted to about £4 billion. Insurance experts state that a modern semi-detached house affected by 500mm of floodwater will require between £15,000 and £30,000 for repairs, plus some £9,000 for replacing contents.

Geographically, the western parts of the British Isles are the wettest as the prevailing weather is from the Atlantic, some of the higher ground being among the wettest areas in Europe; on the other hand, parts of eastern England are among the driest in Europe. Average annual rainfall in England is 838mm, although parts of Cumbria receive over 1600mm. Average annual rainfall in Wales is 1433mm, and the wettest place in the UK is in Snowdonia with an average rainfall of 4473mm in recent years.

The most extreme examples of flooding in the UK in the twentieth century, before the effects of climate change were felt, were the 1952 Lynmouth flood (34 killed, 100+ buildings destroyed, 400+ people homeless) and the 1953 North Sea flood (in which 307 people were killed and 24,000 buildings were destroyed/damaged). Recent climate change has resulted in short periods of extreme rainfall that far exceed the norm or previous experience. For example, in the twenty-first century, Borrowdale in Cumbria received 314mm of rain in a single 24-hour period in November 2009, leading to floods 2.4m deep in Cockermouth; in June 2012, parts of Wales received 81mm of rain over 24 hours.

A property does not need to be near a river or on the coast to be at risk. Most buildings are at a higher risk of pluvial (i.e. surface water) flooding rather than fluvial flooding (i.e. river or coastal flooding):

- **Pluvial flooding** *occurs when: (1) a short period of intense rainfall overwhelms the existing drainage systems; (2) a blockage occurs; or (3) poor maintenance creates a problem. This is commonly called a flash flood and is more likely to occur than fluvial flooding.*
- **Fluvial flooding** *is as a result of: (1) heavy rainfall causing a river or stream to overflow and burst its bank; or (2) a high tide or severe weather overwhelming sea defences. Often, a high tide and severe weather coincide to create a storm or tidal surge. Rising sea levels may increase the likelihood of coastal flooding in the future.*

Over the last weekend of November 2012, large areas of England and Wales were severely affected by flooding due to heavy and prolonged rain falling on previously saturated ground as the result of the wettest summer since records began. The flooding caused damage to many hundreds of houses. In the photograph, flood water coming off higher ground is threatening a cottage on the Blenheim Estate, Woodstock, Oxfordshire.

Since the 1980s, the number of properties that are potentially at risk from flooding has increased considerably because the relaxation of planning controls at that time has allowed the subsequent development of flood plains, particularly for residential buildings. More than 200,000 properties were built on flood plains in the period 2001–2011, of which some 40,000 properties have been constructed without appropriate community flood defences, a significant percentage being residential buildings.

The reasons for flooding are wide-ranging. Left: These houses are well above the stream which drains into a culvert. However, just after the houses were built, about 10 years or so ago, the culvert became blocked as a result of road widening and the houses were flooded. Right: The culvert blockage was cleared and an audible warning device was installed on a near-by lamp-post in case of any future problems. Flooding has not occurred again, but the houses are now identified as 'at risk'.

Sudden, intense periods of rainfall as a result of climate change which are hard to protect against is another reason for flooding. The hard-surfacing of gardens for car parking and other purposes has also contributed significantly to the possibility of urban surface water drainage problems. Over the last 10 years, the area covered by hard surfaces has increased from 28 per cent to 48 per cent of all garden space. Such areas frequently contribute to so-called flash flooding.

Past under-investment in flood defence works by government and other agencies means that some 500,000 homes and businesses are currently at significant risk from flooding. There is, at present, a funding gap of almost £1 billion per annum between actual spend and necessary financing of flood defence in the UK as a whole. Although the government currently spends more than £400 million per year on flood defence, some 610,000 properties will be at significant risk from flooding by 2035 if no further increase to investment is undertaken.

The Environment Agency, via its web-site, provides information on flooding including flood maps that identify areas at increased risk of flooding from rivers and streams. Each flood map offers general guidance on: main rivers; floodplains; river and sea flood defences; the 'significant', 'moderate' or 'low' likelihood of flooding; flood warnings. The Agency will also give a flood risk factor for an individual property based on its postcode, i.e. 'low', 'moderate' or 'significant' risk and it provides a 24/7 on-line flood warning service for England and Wales. There are also a number of specialist search organisations which, for a small fee, will provide a more detailed property-specific report.

An example of a flood warning area map, in this case for an area of low-lying land in the Severn Valley just north of Gloucester (map courtesy of the Environmental Agency).

The effects of flooding

With all three types of flooding (pluvial, river and coastal), the resultant floodwater is normally a mixture of storm water and foul water, as both types of drainage system are overwhelmed. This mixture results in physical damage to the fabric of a building as well as an obnoxious residue of sewage and other organic and inorganic matters affecting buildings and ground surfaces.

Once the flood water has retreated, and only then because of the health challenges, an inspection of the property can be undertaken There is normally a long period of drying out of a building; where the building is under water for any length of time, the structural and other elements may take many weeks, sometimes months, to dry out fully. For example, the normal drying-out allowance for a solid wall is 25mm per month, but it can take a lot longer. Minor flood damage may take only a few weeks to dry out, but older and thicker walls may take many months.

Once access is available, drying-out work commences. This includes the drainage of any under-floor voids and wall cavities followed by the removal of any accumulated debris and silt (much of which may present a health hazard, so disinfection may also be required). The next step is to check and test services followed by the stripping back of all affected surface finishes to the walls and floors of the building, e.g. water-affected fittings and joinery as well as damaged plaster and impervious floor coverings. All potentially affected structural timberwork needs to be exposed. The work needs to expose the structural elements of the building and enable them to go through a drying-out process. This is often undertaken by specialist drying companies.

The use of a structured drying-out process is essential and dehumidifiers and, in some cases, heaters will normally be a key element of the process. Care is needed throughout the drying-out period to monitor levels of moisture, especially if timberwork has been left in place, as the potential for an outbreak of dry rot is high (see Chapter 15). Remedial building work can only commence when the structural elements are at an acceptable moisture level, i.e. below 14–16% moisture. The specialist should provide a drying certificate that confirms when this is achieved.

Further problems that may occur include moisture-saturated materials may be damaged by frost during cold weather, rust may build up in hidden parts of the service equipment and, occasionally, foundations on sandy sub-soils may be undermined or clay sub-soils may 'heave' (see Chapter 3).

Left: The River Avon in Malmesbury, Wiltshire, has a history of flooding. It flooded again in the winter of 2012–2013. Right: This cottage is next to the bridge, where the water is still running high. The effect of the most recent flood can be seen in the tide mark on the cottage wall. The flood defences include a door guard and sand bags.

Defensive measures

It is possible to provide some defence against flooding. However, if undertaken, flood resistance measures need to defend the whole perimeter of the property if they are to be effective. Such measures include: door and window flood guards (these can hold back floodwater for several hours); air brick guards; toilet bungs and/or one way valves to drain pipes; water-resistant treatments to external walls (and to internal walls if floodwater penetration is likely); pump-and-sump systems to remove water beneath ground floors faster than it can enter any void.

Particularly vulnerable properties, such as timber frame buildings or historic buildings, can have a demountable flood skirt installed. It should be noted that such a measure is expensive. Undertaking partial measures will not be wholly effective as they are unlikely to fully exclude floodwater. The minimum depth of defensive measures should be no less than 500mm, although any measure used should be no higher than a metre. This is because research has indicated that floodwater above this height will have sufficient pressure to break through brickwork which may lead to collapse or the need for structural repair.

Recent guidance from the Department of Communities and Local Government based on a project co-ordinated by the Construction Industry Research and Information Association (CIRIA) lists four basic building design approaches for constructing a building where there is a risk of flooding:

- **Flood avoidance.** *Construction that avoids flooding, e.g. raising a building above any flood level or positioning it to avoid a floodplain.*
- **Flood resistance.** *Construction that prevents floodwater entering a building.*
- **Flood resilience.** *Construction that avoids permanent damage when floodwater enters a building, structural integrity is maintained, drying and cleaning are facilitated.*
- **Flood repairable.** *A form of flood resilience where any damage can be easily repaired or replaced.*

Japanese knotweed

A problem plant

Knotweed comprises a number of different plants of which the most widely spread and problematical in the UK is Japanese knotweed (Fallopia japonica). This large, non-indigenous herbaceous plant was introduced into the UK as an ornamental species in the late nineteenth century. It is a hardy bamboo-like plant that grows in large clumps, sometimes reaching a height of 4 metres, and having white flowers in autumn. It dies back to ground level in the autumn, but from March grows vigorously, quickly achieving a height of 3 metres in 10 weeks or so. It spreads via its underground rhizomes or roots, which can be as long as 20m in length.

Unfortunately, much damage to (mainly) external buildings and infrastructure, but also to the occasional house, can be caused by the sheer vigour and extent of the above-ground and below-ground growth (it can grow up to 100mm a day between April and October).

Apart from Japanese knotweed, there are five other types of non-native knotweed in the UK, all of them invasive plants. They are:

- **Giant knotweed** – *grows up to 5 metres high. Flowers are creamy-white.*
- **Bohemica** – *a hybrid-cross between Japanese and giant knotweed; has creamy-white flowers.*
- **Himalayan knotweed** – *up to 2 metres high, but not common and has white or pink flowers.*
- **Dwarf or lesser knotweed** – *up to 900mm high, but not common. Flowers are pink or white.*

- **Conollyana** – *very rare in the UK, but increasing. Discovered in 1982 in Leicester, it is a cross between Japanese knotweed and Russian Vine. Also known as the railway yard knotweed.*

Left: Japanese knotweed growing right up to the external wall of a house; its bamboo-like appearance and large leaves can be seen (photo courtesy of Samantha Organ). Right: An example of dwarf or lesser knotweed in flower. This variety grows to about a metre in height.

While all non-native knotweeds are considered to be invasive and therefore problem plants, Japanese knotweed, because of its extensive national presence and its aggressiveness, is considered to be the most serious problem plant. As early as the start of the twentieth century, Japanese knotweed began to be recognised as a problem plant. However, because of its ability to grow quickly (up to 40mm per day) and provide good landscape screening, it was (unfortunately) used extensively until the 1930s. At this time, it was fully recognised as a problem plant and it is now one of the World Conservation Union's top 100 invasive plants. In 1900, it could be found at only a few sites in England and Wales, now it is heavily spread throughout mainland Britain.

In 2004, a government Review of Non-native Species Policy gave an estimated cost of £1.56 billion to control knotweed nationally. One of the worst affected areas is Swansea in South Wales where it has recently been estimated that it would cost about £9.5 million to treat the current knotweed problem which is spreading at over 2 hectares per year. Most treatment and removal are normally undertaken as part of the process of building development or re-development, recent costs being as high as £60/m².

Problems caused by Japanese knotweed

The damage that it can cause includes:
- *Growth through paths, roads, patios and surface concrete features including floors where, in particular, knotweed can exploit joints in brick and concrete paving as well as movement joints in concrete and other surfaces. Damage can be so extensive as to result in total replacement of the affected surface after eradication of the knotweed.*
- *The growing roots can undermine garden walls and cause retaining walls to collapse.*
- *Lightweight buildings and conservatories can suffer extensive damage, sometimes collapse, especially if poorly constructed.*
- *Root entry into drains via poor joints and cracks in pipes leads to blockages. The denseness and*

size of the underground growth can distort pipes. It is not uncommon for drainage systems to have to be replaced.

• *Disturbance of other underground services.*

It will be noted from the above list that while Japanese knotweed is an aggressive plant, it tends to damage lightweight buildings, external features and services rather than buildings of a heavier structure, though there are occasional cases of damage to housing. However, once identified, apart from any actual damage caused already or potential damage likely to occur in the future, its presence can adversely affect the value and mortgageability of a residential property. Unfortunately, its propensity for re-growth is such that it can remain dormant in the ground for up to 20 years if ineffectively treated. However, unless disturbed, its root radius is normally no greater than 7 metres and any land or building outside of that zone is unlikely to be affected by the plant.

Eradication of Japanese knotweed

Treatment involves one or more of several methods, as set out below. The key feature of all of the methods is that absolutely no trace of the crown, stem or rhizome (root) can be left as it will re-grow from the smallest scrap:

• *Excavation and removal of the plant, its roots and the soil surrounding it to a depth of 3 metres and a distance of 7 metres horizontally from the clump as that is the extent of its potential spread. It should be noted that, under the Environmental Protection Act 1990, Japanese knotweed-contaminated soil is designated a controlled waste and only properly licensed organisations can remove it. It can only be taken to an appropriately licensed landfill site. Any breach of the Act is a criminal offence.*

• *Excavation of the plant and its roots followed by burial on site at a depth of at least 5 metres plus encapsulation by a specialist root barrier.*

• *Encapsulation by a root barrier where it cannot be buried at depth.*

• *The installation of a vertical root barrier along a boundary line to prevent growth onto adjoining land.*

• *Chemical control involving the application of specialist insecticides and herbicides. This will require applications over several growing seasons to be successful. Once treated, however, Japanese knotweed becomes a hazardous waste because of the treatment chemical(s) used and is subject to statutory control under the Hazardous Waste Regulations 2005.*

• *Biological control is now under trial in the UK, but is unlikely to be approved for general use in the foreseeable future. It involves the use of natural enemies of the plant from its native homeland.*

Treated soil from a Japanese knotweed infested site can be re-used for landscaping, but only on the cleaned site itself. It must not be taken off-site.

In overall terms, Japanese knotweed is a plant with extremely vigorous growth that is very difficult to eradicate. Only rarely does it cause structural damage to buildings, but it can be deleterious towards external paved and tarmacadam surfaces and drainage systems. Unfortunately, its presence can cause property blight, with funding bodies such as mortgage lenders becoming very concerned about its presence. The Environment Agency publishes detailed guidance for property owners and developers on its control and disposal in its Code 'Managing Japanese Knotweed on Development Sites: The Knotweed Code of Practice'.

Mundic

Mundic is a problem associated with the south-west of England, i.e. Devon and Cornwall. During the first half of the twentieth century, it was common practice to manufacture concrete blocks from local materials. Similarly, local aggregates were used for concrete foundations. In Cornwall and Devon, the coarse aggregate was frequently mine waste and the fine aggregate was beach sand, china clay waste or mine-processing residue.

Unfortunately, much of the mine waste contained sulfide minerals. These can oxidise in damp, atmospheric conditions and produce sulfuric acid which attacks the cement in concrete, leading to weakness and expansion. Known as 'mundic decay', it results in any affected concrete cracking and, ultimately, in its disintegration. The result is that many properties in the south-west (more specifically all of Cornwall and an area within 15km of Tavistock in Devon) built with concrete blockwork and/or concrete foundations between 1900 and the 1960s may be affected.

The problem was first properly recognised in the 1980s and such properties were blighted for a short period in the 1990s because the Council of Mortgage Lenders banned lending on suspected buildings due to a lack of any suitable test. However, in 1994, the Building Research Establishment (BRE) and the Royal Institution of Chartered Surveyors (RICS) introduced a screening test based on taking a small number of 50mm diameter core samples of the suspected concrete/concrete blocks and subjecting them to petrographic analysis in a laboratory. Subsequently, the screening test has been refined further and results in a number of classifications ranging from A (sound; property is mortgageable) to C (clearly unsound; property is unmortgageable).

Left: This house in Cornwall was suspected of having a mundic problem. Middle: Investigation revealed this mundic-affected concrete block. (Left and middle photos courtesy of Gregg Kerry.) Right: Mundic-affected concrete is so 'rotten' that it can be crumbled by hand.

Sometimes, there is visual evidence that the concrete/concrete block is suffering from mundic deterioration, e.g. crumbling concrete, cracking of surface render. However, for many properties this is not the case (see photograph above) and any suspected property will need to be tested. Where identified, affected components must be replaced. Unfortunately, in some cases, severely affected properties have to be demolished.

Further advice on mundic investigation is given in RICS Guidance Note: The Mundic Problem (1997), Second Edition.

Older buildings: the challenges of preventing fire spread and ensuring adequate sound insulation

Introduction: structural implications

The prevention of fire spread and noise transmittance in domestic buildings are complex topics; far too complex to be tackled in any detail in a book of this nature. This section summarises some of the most common problems relating to partitions, separating walls and floors in older buildings. Where there is any doubt about the provision of fire protection and acceptable sound insulation, appropriate specialist advice should be sought. This is particularly important in multi-storey housing and for HMOs (i.e. houses in multiple occupation such as flats and hostels).

Many large Georgian, Victorian and Edwardian buildings have been converted into flats/apartments. In many cases, these conversions will have been carried out in the past when Building Regulation requirements were less strict than today.

In structural terms, it is the higher floors that are normally most at risk. In many of these early properties, joist size varied from floor to floor. The top floor, for example, was possibly built as servant accommodation and would contain joists of modest size to reflect the light loading. Change of use can result in these floors being overloaded; this is a particular problem where layout changes have required the installation of additional partitions. The situation is compounded by the need for adequate fire protection and sound insulation, both of which increase the load on the floor. Victorian and earlier buildings often had floors formed with joists of various widths and varying centres – sometimes the centres were wider than current practice requires. Such floors are easily overloaded unless extra joists have been added.

Protection against the spread of fire

Walls and partitions

Lack of adequate fire protection is a common potential hazard, particularly in older properties and those which have been converted from a previous use in the past. Nowadays, all conversion work is subject to scrutiny by local authorities, but it should be remembered that not all alteration works are 'approved', and that older conversions may have been subject to less stringent scrutiny.

Walls separating dwellings and walls lining routes of escape are particularly at risk. Although block and brick partitions provide good fire protection, timber studding depends on (1) the size of its framing members; and (2) the thickness and fixing of its plasterboard covering for its resistance to fire. Older stud partitions with lath and plaster facings may need a further plasterboard layer applied in order to meet an acceptable level of fire protection. Some timber partitions, especially in older conversions, may be constructed with a facing of hardboard, ply or fibre board and these materials will not provide acceptable levels of fire protection.

Timber studding covered with lath and plaster offers very little resistance to the spread of fire. Similarly, ceilings finished with lath and plaster will readily allow the spread of fire even if any timber stud partition below is correctly constructed. Fire can also spread from one dwelling, or room, to the next through the floor void immediately below the partition.

Up until the 1970s, buildings that required increased fire resistance were commonly lined out with asbestos sheeting or asbestos-based products. While effective in terms of fire protection, asbestos is now recognised as dangerous to health if in an unstable state, when drilled or sanded or when removed (see earlier section for a more detailed discussion of asbestos and its problems).

Walls separating semi-detached houses or terraced houses (i.e. party walls) should always be built up to the underside of the roof covering. In older housing, this is not always the case. Depending on when the houses were first constructed, there may have been less onerous requirements than now regarding fire-stopping to prevent fire from spreading behind the fascia board or over the top of the separating wall.

The Building Regulations specify minimum periods of fire resistance for separating walls between compartments and for any wall adjacent to a means of escape. These periods range from 30 to 120 minutes depending on the use of the building. In addition, where a partition supports a floor with a specified level of fire protection, the supporting wall should provide the same level of protection. The Building Regulations, and a number of companies (such as British

Gypsum for the use of plaster and plasterboards), provide more detailed information on specific requirements for periods of fire resistance together with appropriate construction guidance.

In timber frame housing, fire is obviously a major risk. (More detailed guidance on fire-related defects in this type of construction can be found in Chapter 17.)

Floors

In a modern two-storey house, the use of plasterboard ceilings and tongued and grooved floor or chipboarding both contribute to the floor's resistance to the spread of smoke and fire. In many older properties, lath and plaster ceilings together with square-edged floor boarding are more likely to be found. Square-edged boards are not effective in limiting the spread of smoke (the main risk to life) and lath and plaster ceilings may fail quickly if a fire starts. In some older conversions, the square-edged floor boards were overlaid with hardboard which acts as a smoke stop.

Earlier construction does not necessarily require renewal, although it should be recognised that the performance of a pre-Second World War floor may be less than satisfactory if a fire breaks out. The risks are more significant where properties have been converted into flats. In these cases the performance requirements of the floor are much more onerous; the higher the building, the greater the level of performance required. The purpose of the Building Regulations is not to make the floor non-combustible, but to provide a reasonably 'safe time' to ensure that occupants can escape. In converted buildings, there may be some 'trade-off' in standards of fire protection depending on the level of fire alarm equipment and emergency lighting.

150 × 50mm joists at 400mm centres can safely span about 3.30 metres

150 × 50mm joists @ 400mm centres

In good construction joists will be 'cross bonded' to provide walls with adequate restraint.

225 × 50mm joists @ 400mm centres

With the added weight of fire protection their safe span drops to about 2.5 metres

Floating floor or smoke resistant layer

Pugging or quilt

Additional layer of plasterboard

If partitions are added to the floor the safe span is reduced even further.

To prevent smoke from reaching the room above tongued and grooved floor coverings are necessary.
An alternative approach is to cover square edged boards with a material such as 6mm plywood.

Additional layer(s) of plasterboard can be fixed to the underside of the joists to improve fire protection.

Airborne noise

Impact noise

Rubber strip

Resilient layer bonded to chipboard

Airborne sound resistance can be improved by adding quilts and/or plasterboard to the void. Impact sound is improved by adding resiliebt layer over floor.

NB: fire and sound protection measures will increase the load on the floor.

Noise prevention: inadequacy of sound insulation

In houses, the occupants should have some degree of control over levels of internal noise. In flats, the situation is very different as individual units of accommodation are under separate occupation and use. In design terms, the challenge is to provide (1) resistance to airborne sound; and (2) resistance to impact sound in compartment walls and floors. Airborne sound can be very stressful and is a particular problem in properties which have been converted from a previous use.

Walls and partitions

Walls separating dwellings can allow unacceptable levels of sound transmission if they are of inadequate thickness and mass, if the bed and perpend joints are not filled, and/or if joists or services penetrate the separating wall. Separating walls should also be built up to the underside of the roof covering.

In converted properties, problems of sound insulation between separate units of accommodation are a common problem particularly in older conversions where the Building Regulation requirements for noise reduction were much lower than current requirements. Problems also often occur where living accommodation is next to common circulation space, e.g. communal landings, corridors and stairwells. Stud partitions, which are commonly used in this type of conversion, often provide inadequate sound insulation. They may be too thin so that they readily flex and reverberate; they may have inadequate thickness of plasterboard either side; or they may not include insulation mats or quilts within their construction. Also, lightweight partitions are inherently less effective than dense, solid partitions with regard to reducing airborne sound. Walls, unlike floors, are not required to provide resistance to impact sound.

Even within houses and flats, partitions may not provide adequate sound insulation between separate rooms. Although most forms of block partition perform adequately, stud partitions can often be found to perform poorly. Occasionally, early stud partitions may be found with hardboard as opposed to plasterboard linings. In stud partitions and cellular core partitions, the addition of extra power sockets that are recessed rather than surface-fitted necessitates cutting part of the plasterboard away and this forms a path for the transfer of sound.

Floors

It is only in recent years that converted properties have had to conform to the Building Regulations in terms of sound insulation. Buildings altered before this date (and even after), however, may still fall within the requirements of the Environmental Protection Act 1990. Many local authorities, housing associations and private landlords are struggling to cope with the financial implications of this Act which can require them to provide adequate sound insulation even if the property conformed to the relevant legal requirements at the time of conversion.

As mentioned earlier, there are two discrete problems to overcome: providing resistance to airborne sound and providing resistance to impact sound.

Meeting these requirements can dramatically increase the load on the floor. In many cases, existing joists will not cope with the increased load and this may require additional joists, new joists of deeper section or intermediate support to convert the floor structure to a double floor. In the latter option, fire protection of the supporting beam will need to be considered.

Nowadays, sound insulation systems depend, to a large extent, on plasterboard. Prior to this, it was common practice to use sand pugging spread between the joists. This was a heavy material, even when dry – if it becomes wet, through, say, a plumbing leak, the ceiling may collapse. Unfortunately, however good the level of sound insulation in a floor, the transfer of sound may still be a problem and this is usually caused by flanking transmission, i.e. the transfer of sound from a room to another room that is either above or below it is achieved via the walls and partitions rather than the floor between.

Services penetrating separating floors

Where drainage or water services penetrate a separating floor in flats, there are strict rules to ensure adequate sound insulation and fire protection. These rules are quite complex and are, again, beyond the scope of this book. However, there are some general defects, which, if found may require more specialist investigation. These include airborne sound transfer between flats where the pipes penetrate the floor, annoying sound from waste water in soil vent pipes and lack of adequate fire protection.

In modern construction, the communal pipework on either side of a compartment floor should be protected by some form of ducting or casing which will provide a minimum of one hour's fire resistance. Access panels fitted in the ducting or casing should not be located in bedrooms or circulation space. The compartment floor should be sealed or fire stopped where the pipe penetrates it. No services other than water pipework or drainage should share this duct. At the same time, to reduce the risk of sound transfer, the pipe should be wrapped in mineral wool (or similar) and the casing boarding should have a minimum density as set out in the Building Regulations.

Radon

Radon is a naturally-occurring colourless gas that is emitted by some rocks and soils and is breathed in by humans throughout their life span. It is radioactive and, on average, accounts for half of a human's total annual radiation dose. Radon is the second most important cause of lung cancer after smoking and the lung cancer risk it presents is proportional to long-term radon exposure. Outdoors, radon presents no danger to humans, but its presence within a building does present a significant health hazard and there are more than 1,000 lung cancer deaths in the UK from radon each year. Unfortunately, radon tends to seep into buildings from the sub-soil.

When first identified as a health hazard, it was thought that the south-west of England was the region most affected as the gas is commonly released from the granite rock strata which are found in Devon and Cornwall. However, research over the past 20 years has discovered high radon levels in buildings located over a number of other types of geology including limestone and some granites that are found elsewhere in the UK. Apart from the south-west of England, large parts of Wales and the Midlands as well as some areas in the north of England have high levels of radon.

It has, therefore, become apparent over a number of years that varying levels of radon can be found across many areas throughout the UK. Maps have been produced to show where high levels of radon are more likely. In these areas, more homes and work places are likely to have a high level of radon that presents a significant hazard to health, resulting in the need for specific protection measures to be undertaken, either during the initial construction of a building, or subsequently if not already provided.

The colours show the maximum percentage band within each 1-km grid square of the national grid. The best estimate for an individual property in a coloured square can be obtained for a small charge from www.ukradon.org.

Percentage of homes at or above the Action Level	Settlements	Transport Network	National Grid
0 - 1	• Gloucester	──── Motorways	100-km 10-km
1 - 3	• Cirencester	──── A Roads	**Other Features**
3 - 5	Ross-on-Wye	──── B Roads	LOCAL ADMINISTRATIVE DISTRICT
5 - 10	Cleverton	┼─┼─┼ Railway Lines	Water features
10 - 30			
Greater than 30			

The Health Protection Agency (now part of Public Health England) publishes an 'Indicative Atlas of Radon in England & Wales'. In the overall map of England and Wales above, the darker the colour, the higher the background level of radon. The atlas also includes much more detailed maps (see example) showing levels of radon for the various areas within both countries. (Maps courtesy of Public Health England: contain public sector information licensed under the Open Government Licence v1.0)

In the UK, much of the research and advice on radon and its hazards has been undertaken by the Centre for Radiation, Chemical and Environmental Hazards (CRCE) which has been advising, researching and providing radon services for over 30 years. CRCE (part of the Health Protection Agency (HPA) until 1 April 2013 and now within Public Health England) publishes guidance on radon including:

- *detailed national and area maps (developed with the British Geological Survey) setting out which locations within the UK are at risk;*
- *the varying levels of risk of each location;*
- *advice on how radon gas can be detected and measured;*
- *guidance on the steps to be taken to safeguard a building and its occupants against radon gas;*
- *a website dedicated to radon (www.ukradon.org).*

The concentration, or activity, of radon in the atmosphere is measured in becquerels (Bq) in each cubic metre of air. The becquerel is the unit of radioactive intensity. Most homes in the UK have a fairly low radon level with an average of 20 becquerels per cubic metre (Bq m^{-3}). The PHE recommendation is that, where the indoor radon level is at or above an Action Level (AL) of 200 Bq m^{-3}, it should be reduced.

Where an existing house or workplace needs to be assessed, this can be undertaken by means of passive monitors. The period of monitoring is recommended to be three months in order to average out short-term variations in radon levels. Once the radon level has been determined, a written report is issued to the householder who, where appropriate, can then decide on the remedial works to be undertaken to reduce radon to a safe level. New buildings in a known area of high radon level are designed and constructed with appropriate safety measures incorporated.

A more detailed discussion of the construction measures needed to safeguard the occupants of a building against the effects of radon is included in Chapter 7.

Water and heating 19

Cold water supply

Incoming mains

In modern houses, the incoming water main is made from plastic; in older properties it is more likely to be lead or copper. Lead pipes, like copper, are easily deformed, thus restricting the flow of water in them. Some soils can cause corrosion in metal pipes and they should, therefore, be protected with a plastic coating or some form of bituminous paint. Modern incoming mains should be at least 750mm below ground level to avoid the effect of frost, but in older properties the main is often very close to the surface. If the rising main within the property is situated on the inner face of an external wall, there is the consequential risk of freezing. Where the main enters the property, there should be a stop tap to isolate the distribution pipework. In older properties, these are often missing or they may have corroded to such an extent that they are no longer operational. Temporarily freezing the pipe (which is not as difficult as it sounds) may be the only option where internal maintenance or isolation is required.

Direct cold water systems

These systems are the most cost effective, providing there is a reliable and continuous water supply and the building has no need of a stored water supply. As water enters the property directly from the mains, it is at a high pressure and the fittings throughout the system must be able to cope with this high pressure. This applies especially to the ball valves feeding cisterns

Direct supply is often used where mains pressure is constant. No water is stored, so there is no need for a tank in the roof space.

However, where hot water is supplied through a cylinder rather than a multipoint, the cylinder will require an indirect feed and a storage tank. Alternatively, a special pressurised cylinder can be used. In practice, many houses like this use a combination boiler to supply the heating and the hot water; this does away with the need to store hot water.

and tanks for heating, hot water and toilet facilities. Water hammer is a problem experienced with variations in water pressure and this can be caused by incorrectly specified valves and fittings. Worn fittings can also start vibrating and this noise may be amplified in other parts of the building.

Where pipework changes direction using bends, there is a tendency for the high water pressure to exert forces on the pipework in its efforts to straighten the run in the same way it will in a garden hose. Fixing brackets must be provided at appropriate distances apart to eliminate this tendency and to reduce the forces acting upon the bends. (Note: With the water company's mains, the forces can be so high as to warrant special concrete securing blocks known as 'thrust blocks' to prevent the water pressure lifting the water mains out of the ground.)

In a direct supply, double check valves are required where there is the risk of contaminated water being sucked back into the mains. A typical example is where one end of a hose is connected to the mains and the other end is inadvertently dropped into a drainage gully. Any sudden drop in the mains pressure will suck the drain contents into the main, resulting in contamination.

Indirect cold water

This system is necessary where the mains water supply is unreliable for any reason. The internal supply becomes indirect when it is fed from a storage facility within the building. It is often necessary in a very densely populated area where, if a direct system were provided, the many people calling for water at any one time would warrant the size of larger, uneconomical main.

Drinking water should never be consumed from a tank supply unless the tank has been manufactured specifically for the purpose. Most houses with an indirect supply have a direct supply to a downstairs tap (in the past it was not uncommon to see a tap specifically labelled 'drinking water'). Modern water regulations require all cold tap water to be potable.

INDIRECT

Overflow

Expansion pipe

Hot supply

Indirect supply to bathroom

Cold – direct to kitchen; for 'clean' drinking supply

Stored water, i.e. indirect supply, allows for intermittent supply. A storage tank in the roof supplies all taps except the kitchen and also feeds the hot water cylinder. The cylinder is vented to allow for expansion of heated water and in case of possible boiling due to thermostat failure.

The pressure in the hot water circuit is determined by the height between the taps and the cold water storage tank. The distribution pipes from the tank are normally of slightly larger diameter than the rising main due to the decrease in pressure.

Problems common to cisterns, tanks and pipework, etc.

Lead pipework was a very common material until the 1950s, but it has been found to be detrimental to health as lead may be absorbed into the water, a particular problem in soft water areas. The skills involved in making lead joints are also no longer readily available.

Forming bends in copper pipes inevitably stretches the copper so fittings of capillary and compression types are often preferred. The thickness of copper pipe has reduced in recent years and bending copper pipes can result in failure of the pipework.

Plastic pipework is now becoming more common, but the support distances need to be matched to the particular material specification. Some water has been found to absorb the taste of plastic from certain grades (usually earlier ones). The expansion rates of plastics are very high and the surrounding temperature will also affect the composition of the pipes. Earlier pipes could become very brittle, developing hairline cracks if subjected to shock loads when very cold, and becoming quite soft and malleable when hot. Some grades can even acquire a sticky surface under adverse conditions and certain colours are affected by ultra-violet light.

The materials used in the manufacture of cisterns and tanks have changed with time, and, while old relics can still be found in use, made from timber, slate, steel, ceramics and copper, the most widely used materials now are either plastic or fibreglass. While the older materials were quite rigid and could safely be mounted on beams or joists, it is now essential that plastic is uniformly supported on suitable load-bearing boarding, capable of withstanding possible water penetration.

All tanks and cisterns should have tight-fitting lids which prevent dust, dirt, bird droppings and even rodent and bird carcasses from polluting the water.

The correct specification of ball valve is also essential to prevent back siphonage, noise and water hammer, and should be matched to the required inlet water pressure.

Well-supported overflow pipes should run to their point of discharge without sagging, which could become a problem in freezing conditions. These overflows are essential as a safety precaution should valve washers, seals or floats cease to function properly and water overfills the storage vessel. The point of discharge should be in a very prominent position; this is so that any overflow will be noticed quickly and steps taken to avoid water wastage.

Any leaking overflows should be fixed promptly to prevent waste, unsightly staining and to resolve any float and valve problems before they become serious. Lead overflows, unfortunately, have a high scrap value. Storage cistern (right) should be insulated.

If cover is missing, water can become contaminated.

Screened air inlet

Ball valve and float

If valves are missing or worn, it will be difficult to isolate the tank for maintenance.

Copper floats are prone to perforation.

If pipes are not lagged, they may freeze.

25mm

If overflow is undersized, it will not discharge safely. If it sags, ponding water can freeze in cold weather.

Warning or overflow pipe with insect screen, bottom of pipe 25mm clear of water level.

If pipe is not turned down into the water, cold air can enter the cistern.

If tank is not insulated, the water may freeze in cold weather.

There should be no insulation below the tank.

Water temperature should not rise above 20 degrees Celsius to limit the growth of bacteria.

If the distribution pipe is taken from the side of the cistern, there is more chance of sediment building up in the bottom of the tank. It is usually harmless but, if it settles, can provide nutrients which encourage the growth of bacteria and other organisms.

Nowadays, a wide range of outlet taps is available – some are more complex than others. Generally, tap washers will need more attention on high pressure supplies; even the ceramic types have a limited life expectancy. Dripping taps and ball valves will ultimately suffer internal damage if worn washers are not replaced at the first sign of seepage. It is much cheaper to replace the washer than the whole tap or valve. It is also cost-effective to have isolation valves provided to cover sections of the installation and avoid total system drain down with its resulting water wastage or excessive time involved when washer replacement is necessary.

Many modern taps (usually shower mixers and monobloc taps) will not work properly if the water pressure is low – they may not be suitable, therefore, for indirect water supplies. Both the hot and cold water pressures must normally be above 1.0 bar if they are to supply a high pressure tap. Pressure can be calculated by measuring the vertical distance in metres between the bottom of the cold water storage tank and the tap (or shower) and dividing the answer by 10 – thus 1 metre of Head = 0.1 bar.

Hot water supply

Direct hot water systems

The main problems which occur in direct hot water systems are water discoloration, possible scale formation, poor temperature control and noise. Older systems are usually heated with coal- or gas-fired back boilers; more modern assemblies are single/multi-point gas and electrical water heaters of both storage and non-storage types. A typical coal-fired back boiler system is shown below.

DIRECT HOT WATER

Expansion pipe

Hot supply

Backboiler

'Furred' pipe

Direct hot water supply was quite common before the 1960s. The water from the cylinder was heated by a cast-iron heat exchanger situated behind the open fire. As the water heated up, it gravitated up the pipework and heated the water in the cylinder. Thus, all water heated ran through the heat exchanger behind the fire.

Rust in the heat exchanger could contaminate the hot water and eventually come out of the taps. In addition, in hard water areas, the pipework coming from the back boiler could quickly get blocked by limescale deposits.

Early back boilers were made from either copper or cast iron. Providing pipes were made of the same materials, the only problems experienced with copper were in hard water areas, resulting in scale build-up. Cast iron, however, would oxidise and the resulting rust deposits would cause water discoloration, being even more of a problem in soft water areas. The lack of thermal control, however, sometimes leads to the water boiling, a problem that causes noise, vibration, heat wastage and, in extreme cases, scale build-up with a risk of explosion.

Scale formation (i.e. limescale) remains the major problem with modern direct hot water systems as the alkaline deposits, through which ground water passes, are drawn into solution.

On heating, these precipitate onto the higher temperature surfaces forming levels of hardness of the scale, depending on temperature. The only way to eliminate the problem is to provide some form of water treatment either by softening or holding the troublesome salts in suspension; the latter is achieved by installing a magnetic-type treatment unit.

Most modern direct hot water systems are made with compatible materials so that electrolytic action and water discoloration problems have been eliminated. However, with many of the modern heating boilers, the numerous electronic controls, safety devices, fans, heating elements and burners, still require regular maintenance, together with care to ensure adequate ventilation rates are fully provided. The more refined and efficient modern technology becomes, the more there is potentially to go wrong and regular servicing has become essential.

Indirect hot water systems

This has been the most popular means of providing hot water domestically, because it is simple and eliminates most of the potential problems associated with direct systems. In an indirect system there are two discrete volumes of water. The primary system circulates through the boiler and through a heat exchanger in the hot water cylinder. It will also circulate through radiators, where they are fitted. A secondary system feeds the cylinder and serves the hot water taps. The two systems must be kept separate to prevent contamination of the water which comes out of the taps.

INDIRECT HOT WATER

Feed and expansion cistern

Vent pipe

Vent pipe

Cold water cistern

To taps

Cold water feed

Indirect cylinder

Boiler

Nowadays, most hot water is heated indirectly. The water which passes through the boiler does not come into direct contact with the water in the cylinder. Instead, it passes through a coiled tube in the cylinder and then returns to the boiler. The boiler usually has the additional function of providing hot water for radiators.

There are, therefore, two separate volumes of water in this system, each with its own feed and expansion tank. The risk of contamination and limescale is reduced in the indirect system.

Scale formation is still seen as a problem, but at a much reduced level. This is because the primary hot water medium, which is the means of indirectly heating the domestic hot water, does so with much lower temperatures on the secondary heat exchanger surfaces (inside the indirect cylinder) than are experienced within a boiler or on the surface of an electrical element.

The primary water, which is first heated in the boiler, is in what is described as a sealed system, so that the water circulating is being continually recycled without replacement, giving off its heat to the secondary system and returning to the boiler for reheating. Once the salts in the original supply water are in this primary circuit, they will be precipitated out at their first heating and, with no further fresh water being introduced, the primary circuit stays free of further scale.

The secondary water flowing into the cylinder and on to the tap outlets is from a different supply and is heated by surfaces at temperatures not usually more than 85°C. This lower temperature means that only some of the salts precipitate out on heating, the remainder staying in solution, passing out of the taps without coating the surfaces.

In small housing developments, the pipework lengths and sizes are usually quite small and so the amount of heat which is lost from these pipes is minimal. Single-run pipes, in which the hot water can cool down when the water is stationary, are called 'dead legs'. These must be kept to a minimum in order to avoid heat loss and wastage of water (which can occur when cold water runs to waste awaiting the arrival of the hot through the tap). On larger installations, a water circulation system is incorporated into the system to ensure the hot water is continually pumped around to avoid this wastage.

Certain bacteria are known to multiply in water-based systems and can cause problems of ill health to building occupants. Legionnaires' Disease is one of these. The most straightforward way of preventing such problems is to ensure water services run either hot or cold and never at tepid or luke-warm temperatures. Hot water should always be available at above 65°C and the cold water below 15°C. Should an intermediate temperature be required, the hot and cold should be mixed at the point of use. If required, mixing valves are available for this purpose. Thermal insulation is used to prevent heat loss or gain, and pipework carrying water at different temperatures should take separate routes through the building to avoid problems.

It is not always possible to incorporate indirect systems within our buildings, especially those which have low floor-to-floor dimensions or lack of roof spaces for location of storage tanks. Some manufacturers will supply combined tank and cylinder components for mounting within cupboards, but the lack of height above the likes of a shower outlet will then make the indirect system less attractive. For very low-use occupiers, the indirect hot water system may also be expensive.

A rather alarming problem can occur with hot water cylinders if the roof storage tank and ball valve become frozen. If a coating of ice seals the feed tank and the user draws water from the cylinder, a potential vacuum can occur within the cylinder. This, in extreme cases, may cause the cylinder to implode in a similar fashion to an empty drinks can. Adequate thermal insulation is obviously essential around the roof-mounted tanks.

Primatic cylinders

During the 1960s and 1970s efforts were made to reduce capital costs for indirect systems by introducing an alternative type of indirect cylinder. These were called 'primatic' cylinders and were used extensively in domestic property.

PRIMATIC CYLINDERS

Vent pipe

Cold water cistern

To taps

Primatic cylinder

Boiler

A Primatic cylinder is installed as a direct cylinder, but functions as an indirect one. When the cylinder is first filled, the primary circuit is charged with water. When the cylinder is full, air seals separate the primary and secondary circuits. When the water in the primary circuit is heated, expansion forces some of the air from the upper chamber into the lower chamber. The primary system vents air into the secondary system which also tops-up the primary system as, and when, necessary.

The design of the heat exchanger within the primatic cylinder enabled the cold feed to the boiler and central heating circuit to be fed from the domestic hot water side of the installation, thus eliminating the need to provide the separate feed and expansion tank and pipework necessary in the more conventional system. This system worked quite well, but there is a possibility of the boiler water (primary circuit) mixing with the domestic hot water (secondary circuit) and, as some of the more recent additives now used in central heating systems to reduce corrosion are toxic, it is no longer acceptable to allow this to happen. The result is that, once identified, primatic cylinders are usually replaced by conventional cylinders, with additional tanks and pipework being incorporated during building or system refurbishment.

Pressurised cylinders

In a pressurised cylinder, the expanding water is contained within a pressurised expansion vessel. This is unlike a vented system where the water expands back into the cistern and up the vent pipe. Pressurised cylinders have to be stronger than vented ones and are often made from stainless steel or glass enamelled steel.

There are usually three main 'functional' controls to protect mains water quality and prevent damage to the cylinder:

1 *a pressure-reducing valve reduces the mains pressure to the safe working pressures of the cylinder;*
2 *a single-check valve prevents contamination of the main (i.e. prevents backflow);*
3 *an expansion valve protects the cylinder from over-pressure caused by failure of the pressure-reducing valve.*

In addition, there are normally three safety controls to protect the householder:

1 *a control thermostat usually set between 60–65°C – to close a motorised valve on the primary flow or switch off the immersion heater;*
2 *one or two energy cut-out devices with manual reset, usually set between 85–89°C – to close down the primary flow and to switch off an immersion;*
3 *a temperature and pressure relief valve set to 90°C or so – to allow water to escape if pressure or temperature gets too high.*

These installations require local authority approval and should be serviced every year.

PRESSURISED CYLINDER (INDIRECT)

mixer taps

NB: water heating circuit not shown

Pressure-reducing valve

Single-check valve

Expansion vessel

Expansion valve and tun dish

Temperature and pressure relief valve and tun dish

Immersion heaters

In most of the above systems, an electric immersion heater can provide a secondary form of water heating. A thermostat prevents the water from over-heating. There have been examples of thermostats failing, causing scalding hot water to flow into the feed and expansion tank. If the overflow is blocked or if the tank is old and made from plastic, it can split. Thermostats should be of the fail-safe type – if it fails, the power is automatically cut off.

Central heating systems

Boiler types

Most modern boilers are of three types (see below). However, they work on broadly similar principles – water is heated in a heat exchanger and is then circulated around a series of radiators. In a 'regular' boiler and a 'system' boiler, the water from the heat exchanger is also fed through a coil (another heat exchanger) in a separate cylinder where it heats up water for washing. In a combination (or 'combi') boiler, water for washing is heated by a small heat exchanger located inside the boiler.

Regular (or vented) boilers

These 'traditional' vented boilers have been around, in one form or another for nearly a hundred years. Modern ones are much more efficient than their predecessors. Regular boilers are designed to provide hot water for radiators and to heat water in a separate cylinder (usually located in an airing cupboard). The boiler is fed by a storage (feed and expansion) cistern usually located in the roof void. This cistern accommodates the expansion of hot water (its level rises slightly) and a pipe from the boiler vents back over, and into, the cistern if the boiler controls go wrong and the water boils. Nowadays, most regular boilers are wall-mounted; early ones were floor-standing.

System boilers

These are designed along similar lines to a regular boiler, but for use with an expansion vessel rather than a feed and expansion cistern. In some boilers, the expanding water is accommodated in an expansion vessel fitted inside the boiler. In others, it is not part of the boiler, but is usually located near to it. To provide hot water for washing, a system boiler heats water in a cylinder. This can be either a vented (traditional) cylinder or a modern pressurised one.

⋈	Pressure-reducing valve
⋉	Servicing valve
▷	Single-check valve
▷▷	Double-check valve
🛢	Expansion vessel
⋈	Expansion valve and tun dish
⋈ TPRV	Temperature and pressure relief valve and tun dish

Combination (combi) boilers

Combination or combi boilers (so-called because they deliver hot water for washing and heating) now account for over 60 per cent of all the new domestic boilers installed in the UK every year. They were originally developed for flats to save having lots of tanks in the roof space. A combi eliminates the need to store hot water – a hot water cylinder is not required. Hot water for washing is heated and delivered via a water-to-water 'plate' exchanger as and when required – not always at very high flow rates, but at mains pressure. The savings in installation cost (no cylinder and no tank) are probably offset by the slightly higher cost of the boiler. These boilers do have minimum requirements in terms of mains water pressure. They also need a good gas supply – at times, these boilers have to work quite hard.

In most modern combis hot water for the taps is heated in a 'plate' exchanger – it is therefore indirect.

Hot water is supplied on demand.

Some older combi systems are likely to have a thermostatic radiator valve (TRV) on every radiator, but no room thermostat. It is better practice to have a thermostat in the lounge (for example, one radiator in the lounge without a TRV) and TRVs elsewhere.

Filling loop – normally closed

TRV

Condensing boilers (regular, system or combination)

These are available in regular, system and combi versions. In a modern (non-condensing) oil- or gas-fired boiler about 70–80 per cent of the fuel is converted into useful heat. The remainder is lost through the flue – the flue gases can be as hot as 250°C. In condensing boilers, more heat is extracted from the fuel; their efficiency can exceed 90 per cent. In fact, so much heat is extracted that the flue gases are only about 50–60°C. In a typical condensing boiler, there are two heat exchangers. After crossing the first heat exchanger, the combustion gases are directed over the second heat exchanger to pre-heat the water returning from the radiators. If the water in the radiator return pipe is cool enough, the exhaust gases will condense on the heat exchanger and release all their latent heat. The secondary heat exchanger must, therefore, be corrosion-resistant and a drainpipe must be fitted to remove the condensate. The exhaust gases of condensing boilers 'plume' – this is quite normal and reflects their lower temperature. Nearly all new boilers are condensing boilers.

General boiler problems

Some boiler problems are unique to specific models; others can occur in all boilers. This section describes some of the more general problems.

Most modern boilers are designed with sealed combustion chambers, so that the poisonous gases produced during combustion of the fuel cannot pass into the building. As part of the regular servicing arrangements, this must be checked. Adequate ventilation of the room housing the boiler may be required for certain types of boiler in case there should be a failure of the combustion chamber seals. Most boilers will not operate if they sense a leak from the combustion chamber.

Some of the most common failures within boilers relate to the controls and any fans or pumps within the unit. High limit thermostats (which switch off the boiler when overheating occurs) are usually incorporated to prevent the water boiling should the boiler temperature thermostat fail. If the fans or pumps stop working, the unit should 'fail-safe'.

The air to fuel mixture controls the efficiency of the boiler and the flame colour gives a very good indication of its efficient operation. Blue/purple flames indicate correct settings, but any indication of yellow means the fuel is not being burnt properly. This may be caused by contaminated fuel, dirt particles on the burner jets causing some form of blockage or a restriction of the air supply/flue gas discharge. Unburnt fuel results in a carbon/soot build-up which will occur rapidly, requiring additional cleaning and maintenance and causing possible damage to other components.

During summer shutdown periods when the boiler has totally cooled down, it is often found that a water leak occurs from within the boiler casings. This is often cured as soon as the boiler is returned to its normal working temperature as the gaskets and boiler sections expand to fill the gaps which occur during the cooling down processes. It is often more advisable to leave the pilot lights on (where they exist) or retain the boiler for heating the domestic hot water, unless total shutdown is required for maintenance purposes.

Shutdown periods are also the most common times of pump and other component failure, as many will continue to work when quite badly worn out right up to the time they are switched off.

There are a number of safety systems associated with fuel-burning appliances and these are provided to ensure fail safe or warning is given to the user. Safety and vacuum release valves protect the structural integrity of appliances and reduce risks of explosion. Core plugs are also sometimes fitted for the same reason. Pilot lights (common in early boilers) impinge on flame failure devices which automatically switch off the gas supply if the pilot light is extinguished. Other controls sense when the oil has not ignited in a burner.

Typical fan

Circuit board – often fitted at bottom of boiler

The boiler shown above dates from about 2000 – it is a combination boiler (non-condensing). The arrows identify the main components (many of which can be found in other boiler types).

According to Which? Magazine, the most common causes of boiler breakdown are failure (in descending order) of:

1 *circuit board*
2 *automatic air vent*
3 *pump*
4 *overheat thermostat*
5 *heat exchanger*
6 *expansion vessel (not present in a regular boiler)*
7 *gas valve*
8 *burner.*

Flue fans can also fail quite often – sometimes within a couple of years. Most boilers will not start if the fan is not working properly.

In older boilers, thermocouple failure is quite common. This is the sensor that sits in the pilot light flame and delivers an electrical current when hot, holding a solenoid valve open and allowing gas to flow. If the pilot goes out, the thermocouple cools, the electricity stops and the gas gets shut off.

Boiler flues

There are a number of regulations regarding the siting of boiler flue terminals. These are quite detailed and well beyond the scope of this chapter. A fan-assisted balanced-flue gas boiler, for example, should have its terminal sited at least:

• *300mm away from an opening*
• *200mm below the eaves*

- *150mm away from a vertical soil or drainpipe*
- *600mm away from a boundary facing the terminal (1200mm if it is opposite another terminal)*
- *300mm away horizontally from a terminal on the same wall (1500mm vertically).*

Carbon monoxide poisoning

Carbon monoxide (CO) is a colourless, odourless, tasteless toxic gas produced by fuel-burning appliances or vehicle engines. Mild exposure can cause headaches, nausea and flu-like symptoms while higher doses can lead to collapse and death within minutes. Incorrectly installed and/ or poorly maintained heating appliances as well as blocked or cracked chimneys and flues are frequently the culprit.

Sensors are available connected to alarms which measure excess quantities of carbon monoxide gas as it can cause loss of consciousness without smell or other warning. An alarmed sensor should be installed in or near every room that contains a fuel-burning appliance. Automatic fire fighting systems may also be required where risks warrant them.

A typical carbon monoxide detector and alarm – this one is battery-operated and is based on an electrochemical cell type gas sensor. The higher the level of carbon monoxide, the faster the alarm responds.

Problems with whole heating installations

Although most new boilers are combis, the majority of existing central heating systems are vented, i.e. the boiler is fed by a storage tank and the water in the boiler can discharge back into this tank if the water overheats and reaches boiling point. Because of the risk of this discharge, the storage tank for the boiler must be separate from the cold water storage tank which serves the cylinder.

Most modern central systems also heat the domestic hot water via a heat exchanger in the indirect storage cylinder. Modern control systems allow the user to select heating, hot water or both, switching on and off the systems at variable times according to need. A summertime problem is sometimes experienced when the radiators on upper floors continue to get hot even when the heating is apparently switched off. This is due to what is called gravity circulation. The only way to cure this problem is to insert additional isolation valves into the upper floor circuits, making sure that the vent pipes are not affected.

Other problems with older systems include:

- *gravity circulation (i.e. no pump) to the hot water cylinder – results in stored water being slow to re-heat*
- *lack of hot water cylinder thermostat – results in excessive stored water temperature*
- *lack of room thermostat – rooms get too hot as system depends solely on boiler temperature*
- *no thermostatic radiator valves (TRVs) – causes excessive room temperatures and poor system balancing*
- *a TRV in the same location (i.e. room or space) as the room thermostat*
- *no boiler interlock – causes the boiler to stay hot and to cycle unnecessarily during programmed periods*
- *incorrect positioning of the pump in the system – can cause a number of problems leading to water 'venting' into the feed tanks or air being drawn into the system through the bleed valves of the radiators.*

Sometimes, the float valve in the feed and expansion tank can seize (usually in the closed position). Over a long period, the water level in the tank and heating pipework drops. If, due to minor leaks, etc. the water level falls too much, the pump will only operate intermittently, if at all. Sometimes, this will be accompanied by strange (and loud) banging noises from the boiler. The boiler can also stop working because the boiler thermostat, sensing a rise in water temperature, turns the boiler off.

Another problem can be caused by a build-up of deposits in the cold feed to the boiler. These deposits can slowly block the pipe and starve the system of fresh water. It can be difficult to diagnose because the supply cistern shows the correct water level.

Deposits or sediment in the pipework or a badly sited pump can cause 'over-pumping'. If this occurs, water will discharge into the feed and expansion tank. If this is not addressed promptly, the constant replenishing of the water will add oxygen to the system which will hasten the rusting of the radiators. It can also split a tank or lead to its collapse.

Specific problems of sealed installations (combi and system boilers)

Most combi and sealed system manufacturers recommend a pressure of about 1.0 to 1.5 bar (when the boiler is cold). Over a period of months, some heating systems steadily lose water; this is usually from tiny leaks which can be very difficult to trace. Water and, therefore, pressure will also be lost if the radiators are bled. If left unattended, the pressure will drop to zero (many boilers have a dial on the front of the casing showing the pressure). At low pressure, air can enter the system and speed up internal corrosion of the radiators. Most boilers have an internal sensor which prevents the boiler firing if the pressure is too low – this can be mistaken for boiler failure.

An expansion vessel accommodates the increased volume of expanding water. If it fails, possibly because of a perished or torn diaphragm or due to leakage of its air charge, the increased pressure should trigger an automatic pressure relief valve; this valve usually opens at about 3 bar and ejects water to the outside. The ejection of water will lower the pressure in the system and, after the boiler has been switched off for a while (with a further lowering of pressure), the boiler will not restart because the pressure switch senses the low pressure.

In combi and system boilers, the water pressure is adjusted by opening the filler loop valve. When this valve is not turned off properly, the pressure in the system will slowly build up until the pressure relief valve opens. If this problem is not rectified, the slow but constant supply of fresh water will lead to corrosion (and a steady stream of water being discharged from the pressure relief valve). If mains pressure water leaks into the heating pipework from a faulty combi water-to-water plate exchanger, the heating circuit pressure will rise and cause similar problems.

Radiators, pipes and thermostats

The majority of existing installations utilise copper for the pipework and steel for the radiators. Using these dissimilar metals can cause problems as an electrolytic action takes place due to the water acting as an electrolyte between them. Over a period of time, varying between 2 and 30 years, pin holes may appear in the radiators requiring their replacement. This problem is generally regarded as acceptable because of the initial high capital cost of alternatively supplying radiators manufactured in copper. The problem is even more severe when the steel pipework has been galvanised; the pin holes can appear quickly and the steel looks as if it is infected by woodworm.

This radiator has surface rust. Whilst unsightly, it is not usually a serious problem and can normally be treated with an anti-rust paint. Note the small bore heating pipe running into and then down behind the plasterboard finish to the wall.

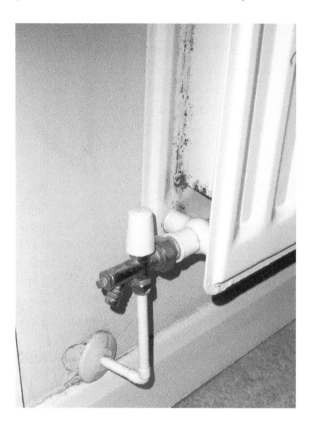

Concealing pipe runs within walls (i.e. within the actual structure – not just behind plasterboard) and solid floors was considered acceptable a few years ago. Damage can occur by nail penetration when securing future fixtures, or chemical reaction may occur over time with the in-situ building materials. A nail driven through a water pipe may not leak immediately and so lead to an unexpected problem months or even years later. Some copper pipework embedded in floors in the early 1970s was of a specification inferior to the normal British Standard, being imported from overseas during a time of severe shortage. This was expected to cause problems of interaction with normal in-situ building materials unless it was suitably wrapped by protective tape at the time of installation. Further problems have occurred with some cheaper pipework manufactured in the UK in the early 1980s, resulting in pitting and premature leaks.

The route chosen for the pipework may, in some buildings, have to rise to a higher level, e.g. to pass over a doorway. The resulting high point is where air within the system becomes trapped and will eventually build up, stopping the circulation of the water. Air-release valves need to be incorporated, but are often hidden away and forgotten. Should the system not be working, this possibility should be investigated after faults with boiler operation and pumps have been checked. Note that where pipes sit in notches in floor joists, they should be wrapped in felt to prevent them from 'knocking'.

RADIATOR PROBLEMS

If the radiators are hot or warm even though the heating is turned off (hot water on), there may be a problem with two or three port valves.

If the radiator is cold at the top and warm at the bottom, it probably needs 'bleeding' to remove air.

If the radiator is hot at the top and cold/warm at the bottom, it could be blocked with sludge.

If the radiators near the boiler are hot and the others are cool, the radiators may not be balanced properly.

Thermostatic valves sometimes stick if heating is left 'off' for long periods.

If the radiators upstairs only are cold, it could mean that the feed and expansion tank in the roof is empty or, on a combi system, that the boiler pressure is too low.

If the radiators downstairs are cold, there could be a problem with the pump.

RETURN

FLOW

In a single-pipe system, the radiators at the end of the circuit are likely to be cooler.

In micro-bore systems, the pipes are easily damaged and small 'blobs' of solder are more likely to block the pipework.

The first source of leaks is likely to be the joint between the copper pipe and the steel flange of the radiator. Electrolytic action occurs here (where different metals are in contact).

Air trapped in heating systems reduces water flow and, if lodged in radiators, will reduce their heating efficiency. After a system has been refilled, it may take the air some days to arrive at radiators and other high points. Regular checks may need to be carried out in order that the air can be located and removed through the air-release valves. If the problem persists, then the position in the system of the pump (as mentioned earlier) should be examined together with possibly system leaks, where air could be sucked in.

Some heating systems may have had inexpensive valves installed, especially on radiator connections. The process of opening and closing the valves may have the effect of loosening the gland, resulting in the fibrous washers or packing becoming loose. This means water can leak out and damage ceiling and floor coverings, especially when the system has cooled down. Repacking and tightening the glands of the valves should be carried out on a regular basis or it may be considered more economical in the long term to replace the valves with those of a higher specification. Most modern heating systems tend to have thermostatic radiator valves (TRVs) controlling individual room temperatures. Note that some early valves only function correctly when they are installed on the flow pipe to the radiator (not the return pipe).

Many heating systems (including those with TRVs) are controlled by a room thermostat switching the pump on and off, so regulating the heat source to all the radiators. The positioning of the thermostat is important and some improvement in comfort levels can be made by moving the thermostat away from ancillary heat sources such as the back of a television or other heat-generating appliances. The room in which the room thermostat is installed may also affect the overall economy of the heating system and the quantity of fuel burnt; any radiator in that room should not have a TRV fitted as this will 'fight' with the room thermostat and upset its efficiency.

Micro-bore pipework systems utilise very small diameter tubes of a soft copper specification. If bumped, e.g. by a vacuum cleaner during room cleaning, the resulting damage can reduce water flow or, in the worst case, cause a major leak. Normally installed in floors and ducts, some form of protection may be necessary where micro-bore pipework is exposed.

In modern installations, plastic pipework has become common. It is easy to cut and join and ideal for use where the floors are made from metal web joists (in coil form, it is flexible enough to be fed through the webs). If it is not properly supported, it tends to sag. In some cases, sagging pipes plus plastic's high thermal expansion and contraction rates can actually pull joints apart. Sagging pipes are also difficult to drain down and they look unsightly – one reason why, in modern houses, the pipework is often hidden behind plasterboard. On the other hand, unlike copper, it does not need to be bonded (earthed). Also, its scrap value is nil, so there is no risk of the pipes being removed while building occupants are away on holiday.

Drainage 20

Introduction

In the Georgian period, sanitation was, when compared to modern standards, primitive. Better-quality houses had rainwater pipes, made from lead, or in some cases timber. A crude system of brick sewers, often built around natural water courses, carried rainwater away to streams and rivers. Collection and disposal of foul matter, a much more hazardous material in terms of health, were dealt with in a variety of ways. Perhaps the most common was the cesspool. This comprised a pit dug in the back yard, possibly lined with brick, which was emptied periodically by nightmen (so called because they were compelled by law to operate during the early hours before dawn). Some cesspools were fitted with overflows; surplus liquid matter would percolate into the surrounding ground where it often contaminated local water courses and water supplies. A privy was often sited directly over the cesspool; few houses had flushing water closets until the early nineteenth century. In affluent households, although servants would be expected to use the privy, the house owners would have used commodes, often discreetly hidden in elegant furniture. Foul matter from commodes and waste water from washing would have been carried down to the cesspool in buckets.

Following the Industrial Revolution there were a number of technological developments which improved sanitation; for the wealthy at least. The first was the steam engine which enabled water to be pumped to reasonable heights. The second was the introduction of machine-made lead piping to carry the mains system. This, together with the development of relatively cheap cast-iron drainage pipes, led to the introduction of water closets and, eventually wash basins, on upper floors.

By the middle of the Victorian period most prosperous town houses had a system of mains water supply and a system of mains drainage. Although a sewer would contain foul, waste and rainwater, the above ground drainage system was usually a two pipe system. The foul pipe served the WCs; the waste pipe served the sinks, baths and rainwater guttering.

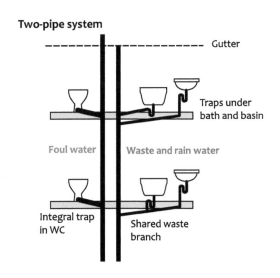

Two-pipe system

Gutter

Traps under bath and basin

Foul water   Waste and rain water

Integral trap in WC

Shared waste branch

Mid Victorian   1930s

Three examples of two-pipe systems. The foul pipe extends beyond the eaves to act as a vent pipe for the drain and sewer.

Above-ground drainage

Traps

To prevent foul smells from the sewer entering the house, traps were fitted below each sanitary appliance. A trap contains a small volume of water 'trapped' in the 'U' bend. Unfortunately, there are a number of ways in which traps can lose their seals. This can be caused by siphonage, or by back pressure created when a large volume of water running down the discharge pipe slows at a tight bend. Failure of traps (or no traps at all) is probably the most significant defect in early above-ground drainage systems.

Self siphonage

Steep gradients most at risk

S traps are very prone to problems of self siphonage.

Induced siphonage

Induced siphonage can also be caused by other fittings sharing the same branch.

Water level before discharge

If siphonage reduces level of water below this line, trap will lose its seal.

If the main stack runs full bore, it can suck the water out of the trap.

Lavatory Traps and Wastes

No. 6120. "S."   No. 6119. "P."

Bottle traps are less prone to siphonage but more likely to block.

Back pressure

If branch is near foot of stack back pressure caused by water running down stack and slowing down at bend can 'break' seals.

'Tight' bend

FIG. 2

FIG. 3.

Anti-siphon traps became available at the end of the nineteenth century. They were bulky, noisy and expensive.

The risk of siphonage, however, could be reduced by ventilating each branch pipe as explained in the following diagram. This, although effective, created a complex system of pipework which was both expensive to install and expensive to maintain (see following photographs).

As water runs down the branch, any siphonage action pulls air down the vent pipe, thus maintaining the seal in the trap.

Traps can also become blocked. The three main 'culprits' are: disposable nappies blocking WCs, grease blocking kitchen sinks, and hair blocking wash-hand basins.

Typical installations of the early twentieth century

During the first 50 years of the twentieth century, millions of low-rise houses were built as towns expanded and the suburbs were created. Most of these houses adopted the two-pipe system of above-ground drainage. Discharge pipes were usually made from cast iron, with yarn and lead joints. From the 1940s asbestos cement pipes with cement mortar joints became common. The discharge stacks were connected to the mains drainage by salt glazed clay pipes (salt glazing provides a smooth surface and improves impermeability). An interceptor, or rat trap, was often required by the local authority to prevent smells (and rats) from entering the house drainage system. There were a number of proprietary interceptor traps, although they all worked on the same principle, i.e. a 'U' bend in the trap held a small volume of effluent. A short length of pipe fixed with a cover allowed access for rodding down the drain should blockages occur. To prevent the build-up of gases, a one-way vent was normally provided just above ground level. This vent, when operating correctly, allowed air to flow into the chamber to ventilate the house drainage system but prevented foul air from escaping into the atmosphere. Unfortunately, these vents, known as Mica valves, often became blocked with paint, dust and debris, which affected their intended function. Interceptor traps, in most cases, were unnecessary and in modern construction they are rarely found.

Some common problems found in early twentieth-century installations are shown in the following drawing.

Typical arrangement from the 1920s and 1930s – and some of the common problems

Cage to prevent birds' nests or leaves blocking air flow in drain

Cast-iron soil pipe serves WC only (Foul)

Branch discharge pipes in lead or copper – can freeze and cause blockages in cold weather if taps drip.

Hopper heads prone to blockage from leaves etc.

Interceptor traps were common in the 1920s and 1930s. They were provided to prevent foul air and rats from the sewer entering the house drainage system. In fact, they were often unnecessary if traps were provided at the base of the waste/rain stack. Interceptor traps are prone to blockage.

Cast-iron combined rainwater and waste pipe

Rust is most likely to occur at the back of the pipes (difficult to paint).

One-way vent to maintain air in house drainage system at atmospheric pressure.

Waste from kitchen runs direct into gully.

Rodding Eye

Interceptor trap in manhole/inspection chamber

Slow bend at foot of soil stack (not trapped)

Trapped gully

This shows the above-ground vent to the interceptor trap located within the manhole. The grill to the vent is missing in this case.

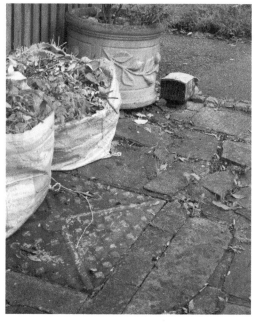

The vent pipe to an interceptor trap in the manhole of a 1930s house.

In the previous example the waste stack runs into a trapped gully. The trap is required to prevent foul smells escaping from the open hopper head. Cast iron pipes, which were the norm until the 1940s, are very durable. However, they can leak at defective joints, sometimes caused by settlement of the property over many years. The back of the pipes can also develop leaks; because this is the surface that is hardest to paint, it is therefore the first part to rust. In better-quality installations the stack is supported on spacer blocks to ensure it is well clear of the wall. This makes it easier to paint.

This is a page from an old manufacturer's catalogue showing an interceptor trap.

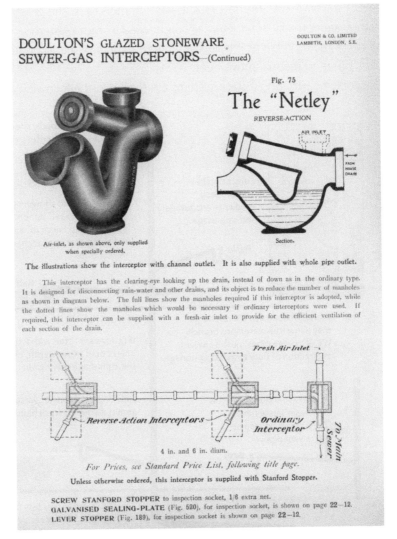

The rust stains show the extent of leakage from this rainwater downpipe.

In the post-war period the two-pipe system of drainage was abandoned and replaced with the single-stack system. In this system, foul water and waste water are collected in the same discharge stack (rainwater is collected separately), which usually ventilates the system as well. The Building Regulations specify a number of rules to ensure that these systems work effectively. These, in the main, control the depth of traps, the size and gradients of pipes, and the number of fittings which can be served by branches.

Modern single-stack systems in low-rise housing are invariably constructed with plastic pipes. The discharge stacks have flexible 'O' ring joints; discharge branches can be jointed with 'O' ring seals or solvent welds. In the 1960s it was common to fix the discharge stacks externally; nowadays, they are normally fitted internally and boxed in with plasterboard. Although internal stacks may cause more problems if leaks occur, they are less prone to degradation by ultra-violet light and less prone to thermal movement.

Typical problems with single-stack systems are shown in the following drawing.

Top of discharge, if badly sited, stack allows foul smells to enter property.

Long branches and/or steep gradients encourage self siphonage.

Baths, sinks and showers are less susceptible to self siphonage as trailing discharge of water (due to flat bottom of appliance) will help re-seal trap.

Basin

WC

Bath

Branches opposite WC discharge may allow foul water into branch. WC discharge pipes must be swept in direction of flow to avoid siphonage.

If fittings are fitted with P traps of less than 75mm depth, or S traps, self-siphonage may occur.

To avoid siphonage this basin waste needs a resealing trap or a vent pipe.

Basin

Bath

Sink, bath and shower waste minimum 40mm diameter, wash hand basin 32mm.

Sink

If this distance is less than 450mm, back pressure in stack may cause loss of trap seal in ground-floor appliances.

Long runs, as with this soil pipe, tend to encourage self-siphonage.

Appliances at ground-floor level are often connected direct to the drainage system. This may be because the branch length exceeds the rules set out in the Building Regulations. It may also be necessary because of problems of back pressure.

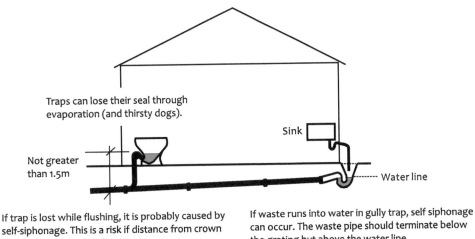

Traps can lose their seal through evaporation (and thirsty dogs).

Sink

Not greater than 1.5m

Water line

If trap is lost while flushing, it is probably caused by self-siphonage. This is a risk if distance from crown of trap to invert exceeds 1.5m.

If waste runs into water in gully trap, self siphonage can occur. The waste pipe should terminate below the grating but above the water line.

The diagrams (previous and following) highlight a number of design criteria which affect the correct operation of single-stack systems. There are also a number of practical problems related to good construction practice.

In very exposed situations wind blowing across the top of the stack can 'break' seals/traps in waste fittings.

Missing cage can result in lack of ventilation if birds nest in pipe.

Leaking joints

100mm soil vent pipes will bend if not supported at 1.8m centres (1.2m where running horizontally).

If 'push fit' couplings are not provided with correct lubricant their ability to accommodate thermal movement will be reduced. Washing-up liquid will aid assembly but will not work in the long term.

Lack of movement joint where two pipes join may lead to distortion and leaks.

Wastes will sag unless supported at correct intervals. If wastes sag joints can be pulled apart or sumps are created where solids will collect. Where wastes pass through walls joints can part if sleeves are not provided.

Kinks or off-sets in SVP can produce sufficient back pressure to 'break' seals in traps.

Polypropylene wastes should not be used externally unless painted. They quickly break down if exposed to sunlight.

Wastes with solvent joints cannot be used in long runs. Provision for thermal movement must be provided with additional 'push fit' couplings at 1.8m intervals.

This waste pipe is too long, and therefore prone to self-siphonage. The steep gradients will also add to this risk. It is also very poorly supported, and therefore sags.

Rainwater systems

Until the 1950s most rainwater goods were formed from cast iron, asbestos cement and even timber – and in lead.

A lead rainwater downpipe taking water from a bay roof and, via a hopper from a main roof.

A cast iron gutter.

Since the late 1960s plastic has been the norm. It is available in a range of colours and profiles. The Building Regulations contain guidance on the size of guttering, and the size and number of downpipes for given roof areas. There are a number of common defects, most of which are shown in the drawings below.

If gutter is too low rain can discharge over outer face.

Sarking felt should run into gutter – otherwise water leaking through tiles can run down wall face.

Where rainwater pipes discharge over lower roofs the water cannot 'fan out' if profiled tiles are used. This will result in rainwater over shooting the lower gutter in heavy rain.

Blockages in pipe caused by leaves, balls, etc will cause leaks at joints.

Ladders leaning against gutter can crack it.

Gutters can sag if not adequately supported.

Overflowing gutters are often due to build up of dust, leaves, etc. or balls which obstruct the outlet. Gutters should be cleared annually.

Lack of movement joint where two pipes join may lead to distortion and leaks.

Downpipes can bend if not adequately supported.

If downpipes do not discharge directly into gullies or rainwater shoes, splashing can bridge DPC and wet pavings. Splashing can also lead to problems of ice (and injury) in freezing weather conditions.

This explains why the ground is wet in the previous photo.

This incorrect slope on the gutter has led to rainwater failing to be removed adequately – and has encouraged growth of grass which has exacerbated the problem.

Leaves blocking the gutter will impede the flow of water and may cause the gutter to overflow – wetting the wall below.

Clearly the relationship between the gutter and the rainwater hopper is never going to work.

Left: An odd arrangement for disposing of rainwater from the roof of the house. Right: A section of the sarking felt under the tiles has been damaged and does not fold into the gutter, causing water to run down the face of the wall.

Gutters may overflow because they are full of debris, undersized or the outlet is blocked.

It can also occur where downpipes are incorrectly positioned. If downpipes are sited at the end of the gutter, the lower parts of the gutter may be carrying excess water. Siting the downpipe in the middle avoids this problem.

This gutter falls the wrong way. It also has to carry water from another roof above.

Below-ground drainage

Combined system

Mains drainage is comparatively recent. Although a number of Georgian buildings had a rudimentary form of drainage, it was not until the end of the nineteenth century that it became an automatic requirement to provide a comprehensive system of mains drainage for all houses in urban areas. Early systems were usually 'combined'; in other words, rainwater and foul water (from WCs and wastes) shared the same drains.

A typical terraced house from about 1900 is shown in the following diagram. At this date, bathrooms, as we know them today, did not exist in most modest houses. People washed themselves over a sink in the scullery. Hot water would have to be heated on a coal-fired stove; galvanised steel or tin baths were filled (and emptied) by hand.

The drawings above and right show a typical drainage installation from the 1900s. The pipes above ground level were usually cast iron, those below, salt-glazed clay. Textbooks of the time refer to unscrupulous builders using reject pipes ('seconds').

To prevent smells from the drainage system rising into the air, traps were provided just below ground level (see earlier drawing). Some local authorities required a trap at the junction between drain and sewer (interceptor traps).

Most early pipework was made from clay; usually glazed with salt to provide a smooth surface. Pipes were laid in trenches, often in the bare earth, although some local authorities required concrete beds below the pipes. Drainage pipes had a socket at one end and a spigot at the other. Joints were formed in mortar alone or by ramming a tarred rope (caulking) around the junction of the socket and spigot; the joint was finished with a mortar fillet. Any minor ground movement could quickly crack the joint and allow effluent to leak into the surrounding ground.

Rigid joists are not necessarily unreliable. Where pipes were laid (and sometimes surrounded) on a concrete bed (common practice in the 1920s to 1950s), there was little movement in the pipework and therefore limited stress on the joint. However, where pipes were laid straight onto the floor of an excavated trench or on a thin layer of stone, rigid joints often failed quite quickly due to minor movement in the pipe line. In the 1960s, the design of drainage pipework was 're-visited' and it was concluded that it was not necessary to bed pipes on concrete as long as the pipes were strong enough to resist compression forces, and as long as flexible joints could be provided to minimise the risks of leakage in case of minor ground movement.

Doulton & Co. pipes – circa 1910

Until the 1960s most pipes had rigid joints. Since then, flexible joints have become the norm. Pipes can be clay or plastic.

Spigot end   Flow   Socket end

Modern pipes are available with socket and spigot ends or plain ended. A sleeved joint is used with plain end pipes; an O-ring joint is used with socketted pipes. (Photos courtesy of Hepworth Clay Wavin Ltd.)

Nowadays clay and plastic drainpipes nearly always have flexible joints. The only exceptions are likely to be where small runs of pipework are being repaired. There are two types of mechanical push-fit joint applied to modern clay pipes; the sleeve joint and the O-ring joint. The sleeve joint is used with plain end pipes, and the O-ring joint is used with socketted pipes.

For an O-ring jointing, the pipes have polyester fairings cast round the outside of the spigot and the inside of the socket. In the O-ring joint, the sealing action is achieved by a rubber ring, located in the groove in the spigot moulding, and compressed between the spigot and socket fairings.

The coupling for the sleeve joint is made from polypropylene with two rubber gaskets incorporated into the sleeve to give the sealing action between the pipe and the sleeve. The coupling is fitted onto one end, before lowering the pipe into the trench. The pipe is placed on a clean, firm base; lubricant is applied to the chamfered pipe end and the coupling is pressed onto the lubricated end.

With the O-ring type joint, the ring is placed into the groove on the spigot moulding and the socket of the preceding pipe is lubricated. The spigot is then placed up to the socket of the mating pipe and pushed in with a slight side-to-side action. The rubber ring is thereby compressed to make a flexible water-tight seal. Note: drains are normally laid uphill, in other words starting at the lower level and laid with the socket ends at the higher end. With both types of flexible joint it is essential that the mating surfaces are clean and that the lubricant supplied by the manufacturer is used. Any other type of lubricant could have a detrimental effect on the long-term performance of the joints; washing-up liquid should never be used.

Separate system

During the last 60 years or so, there has been a move towards 'separate' drainage systems. This requires extra pipework but does reduce the load on sewage treatment works; the storm, or surface, water being directed into rivers or straight out to sea.

The principles of modern drainage systems have not changed much over the last 100 years; what has changed is the rules regarding their installation. The Building Regulations control the design and installation of drainage systems. They contain a number of rules regarding (among other things):

- *pipe size, bedding material and gradients*
- *protection to pipes under roads and buildings*
- *provision for clearing blockages, siting of access points, size of access points*
- *ventilation.*

A modern separate drainage system

Inspection chamber.

Storm

Foul

Concrete manhole.

A more detailed example of a modern system is shown below.

Shallow access point suitable for garden.

SVPs do not require traps as all fittings in dwelling are trapped.

Back inlet gully

Inspection chamber

Rodding eye allows rodding down drain.

Back inlet gully

SVP

RE

RWP

RWP

Foul Sewer

Road gully

SVP	**soil vent pipe**
WC	**water closet**
RE	**rodding eye**
RWP	**rainwater pipe**

WC

WC contains integral trap.

Drive

Gully

Storm sewer

RWP

RWP

RE

In a combined system, traps are required for RW pipes – not necessary in a separate system.

Gully

Manhole

Manholes (with working space at drain level) can be formed in brick or pre-cast concrete. Iron covers cope with the loads of traffic. A plastic inspection chamber is shown below.

Trenches should not be excavated too far in advance of pipe laying and should always be kept to the specified design width. Excessively wide trenches increase the amount of excavation, bedding and backfilling, and may impose a load on the pipe in excess of the design load. The trench width should not be less than the pipe diameter plus 300mm.

Recommended pipe gradients vary depending on pipe diameter, number of dwellings served, and the pipe contents. 100mm foul pipes are usually laid somewhere between 1:40 and 1:80. 100mm storm pipes can be laid as flat as 1:100. There are no maximum gradients although it should be remembered that steep gradients (i.e. quickly dropping trenches) are expensive. In practice, branch drains (i.e. from dwellings to the main drain run) are likely to be quite steep; the main runs will normally be as shallow as possible to minimise excavation.

Narrow trenches may impede the proper placing and consolidation of bedding material around the pipe and restrict working conditions in the trench. Where a specific trench width is defined for the narrow trench design condition, this must be maintained vertically to a height of at least 300mm above the crown of the pipe. The load-bearing capacity of an installed pipeline depends on the construction of a suitable bedding and surround. The bedding should level out any irregularities in the trench formation and ensure uniform support along the pipe barrel. Clay pipes are high strength rigid units which have been designed to carry applied loads with no deformation. Additional bedding and side-fill can, if necessary, enhance the pipes' load-carrying capacity. Correctly sized granular material should be placed to the required level and extend the full width of the trench. Socket holes (where they exist) should be formed to prevent the weight of the pipe bearing onto the socket. Bricks or blocks must never be placed in the bedding material for setting the pipes to level; at the very least any bricks or blocks must be removed before backfill.

Plastics pipes are flexible and have little resistance to load without the support of granular bedding and sidewall material, which should be correctly sized and compacted. The bedding should normally be placed to at least the full height of the pipe. The side-support provided is essential in keeping deformation to the acceptable maximum of 6 per cent of the vertical pipe diameter. However, although this bedding and surround are expensive, it must be accepted as the trade-off for using pipes which are lighter and easier to handle than their clay counterparts.

The Building Regulations with regard to drainage are straightforward but beyond the scope of this book. Most drainage systems were installed a long time before the present Regulations were written and, of course, the Regulations are not retrospective. The following drawing shows typical problems in a modern installation. Its two main differences from the late Victorian example shown earlier are the increased use of access points (inspection chambers and manholes) and the separation of foul and storm water.

Summary of drainage defects

Many existing houses will have drainage systems which appear to work but do not meet the demands of current practice. Early systems, for example, had few access points so blockages were difficult to clear. Fortunately, modern technology means that drain jetting is a relatively straightforward practice. This short section concentrates, not on some of the general problems highlighted earlier in the text, but on defects or poor practice which can cause blockages and/or leaks.

There are a number of ways in which drains can become blocked. Identifying the cause may take a surveyor minutes, although it may require specialist input and specialist equipment. Nowadays, there are a number of companies who can inspect drains internally using remote controlled cameras. These are relatively cheap when the alternative cost of digging up several sections of road or gardens is considered. Blocked drains can be caused by:

* *inappropriate materials finding their way into the drainage system, e.g. disposable nappies, fat which solidifies as it cools*
* *in new developments, building materials finding their way into the drainage system during construction*
* *pipes crushing (see next section on leaks)*
* *pipes laid at incorrect gradients (excess or inadequate fall, pipes backfalling)*
* *undersized pipes*
* *sharp bends*
* *debris, earth, etc. falling into gullies*
* *ground movement, damage caused by tree and shrub roots*
* *crushed inspection chambers.*

Leaks (but not necessarily blockages) can be caused by:

* *broken pipes generally – damaged by traffic, incorrect bedding, surround or backfill, levelling on bricks during construction, ground movement, tree roots*
* *fractures caused by lack of movement joint where pipes enter buildings, concrete or brick inspection chambers*
* *leaking – as above plus damage to sleeves or 'O' rings during construction.*

Typical Drainage Problems

Foul drains without access points at changes in direction are difficult to rod/clear if they become blocked.

Drains should not kink or bend

Tree roots are always a potential problem. If drains are cracked, there may be subsequent problems of ground heave in clay soils.

Back inlet gully

RWP

Rodding eye missing

Storm Sewer

Foul Sewer

SVP

Downstairs sink or basin

Pavement

Plastic and lightweight steel covers easily damaged by vehicles.

WC (gd floor)

RWP

Branch at wrong angle.

Drive

Gully

Grate

No access to main drain where branch connects. Blockages could be expensive to clear.

RWP

RWP

RE

Manhole

If rainwater gulleys are not trapped, smells from an incorrectly connected drain can be a nuisance.

Drive not laid to falls or gully too high will prevent proper drainage. In cold weather, ice can be a threat to safety.

Foul drain incorrectly connected to storm drain.

Where pipes run through walls they are likely to fracture unless there is provision for movement. Joints should be made either side of the wall. The pipes should run through a clear opening with a lintel over the top.

In roads or under drives deep pipes may not need protection. Shallower pipes can be crushed unless they are covered with a reinforced concrete slab or bedded and surrounded in concrete.

In deep trenches below, and close to, foundations, pipes must be protected by concrete.

Other problems:

- *plastic inspection chambers in roads may collapse if not surrounded in concrete*
- *deep but narrow inspection chambers will be difficult to rod/clear*
- *drain pipes laid too shallow will not run properly; drain pipes laid too steeply will cause unnecessary excavation and are more prone to blockage – water flows but solids get left behind.*

Gulleys can block quite easily – especially if they are taking the 'run-off' from patios, etc.

Two photos taken with a drain camera. Left: A small sewer – the pipe is clean. Right: A sewer which is slowly becoming blocked by solidified fat.

Until the 1980s shallow inspection chambers were normally formed in brickwork. If they exceed 900mm in depth they should be one brick thick. The middle example (mid 1970s) has never been finished (there is no benching). Nowadays inspection chambers are usually glass fibre or plastic. They are easily damaged, both during, and after, construction.

Electrical installations 21

Introduction

The electrical supply to most domestic buildings is currently standardised within the European Union at 230V 50Hz. In England and Wales, the Building Regulations (Part P) require all new domestic electrical installations to be designed and installed in accordance with BS7671. Day-to-day working standards for electricians are set by the Institute of Engineering and Technology in the IET Wiring Regulations, 17th Edition (formerly the IEE Wiring Regulations).

The electrical supply to large buildings is usually supplied with three phases, each with its own colour cable. These are generated out of phase with each other and can be arranged to supply large electrical motors which then operate more effectively and with greater efficiency. The majority of UK local electrical supply is made up of these three-phase distribution systems running in the pavements and roads from the local supply sub-station.

However, most domestic properties do not require a three-phase supply and a single-phase supply is normally installed. This is because the motors on domestic appliances are small, many below a rating of 1kW and, whilst some more powerful cleaners may be rated up to 1.5kW, a single-phase supply is considered acceptable even where highly loaded for a short period of use. In larger domestic buildings, designers will usually provide a three-phase supply for any permanently wired motor over the 1kW rating. To receive a single-phase supply in a house, a connection is taken from one of the colour-coded phase supplies, balancing the load across all three phases by alternating the colour to different houses.

In urban areas, the electrical supply has been moved from the original overhead provision to underground, but overhead supply is still common in rural areas. Generally, exposed (i.e. overhead) electrical supplies are more likely to suffer damage during inclement weather from wind, ice build-up or overgrown and falling trees. Underground supplies may suffer from road and building alterations and, occasionally, subsidence or flooding.

Each unit of accommodation should have a supplier's electric meter installed. This measures and charges for the energy used, but there are many different 'tariffs' available and it is important to arrange for the most economic one to be selected depending on the nature of use. For example, a domestic user may be able to justify a night-time tariff even if there is no permanent electrical heating system, if a washing machine, dishwasher, tumble dryer and hot water immersion heater can be arranged to operate regularly overnight. Commercially, there are many more choices available.

The electrical wiring within the property comprises three wires, live, neutral and earth. The live wire supplies the power when required. The neutral completes the circuit allowing the current to flow when switched on. The earth wire is a protective wire diverting fault current away to achieve circuit isolation.

Left: A supply company's externally-positioned meter box. Right: A modern consumer unit which is installed internally in an easily accessible area of the building.

The distribution of electricity within properties is arranged in circuits for differing needs. These circuits start their distribution from a control panel called a consumer unit that is connected to the meter. The electrical supply is received by the consumer unit from the meter and can be fully isolated (normally within the consumer unit) before splitting into circuits. Lighting circuits are limited to a circuit load of 5A (i.e. 5 amp) and the number of circuits is dictated by the number of lighting fittings, but all properties nowadays have a minimum of two. These are generally arranged to cover separate floor levels in a house or day time/night time use in a bungalow. Socket outlets for power are supplied by a ring main or circuit which loops into all fixed outlets and returns to the consumer unit. There is normally a provision of two ring mains, with more installed as a property increases in size. The number of socket outlets is usually chosen for convenience to give flexibility of room layout and not always for exact number of appliances required. The maximum rating of a socket outlet is usually 13A to avoid overload. Immersion heaters and cookers both have their own individual radial circuit with a 30A rating.

A typical wiring layout is shown below.

In modern houses, there are usually additional radial feeds for showers, burglar alarms and smoke alarms.

Protection

Protective devices are inserted in the consumer unit to prevent overload. These are usually fuses in older installations and miniature circuit breakers (MCB) in more modern applications.

Fuses

Fuses can be one of two types: either rewirable or cartridge. Rewirable fuses have a piece of special fuse wire running between the two terminals. The fuse wire generates heat when a fault current flows through it and it melts when the heat exceeds the acceptable level. This can happen if too many appliances are operated at once or if the current runs to earth due to a faulty connection. The melted fuse breaks the circuit and stops the electricity supply. Replacement fuse wire is available in 5, 15 20, 30 and 45amp ratings. Rewirable fuses were the type used historically in consumer units.

Left: An older consumer unit with re-wirable fuses. Middle: An example of a re-wirable fuse. Right: A cartridge fuse.

Cartridge fuses consist of a silica-filled porcelain tube with metal caps at each end. They work along similar principles to those for re-wirable fuses, but once this type of fuse blows the cartridge has to be renewed. Cartridge fuses may be found in use in some consumer units, but are also used at the point of connection of the appliances, either within the connecting plug or in the socket outlet itself, e.g. a fused spur.

Miniature circuit breakers (MCBs)

MCBs are automatic protection devices fitted in the consumer unit which switch off a circuit if there is a fault. MCBs are similar in size to fuse holders, but give more precise protection than fuses. If a MCB 'trips', it can be re-set using the switch on its face, but this should only be after finding and correcting the fault.

Left: MCB is fitted onto a buzz-bar in a modern consumer unit. Right: Two alternative replacement MCBs that can be installed instead of a fuse in an older consumer unit.

An MCB will offer protection against circuit over-load, but will not work quickly enough to protect humans or animals. The modern consumer unit therefore includes one or more residual current devices (RCDs). An RCD disconnects a circuit whenever it detects that the flow of current is not balanced between the live and neutral conductors. Such an imbalance can be caused by current leakage through someone's body if it is 'grounded' and touching a live part of the circuit. An RCD is designed to disconnect quickly enough to avoid injury or lethal shock. Current IET Wiring Regulations require RCD protection to all lighting and power circuits (with a few exceptions) and all sockets which could have outdoor equipment plugged into them must be protected with an RCD.

Several MCBs and an RCD as well as the metal buzz-bar can be seen in this photograph.

Cables and flexes

Modern installations

Cables and flexes (originally known as flexible cable) are quite different and should never be confused. Cables are usually hidden in the structure and carry the power to the sockets and lights. Flexes connect portable appliances and lamps to the power sockets. Flexes are also used for light pendants and connecting items like storage radiators and immersion heaters to fused connections. Cables and flexes are available in a range of sizes – they are given a rating based on the area of the conductor's cross-section. The cable size depends on the amount of current likely to be carried in the circuit. A cable for a ring main, for example, is normally described as 2.5mm2 (i.e. it contains two conductors each of 2.5mm2). The third conductor, the earth, is usually slightly smaller – it only carries current when there is a fault.

Until fairly recently, the live and neutral conductor insulation sheathing (of PVC) was coloured red and black respectively. The earth conductor is not insulated, although it is normally covered with a green and yellow sleeve where it is exposed (i.e. in a socket). This type of cable is known as two core and earth. In 2004, colour coding changed (due to European harmonisation). The live sheathing is now brown and the neutral is blue – these colours had, in fact been used in flexes for several years previously.

Before 1977   1977–2006   2004 onwards

Flexes are made up of multiple strands. Unlike cable, the earth conductor in a flex (if there is one) is covered with green and yellow PVC insulation (now green and yellow bands). Some appliances only have a two-core flex – these are items where there is no risk of the casing becoming live.

Unlike flexes, cables do not have to move during their working life. This means that the expensive multi-stranded thin wire of flexes (originally known as flexible cables) is not required and they can, therefore, be made from single-strand conductors. However, thick cables, serving a cooker or electric shower for example, would be impossible to bend if they were made from single strands and cables over 2.5mm² tend to made from five or seven strands.

All hidden cables within the walls and partitions of a building should rise vertically and not (under any circumstance) diagonally. Unfortunately, this is not always the case.

Older installations

Diagnosing many electrical problems requires specialist help. However, if nothing else, a competent surveyor should be able to recognise evidence of old installations. The cables will give an immediate visual clue (a socket or switch plate may have to removed to see the cable).

1920s–1940s: Rubber-insulated, two-core (multi-strand) covered with an outer lead, or India rubber and cotton braided, sheath. If there is any of this very old wiring still installed, it should be isolated and replaced as soon as possible. The inner and outer rubber insulation will almost inevitably have degraded; in lead-covered cable, any break down in the insulation can cause the outer lead covering to become live – and an old installation still using this type of wiring is unlikely to have modern trips fitted in the consumer unit or an RCD. In addition, as the rubber deteriorates and starts to flake, sparks between the exposed wires can cause a fire. Lighting pendants and bed switches were made from a rubber-insulated conductor covered with cotton or silk. The conductors were then twin-twisted. For portable appliances the best flex was circular braided, in other words the two conductors (three where an earth was included) were separately insulated, but within one braiding. This was less liable to kinking than the twisted type.

Very occasionally, old fuse boxes are still found – they usually have a wooden back and cast-iron switches.

Late 1940s–early 1960s: Rubber-insulated, two or three-core (multi-strand) covered with an outer rubber sheath. The danger with this type of cable is the degrading of the rubber especially where a terminal has become overheated.

Ceiling rose

Fuse boards

Two and three pin sockets

Bed switch and flex

Brass light switch

Wood block mount

Bulb holder

Bakelite switch plate

Surface mounted light switches

Examples of early electrical fittings.

Left: Rubber-insulated cable is shown. Right: A two-strand drop to a pendant light.

It should also be noted that some houses (mostly from the 1960s) were wired with PVC insulated, three core (multi strand) with an outer PVC sheath. This may still be in good condition although broken strands sometimes occur at the terminals which reduces the current-carrying capacity of the cable and can cause local overheating.

Some common electrical defects

Table 21.1 Some of the more common electrical defects

DEFECT	PROBLEM(S) CAUSED
Plugs become unduly hot and/or scorch	Fuse of incorrect rating
Burn marks found on sockets or plugs.	Loose cable terminals Poor connection between plug and socket Prolonged high loading (3kW or above) Poor-quality (non BS) plugs
Sparks at socket outlet	Loose wires Faulty socket
Repeated blowing of fuses	Incorrect fuse rating Cable fault Accessory fault Overload on circuits
Smell of burning	Old wiring Loose connections Cables overheating

DEFECT	PROBLEM(S) CAUSED
Light bulbs blow repeatedly	Cheap bulbs Bulbs of high wattage in enclosed light fittings Mains voltage fluctuation
General loss of power	Blown main fuse Over-sensitive trip switches Mains failure
Loss of power at individual socket.	Blown fuse Loose connection

Care is required in selecting the routes of cabling within buildings to avoid possible physical damage or overheating. Where embedded in walls and partitions, the routes should always be vertical so the socket outlet or light position is known to indicate the position of wiring. Nasty electrical shocks have been experienced by DIY enthusiasts when fixing pins into walls. It is also not recommended that cables are tightly bunched together, nor should they be tightly enclosed beneath thermal insulation where heat cannot dissipate.

A poor electrical connection or cable joint can be dangerous as the electrical energy attempts to jump the gap. Any plug or fitting which shows signs of discoloration or a smell of burning should be attended to immediately. All electrical installations should be tested regularly to ensure their continued safe operation.

Outbreaks of fire are often linked to electrical faults or misuse. The attack of systems by rodents can be devastating and, where mice or rats are detected in a building, it is essential that they are not only removed as a health hazard, but also from an electrical safety point of view. Finally, existing cabling may overheat if buried below thick roofing insulation – if possible, the insulation should be laid below the cable.

Periodic inspection and more complex problems

General requirements

The IET Working Regulations recommend that, for domestic buildings, an inspection is carried out every 10 years (by a 'competent' person) and also when it is sold or let to identify any deficiencies in an electrical installation. A periodic inspection will:
- *reveal if any of the electrical circuits or equipment is overloaded;*
- *find any potential electrical shock risks and fire hazards in the electrical installation;*
- *highlight any lack of earthing or bonding;*
- *identify any defective DIY electrical work.*

This older electrical installation needs assessment by a qualified electrician. Both light switches and the cooker control unit are of some age, as is the surface casing to the switches.

Tests are also carried out on wiring and associated fixed electrical equipment. A schedule of circuits is also provided.

Testing a circuit.

The electrical installation is assessed against the current requirements of BS7671:2008 and the IET Working Regulations and will include (among other things):

- *adequacy of earthing and bonding;*
- *suitability (or otherwise) of the switchgear and control gear;*
- *serviceability of equipment, e.g. switches, socket-outlets and light fittings;*
- *type of wiring system and its condition;*
- *provision of residual current devices for socket-outlets that may be used to plug in electrical equipment used outdoors;*
- *presence of adequate identification and notices;*
- *extent of any wear and tear, damage or other deterioration.*

The overall condition will be assessed as either 'satisfactory', in which case no immediate remedial work is required, or 'unsatisfactory' which means remedial work is required to make the installation safe to use.

The inspection code

The inspection report includes a section, 'Observations and Recommendations for Actions to be Taken', which highlights any departures from the current edition of BS 7671. Each observation is coded as set out below. Where more than one code applies, the most serious is recorded.

- *Code C1: 'Danger present'. Risk of injury. Immediate remedial action required.*
- *Code C2: 'Potentially dangerous'. Urgent remedial action required.*
- *Code C3: 'Improvement recommended'.*

A few examples from each category are set out below. Most of these are items that can only be identified by a qualified electrician or electrical engineer. The examples are taken from Best Practice Guide No. 4: Electrical Installation Condition Reporting, published by the Electrical Safety Council.

Code C1 (Danger present)

- *Exposed live parts that are accessible to touch, such as where:*
  - *A fuse carrier or circuit-breaker is missing from a consumer unit and a blanking piece is not fitted in its place.*

- • *Terminations or connections have no (or damaged) barriers or enclosures.*
- • *Live conductors have no (or damaged) insulation.*
- • *An accessory is badly damaged.*
- • *Conductive parts that have become live as the result of a fault.*
- • *Incorrect polarity.*

Code C2 (Potentially dangerous)

- • *Absence of a reliable and effective means of earthing for the installation.*
- • *A public utility water pipe being used as the means of earthing for the installation.*
- • *A gas or oil pipe being used as the means of earthing for the installation .*
- • *Evidence of excessive heat (such as charring) from electrical equipment causing damage to the installation or its surroundings.*
- • *A 'borrowed neutral', for example, where a single final circuit neutral is shared by two final circuits (such as an upstairs lighting circuit and a separately protected downstairs lighting circuit).*
- • *Absence of a warning notices indicating the presence of an alternative or second source of electricity, such as a standby generator or microgenerator.*
- • *Immersion heater does not comply with BS EN 60335-2-73 (that is, it does not have a built in cut-out that will operate if the stored water temperature reaches 98 °C if the thermostat fails), and the cold water storage tank is plastic.*
- • *Absence of RCD protection for portable or mobile equipment that may reasonably be expected to be used outdoors.*

Code C3 (Improvement recommended)

- • *Absence of RCD protection for circuits of a location containing a bath or shower where satisfactory supplementary bonding is present.*
- • *Absence of 'Safety Electrical Connection – Do not Remove' notice.*
- • *Socket outlet mounted in such a position as to result in potential damage to socket, plug and/or flex.*
- • *Absence of circuit identification details.*
- • *Bare protective conductor of an insulated and sheathed cable not sleeved with insulation, colour coded to indicate its function.*

Electrical thermal storage heating

During the 1950s, the power supply industries were trying to find means of levelling out the peak demands for electricity and one of the ways introduced was to offer electrical power at a cheaper rate during the daily periods of low demand. The most significant period was overnight, and the introduction of electrical thermal storage heaters proved very effective at that time.

The economies to the user, however, were questionable under certain circumstances and led to a generally poor reputation for this type of heating. The main problem lies in the difficulty of anticipating how much heat is required for the following day and initially the systems introduced were unable to account in any way for this. Many buildings were overheated on warmer days, but other buildings were cold by mid-afternoon as the stored heating capacity ran out in colder weather. The size of building would also dictate how much electrical power was consumed. The fact that the capital cost was very attractive hid the revenue implications of high energy costs despite the discounted prices. In the UK, a thermal storage heater earns a low rating on an Energy Performance Certificate.

Left: A typical older night storage heater (NSH) Right: An NSH has been stripped down for repair. Some heating blocks and the heating elements (coils) remain.

The principle of thermal storage heating relies on a mass of material, such as ceramic, clay or concrete blocks, being heated overnight by electrical elements inserted between them, utilising the off-peak electrical supply. The thermal store is then dissipated during the day to keep the building warm. In cold weather, there is often insufficient stored heat to last the day, so there is a late afternoon boost. However, this is at a cost to the user, as this will be utilising on-peak (i.e. expensive) electricity.

Many different designs of thermal storage heater were tried, including water as the thermal store, and a number of control systems incorporated to make the systems more effective. Heaters can be one of two general types:

A   *individually-controlled room heaters (e.g. night storage heaters), the more efficient being fan-assisted; OR*

B   *a single fan-assisted large-capacity heater controlled by a room thermostat that delivers heat to each room or floor of a house via ducting.*

Modern thermal storage heaters can be effective providing certain criteria are met:

•   *The system should not be installed in large, poorly insulated buildings as the low capital cost cannot justify the high energy costs.*

•   *Ideally, the building should be of a low air volume and the construction should be with materials of high thermal insulation values.*

•   *The heaters should be capable of dissipating the heat under fan control and some form of external optimising controls that are capable of regulating energy input relative to weather conditions should be incorporated.*

•   *The capacity of each thermal heater needs to be carefully considered. If it is too large, its capital cost will be excessive and it will be under-utilised; if too small, the cost of supplemental (i.e. daytime on-peak) electrical heating in cold weather will be excessive.*

If these parameters are adhered to, electrical thermal storage systems can have distinct advantages over gas-fired boiler alternatives, especially in timber-framed housing, as long as the power supply remains reliable

Storage radiators can be very reliable and durable; their obsolescence is often due to changing tastes rather than mechanical failure. However, as they age, the efficiency of the heating elements and the electronic controls reduces. They can become highly inefficient and, often, this results in the storage radiator becoming either too hot or insufficiently hot for the weather conditions being experienced.

Index